Cracking the

CHEMISTRY EXAM

2017 Edition

The Staff of The Princeton Review

PrincetonReview.com

Penguin
Random
House

The Princeton Review, Inc.
24 Prime Parkway, Suite 201
Natick, MA 01760
E-mail: editorialsupport@review.com

Published in the United States by Penguin Random House LLC, New York, and in Canada by Random House of Canada, a division of Penguin Random House Ltd., Toronto.

Terms of Service: The Princeton Review Online Companion Tools ("Student Tools") for retail books are available for only the two most recent editions of that book. Student Tools may be activated only twice per eligible book purchased for two consecutive 12-month periods, for a total of 24 months of access. Activation of Student Tools more than twice per book is in direct violation of these Terms of Service and may result in discontinuation of access to Student Tools Services.

ISBN: 978-1-101-91987-3
eBook ISBN: 978-1-101-92013-8
ISSN: 1092-0102

Editor: Meave Shelton
Production Editors: Lee Elder and Liz Rutzel
Production Artist: Keren Peysakh

Printed in the United States of America on partially recycled paper.

10 9 8 7 6 5 4 3 2 1

2017 Edition

Acknowledgments

The Princeton Review would like to give special thanks to Nick Leonardi for his hand in revising and updating this title to better prepare students for the AP Chemistry exam. Nick would like, in turn, to thank the AP Chemistry teacher community from the College Board forums, and his parents, who, despite not knowing chemistry, have contributed more to his success than anybody else.

Contents

Register Your

1 Go to **PrincetonReview.com/cracking**

2 You'll see a welcome page where you can register your book using the following ISBN: 9781101919873.

3 After placing this free order, you'll either be asked to log in or to answer a few simple questions in order to set up a new Princeton Review account.

4 Finally, click on the "Student Tools" tab located at the top of the screen. It may take an hour or two for your registration to go through, but after that, you're good to go.

If you are experiencing book problems (potential content errors), please contact EditorialSupport@review.com with the full title of the book, its ISBN number (located above), and the page number of the error. Experiencing technical issues? Please e-mail TPRStudentTech@review.com with the following information:

- your full name
- e-mail address used to register the book
- full book title and ISBN
- your computer OS (Mac or PC) and Internet browser (Firefox, Safari, Chrome, etc.)
- description of technical issue

Book Online!

Once you've registered, you can...

- Find any late-breaking information released about the AP Chemistry Exam
- Take a full-length practice PSAT, SAT, and ACT
- Get valuable advice about the college application process, including tips for writing a great essay and where to apply for financial aid
- Sort colleges by whatever you're looking for (such as Best Theater or Dorm), learn more about your top choices, and see how they all rank according to *The Best 380 Colleges*
- Access comprehensive study guides and a variety of printable resources, including answer bubble sheets, the periodic table, and a list of equations to review
- Check to see if there have been any corrections or updates to this edition

Look For These Icons Throughout The Book

 Online Articles

 Proven Techniques

 Online Practice Tests

 Applied Strategies

The Princeton Review®

Part I
Using This Book to Improve Your AP Score

PREVIEW ACTIVITY: YOUR KNOWLEDGE, YOUR EXPECTATIONS

Your route to a high score on the AP Chemistry Exam depends a lot on how you plan to use this book. Please respond to the following questions.

1. Rate your level of confidence about your knowledge of the content tested by the Chemistry AP Exam:

   A. Very confident—I know it all
   B. I'm pretty confident, but there are topics for which I could use help
   C. Not confident—I need quite a bit of support
   D. I'm not sure

2. Circle your goal score for the AP Chemistry Exam.

   5 4 3 2 1 I'm not sure yet

3. What do you expect to learn from this book? Circle all that apply to you.

   A. A general overview of the test and what to expect
   B. Strategies for how to approach the test
   C. The content tested by this exam
   D. I'm not sure yet

YOUR GUIDE TO USING THIS BOOK

This book is organized to provide as much—or as little—support as you need, so you can use this book in whatever way will be most helpful to improving your score on the AP Chemistry Exam.

- The remainder of **Part I** will provide guidance on how to use this book and help you determine your strengths and weaknesses

- **Part II** of this book contains Practice Test 1, its answers and explanations, and a scoring guide. (Bubble sheets can be found in the very back of the book for easy tear-out.) We strongly recommend that you take this test before going any further, in order to realistically determine:

  - your starting point right now

  - which question types you're ready for and which you might need to practice

  - which content topics you are familiar with and which you will want to carefully review

Once you have nailed down your strengths and weaknesses with regard to this exam, you can focus your test preparation, build a study plan, and be efficient with your time.

- **Part III** of this book will
    - provide information about the structure, scoring, and content of the Chemistry Exam
    - help you to make a study plan
    - point you towards additional resources
- **Part IV** of this book will explore
    - how to attack multiple-choice questions
    - how to write high-scoring free-response answers
    - how to manage your time to maximize the number of points available to you
- **Part V** of this book covers the content you need for your exam.
- **Part VI** of this book contains Practice Test 2, its answers and explanations, and a scoring guide. (Bubble sheets can be found in the very back of the book for easy tear-out.) If you skipped Practice Test 1, we recommend that you do both (with at least a day or two between them) so that you can compare your progress between the two. Additionally, this will help to identify any external issues: if you get a certain type of question wrong both times, you probably need to review it. If you only got it wrong once, you may have run out of time or been distracted by something. In either case, this will allow you to focus on the factors that caused the discrepancy in scores and to be as prepared as possible on the day of the test.

You may choose to use some parts of this book over others, or you may work through the entire book. This will depend on your needs and how much time you have. Let's now look how to make this determination.

HOW TO BEGIN

1. Take a Test

Before you can decide how to use this book, you need to take a practice test. Doing so will give you insight into your strengths and weaknesses, and the test will also help you make an effective study plan. If you're feeling test-phobic, remind yourself that a practice test is a tool for diagnosing yourself—it's not how well you do that matters but how you use information gleaned from your performance to guide your preparation.

So, before you read further, take AP Chemistry Practice Test 1 starting at page 7 of this book. Be sure to do so in one sitting, following the instructions that appear before the test.

2. Check Your Answers

Using the answer key on page 35, count how many multiple-choice questions you got right and how many you missed. Don't worry about the explanations for now, and don't worry about why you missed questions. We'll get to that soon.

3. Reflect on the Test

After you take your first test, respond to the following questions:

- How much time did you spend on the multiple-choice questions?
- How much time did you spend on each long form free-response question? What about each short form free-response question?
- How many multiple-choice questions did you miss?
- Do you feel you had the knowledge to address the subject matter of the free-response questions?
- Do you feel you wrote well organized, thoughtful answers to the free-response questions?

- Circle the content areas that were most challenging for you and draw a line through the ones in which you felt confident/did well.
    - **Big Idea #1:** Atoms, Elements, and the Building Blocks of Matter
    - **Big Idea #2:** Chemical and Physical Properties of Matter
    - **Big Idea #3:** Chemical Reactions, Energy Changes, and Redox Reactions
    - **Big Idea #4:** Chemical Reactions and their Rates
    - **Big Idea #5:** Laws of Thermodynamics and Changes in Matter
    - **Big Idea #6:** Equilibrium, Acids and Bases, Titrations and Solubility

4. **Read Part III of this Book and Complete the Self-Evaluation**

   As discussed previously, Part III will provide information on how the test is structured and scored. As you read Part III, re-evaluate your answers to the questions on the previous page. You will then be able to make a study plan, based on your needs and time available, that will allow you to use this book most effectively.

5. **Engage with Parts IV and V as Needed**

   Notice the word *engage*. You'll get more out of this book if you use it intentionally than if you read it passively, hoping for an improved score through osmosis.

   The strategy chapters will help you think about your approach to the question types on this exam. Part IV will open with a reminder to think about how you approach questions now and then close with a reflection section asking you to think about how/whether you will change your approach in the future.

   The content chapters are designed to provide a review of the content tested on the AP Chemistry Exam, including the level of detail you need to know and how the content is tested. You will have the opportunity to assess your mastery of the content of each chapter through test-appropriate questions.

6. Take Test 2 and Assess Your Performance

Once you feel you have developed the strategies you need and gained the knowledge you lacked, you should take Test 2, which starts at page 333 of this book. You should do so in one sitting, following the instructions at the beginning of the test.

When you are done, check your answers to the multiple-choice sections. See if a teacher will read your answers to the free-response questions and provide feedback.

Once you have taken the test, reflect on what areas you still need to work on, and revisit the chapters in this book that address those deficiencies. Through this type of reflection and engagement, you will continue to improve.

7. Keep Working

After you have revisited certain chapters in this book, continue the process of testing, reflecting, and engaging with the second test. Each time, consider what additional work you need to do and how you will change your strategic approach to different parts of the test.

Part II
Practice Test 1

AP® Chemistry Exam

SECTION I: Multiple-Choice Questions

DO NOT OPEN THIS BOOKLET UNTIL YOU ARE TOLD TO DO SO.

At a Glance

Total Time
1 hour and 30 minutes
Number of Questions
60
Percent of Total Grade
50%
Writing Instrument
Pencil required

Instructions

Section I of this examination contains 60 multiple-choice questions. Fill in only the ovals for numbers 1 through 60 on your answer sheet.

CALCULATORS MAY NOT BE USED IN THIS PART OF THE EXAMINATION.

Indicate all of your answers to the multiple-choice questions on the answer sheet. No credit will be given for anything written in this exam booklet, but you may use the booklet for notes or scratch work. After you have decided which of the suggested answers is best, completely fill in the corresponding oval on the answer sheet. Give only one answer to each question. If you change an answer, be sure that the previous mark is erased completely. Here is a sample question and answer.

Sample Question	Sample Answer
Chicago is a (A) state (B) city (C) country (D) continent	Ⓐ ● Ⓒ Ⓓ

Use your time effectively, working as quickly as you can without losing accuracy. Do not spend too much time on any one question. Go on to other questions and come back to the ones you have not answered if you have time. It is not expected that everyone will know the answers to all the multiple-choice questions.

About Guessing

Many candidates wonder whether or not to guess the answers to questions about which they are not certain. Multiple-choice scores are based on the number of questions answered correctly. Points are not deducted for incorrect answers, and no points are awarded for unanswered questions. Because points are not deducted for incorrect answers, you are encouraged to answer all multiple-choice questions. On any questions you do not know the answer to, you should eliminate as many choices as you can, and then select the best answer among the remaining choices.

Disclaimer

This test is an approximation of the test that you will take. For up-to-date information, please remember to check the AP Students website.

CHEMISTRY
SECTION I

Time—1 hour and 30 minutes

INFORMATION IN THE TABLE BELOW AND ON THE FOLLOWING PAGES MAY BE USEFUL IN ANSWERING THE QUESTIONS IN THIS SECTION OF THE EXAMINATION.

DO NOT DETACH FROM BOOK.

PERIODIC TABLE OF THE ELEMENTS

1 **H** 1.008																	2 **He** 4.00
3 **Li** 6.94	4 **Be** 9.01											5 **B** 10.81	6 **C** 12.01	7 **N** 14.01	8 **O** 16.00	9 **F** 19.00	10 **Ne** 20.18
11 **Na** 22.99	12 **Mg** 24.30											13 **Al** 26.98	14 **Si** 28.09	15 **P** 30.97	16 **S** 32.06	17 **Cl** 35.45	18 **Ar** 39.95
19 **K** 39.10	20 **Ca** 40.08	21 **Sc** 44.96	22 **Ti** 47.90	23 **V** 50.94	24 **Cr** 52.00	25 **Mn** 54.94	26 **Fe** 55.85	27 **Co** 58.93	28 **Ni** 58.69	29 **Cu** 63.55	30 **Zn** 65.39	31 **Ga** 69.72	32 **Ge** 72.59	33 **As** 74.92	34 **Se** 78.96	35 **Br** 79.90	36 **Kr** 83.80
37 **Rb** 85.47	38 **Sr** 87.62	39 **Y** 88.91	40 **Zr** 91.22	41 **Nb** 92.91	42 **Mo** 95.94	43 **Tc** (98)	44 **Ru** 101.1	45 **Rh** 102.91	46 **Pd** 106.42	47 **Ag** 107.87	48 **Cd** 112.41	49 **In** 114.82	50 **Sn** 118.71	51 **Sb** 121.75	52 **Te** 127.60	53 **I** 126.91	54 **Xe** 131.29
55 **Cs** 132.91	56 **Ba** 137.33	57 ***La** 138.91	72 **Hf** 178.49	73 **Ta** 180.95	74 **W** 183.85	75 **Re** 186.21	76 **Os** 190.2	77 **Ir** 192.2	78 **Pt** 195.08	79 **Au** 196.97	80 **Hg** 200.59	81 **Tl** 204.38	82 **Pb** 207.2	83 **Bi** 208.98	84 **Po** (209)	85 **At** (210)	86 **Rn** (222)
87 **Fr** (223)	88 **Ra** 226.02	89 **†Ac** 227.03	104 **Rf** (261)	105 **Db** (262)	106 **Sg** (266)	107 **Bh** (264)	108 **Hs** (277)	109 **Mt** (268)	110 **Ds** (271)	111 **Rg** (272)							

*Lanthanide Series	58 **Ce** 140.12	59 **Pr** 140.91	60 **Nd** 144.24	61 **Pm** (145)	62 **Sm** 150.4	63 **Eu** 151.97	64 **Gd** 157.25	65 **Tb** 158.93	66 **Dy** 162.50	67 **Ho** 164.93	68 **Er** 167.26	69 **Tm** 168.93	70 **Yb** 173.04	71 **Lu** 174.97
†Actinide Series	90 **Th** 232.04	91 **Pa** 231.04	92 **U** 238.03	93 **Np** (237)	94 **Pu** (244)	95 **Am** (243)	96 **Cm** (247)	97 **Bk** (247)	98 **Cf** (251)	99 **Es** (252)	100 **Fm** (257)	101 **Md** (258)	102 **No** (259)	103 **Lr** (262)

GO ON TO THE NEXT PAGE.

ADVANCED PLACEMENT CHEMISTRY EQUATIONS AND CONSTANTS

Throughout the test the following symbols have the definitions specified unless otherwise noted.

L, mL	= liter(s), milliliter(s)	mm Hg	= millimeters of mercury
g	= gram(s)	J, kJ	= joule(s), kilojoule(s)
nm	= nanometer(s)	V	= volt(s)
atm	= atmosphere(s)	mol	= mole(s)

ATOMIC STRUCTURE

$E = h\nu$

$c = \lambda\nu$

E = energy
ν = frequency
λ = wavelength

Planck's constant, $h = 6.626 \times 10^{-34}$ J s
Speed of light, $c = 2.998 \times 10^{8}$ m s^{-1}
Avogadro's number = 6.022×10^{23} mol^{-1}
Electron charge, $e = -1.602 \times 10^{-19}$ coulombs

EQUILIBRIUM

$K_c = \frac{[C]^c[D]^d}{[A]^a[B]^b}$, where $a\,A + b\,B \rightleftarrows c\,C + d\,D$

$K_p = \frac{(P_C)^c(P_D)^d}{(P_A)^a(P_B)^b}$

$K_a = \frac{[H^+][A^-]}{[HA]}$

$K_b = \frac{[OH^-][HB^+]}{[B]}$

$K_w = [H^+][OH^-] = 1.0 \times 10^{-14}$ at 25°C

$= K_a \times K_b$

$pH = -\log[H^+]$, $pOH = -\log[OH^-]$

$14 = pH + pOH$

$pH = pK_a + \log\frac{[A^-]}{[HA]}$

$pK_a = -\log K_a$, $pK_b = -\log K_b$

Equilibrium Constants

K_c (molar concentrations)
K_p (gas pressures)
K_a (weak acid)
K_b (weak base)
K_w (water)

KINETICS

$\ln[A]_t - \ln[A]_0 = -kt$

$\frac{1}{[A]_t} - \frac{1}{[A]_0} = kt$

$t_{1/2} = \frac{0.693}{k}$

k = rate constant
t = time
$t_{1/2}$ = half-life

GO ON TO THE NEXT PAGE.

GASES, LIQUIDS, AND SOLUTIONS

$PV = nRT$

$P_A = P_{total} \times X_A$, where $X_A = \frac{\text{moles A}}{\text{total moles}}$

$P_{total} = P_A + P_B + P_C + \ldots$

$n = \frac{m}{M}$

$K = {}^\circ C + 273$

$D = \frac{m}{V}$

KE per molecule $= \frac{1}{2}mv^2$

Molarity, M = moles of solute per liter of solution

$A = abc$

P = pressure
V = volume
T = temperature
n = number of moles
m = mass
M = molar mass
D = density
KE = kinetic energy
v = velocity
A = absorbance
a = molar absorptivity
b = path length
c = concentration

Gas constant, $R = 8.314\ \text{J mol}^{-1}\text{K}^{-1}$
$= 0.08206\ \text{L atm mol}^{-1}\text{K}^{-1}$
$= 62.36\ \text{L torr mol}^{-1}\text{K}^{-1}$
1 atm = 760 mm Hg
= 760 torr
STP = 0.00°C and 1.000 atm

THERMOCHEMISTRY/ ELECTROCHEMISTRY

$q = mc\Delta T$

$\Delta S^\circ = \sum S^\circ$ products $- \sum S^\circ$ reactants

$\Delta H^\circ = \sum \Delta H_f^\circ$ products $- \sum \Delta H_f^\circ$ reactants

$\Delta G^\circ = \sum \Delta G_f^\circ$ products $- \sum \Delta G_f^\circ$ reactants

$\Delta G^\circ = \Delta H^\circ - T\Delta S^\circ$

$= -RT \ln K$

$= -nFE^\circ$

$I = \frac{q}{t}$

q = heat
m = mass
c = specific heat capacity
T = temperature
S° = standard entropy
H° = standard enthalpy
G° = standard free energy
n = number of moles
E° = standard reduction potential
I = current (amperes)
q = charge (coulombs)
t = time (seconds)

Faraday's constant, F = 96,485 coulombs per mole of electrons

$1\ \text{volt} = \frac{1\ \text{joule}}{1\ \text{coulomb}}$

GO ON TO THE NEXT PAGE.

1. $2ClF(g) + O_2(g) \leftrightarrow Cl_2O(g) + F_2O(g)$ $\Delta H = 167$ kJ/mol$_{rxn}$

During the reaction above, the product yield can be increased by increasing the temperature of the reaction. Why is this effective?

(A) The reaction is endothermic; therefore adding heat will shift it to the right.
(B) Increasing the temperature increases the speed of the molecules, meaning there will be more collisions between them.
(C) The reactants are less massive than the products, and an increase in temperature will cause their kinetic energy to increase more than that of the products.
(D) The increase in temperature allows for a higher percentage of molecular collisions to occur with the proper orientation to create the product.

2. $CH_3NH_2(aq) + H_2O(l) \leftrightarrow OH^-(aq) + CH_3NH_3^+(aq)$

The above equation represents the reaction between the base methylamine ($K_b = 4.38 \times 10^{-4}$) and water. Which of the following best represents the concentrations of the various species at equilibrium?

(A) $[OH^-] > [CH_3NH_2] = [CH_3NH_3^+]$
(B) $[OH^-] = [CH_3NH_2] = [CH_3NH_3^+]$
(C) $[CH_3NH_2] > [OH^-] > [CH_3NH_3^+]$
(D) $[CH_3NH_2] > [OH^-] = [CH_3NH_3^+]$

3. The diagram below shows the relative atomic sizes of three different elements from the same period. Which of the following statements must be true?

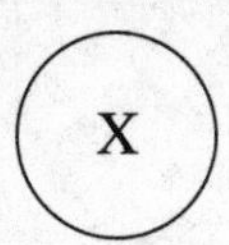

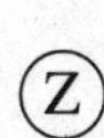

(A) The effective nuclear charge will be the greatest in element X.
(B) The first ionization energy will be greatest in element X.
(C) The electron shielding effect will be greatest in element Z.
(D) The electronegativity value will be greatest in element Z.

4. A sealed, rigid container contains three gases: 28.0 g of nitrogen, 40.0 g of argon, and 36.0 g of water vapor. If the total pressure exerted by the gases is 2.0 atm, what is the partial pressure of the nitrogen?

(A) 0.33 atm
(B) 0.40 atm
(C) 0.50 atm
(D) 2.0 atm

5. A sample of liquid NH_3 is brought to its boiling point. Which of the following occurs during the boiling process?

(A) The N–H bonds within the NH_3 molecules break apart.
(B) The overall temperature of the solution rises as the NH_3 molecules speed up.
(C) The amount of energy within the system remains constant.
(D) The hydrogen bonds holding separate NH_3 molecules together break apart.

GO ON TO THE NEXT PAGE.

Questions 6-10 refer to the following.

Two half-cells are set up as follows:

Half-Cell A: Strip of Cu(*s*) in $CuNO_3(aq)$
Half-Cell B: Strip of Zn(*s*) in $Zn(NO_3)_2(aq)$

When the cells are connected according to the diagram below, the following reaction occurs:

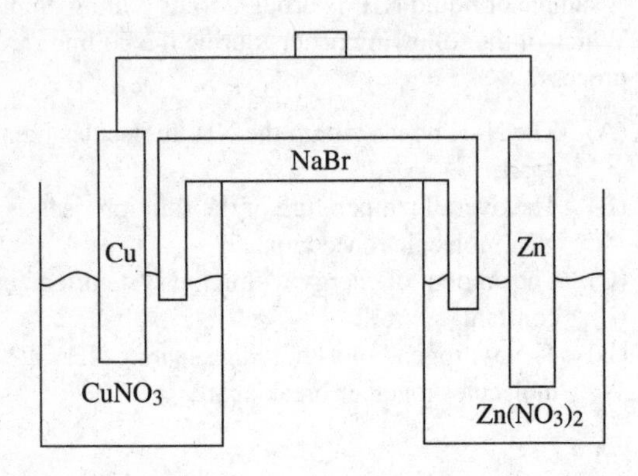

$2Cu^+(aq) + Zn(s) \rightarrow 2Cu(s) + Zn^{2+}(aq)$ $E° = +1.28$ V

6. Correctly identify the anode and cathode in this reaction as well as where oxidation and reduction are taking place.

(A) Cu is the anode where oxidation occurs, and Zn is the cathode where reduction occurs.
(B) Cu is the anode where reduction occurs, and Zn is the cathode where oxidation occurs.
(C) Zn is the anode where oxidation occurs, and Cu is the cathode where reduction occurs.
(D) Zn is the anode where reduction occurs, and Cu is the cathode where oxidation occurs.

7. How many moles of electrons must be transferred to create 127 g of copper?

(A) 1 mole of electrons
(B) 2 moles of electrons
(C) 3 moles of electrons
(D) 4 moles of electrons

8. If the $Cu^+ + e^- \rightarrow Cu(s)$ half-reaction has a standard reduction potential of +0.52 V, what is the standard reduction potential for the $Zn^{2+} + 2e^- \rightarrow Zn(s)$ half-reaction?

(A) +0.76 V
(B) –0.76 V
(C) +0.24 V
(D) –0.24 V

9. As the reaction progresses, what will happen to the overall voltage of the cell?

(A) It will increase as $[Zn^{2+}]$ increases.
(B) It will increase as $[Cu^+]$ increases.
(C) It will decrease as $[Zn^{2+}]$ increases.
(D) The voltage will remain constant.

10. What will happen in the salt bridge as the reaction progresses?

(A) The Na^+ ions will flow to the Cu/Cu^+ half-cell.
(B) The Br^- ions will flow to the Cu/Cu^+ half-cell.
(C) Electrons will transfer from the Cu/Cu^+ half-cell to the Zn/Zn^{2+} half-cell.
(D) Electrons will transfer from the Zn/Zn^{2+} half-cell to the Cu/Cu^+ half-cell.

---

11. For a reaction involving nitrogen monoxide inside a sealed flask, the value for the reaction quotient (*Q*) was found to be 1.1×10^2 at a given point. If, after this point, the amount of NO gas in the flask increased, which reaction is most likely taking place in the flask?

(A) $NOBr(g) \leftrightarrow NO(g) + \frac{1}{2}Br_2(g)$ $K_c = 3.4 \times 10^{-2}$
(B) $2NOCl(g) \leftrightarrow 2NO(g) + Cl_2(g)$ $K_c = 1.6 \times 10^{-5}$
(C) $2NO(g) + 2H_2(g) \leftrightarrow N_2(g) + 2H_2O(g)$ $K_c = 4.0 \times 10^6$
(D) $N_2(g) + O_2(g) \leftrightarrow 2NO(g)$ $K_c = 4.2 \times 10^2$

12. Which of the following substances has an asymmetrical molecular structure?

(A) SF_4
(B) PCl_5
(C) BF_3
(D) CO_2

vs. electron structure

GO ON TO THE NEXT PAGE.

13.

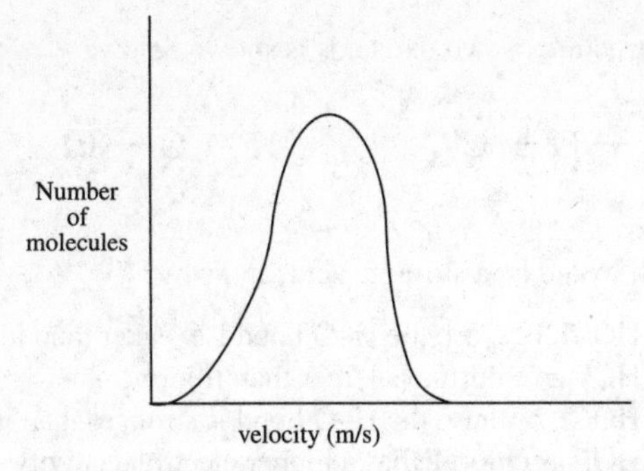

The diagram above shows the speed distribution of molecules in a gas held at 200 K. Which of the following representations would best represent the gas at a higher temperature? (Note: The original line is shown as a dashed line in the answer options.)

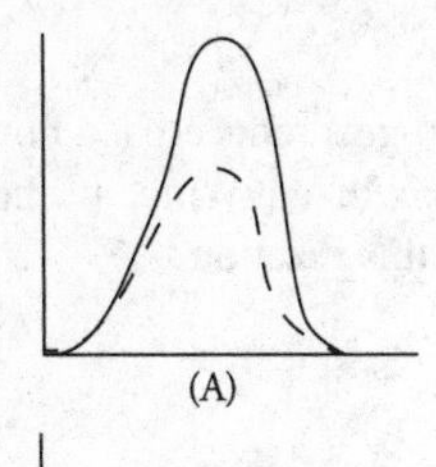
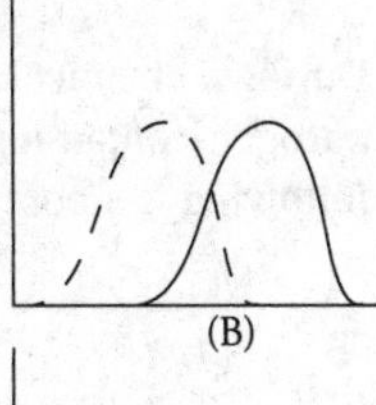

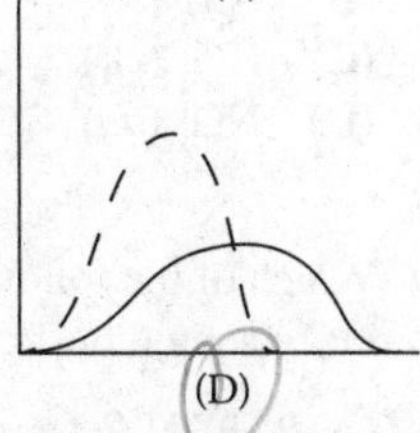

(A) (B) (C) (D)

14. Nitrogen's electronegativity value is between those of phosphorus and oxygen. Which of the following correctly describes the relationship between the three values?

(A) The value for nitrogen is less than that of phosphorus because nitrogen is larger, but greater than that of oxygen because nitrogen has a greater effective nuclear charge.
(B) The value for nitrogen is less than that of phosphorus because nitrogen has fewer protons but greater than that of oxygen because nitrogen has fewer valence electrons.
(C) The value for nitrogen is greater than that of phosphorus because nitrogen has fewer electrons, but less than that of oxygen because nitrogen is smaller.
(D) The value for nitrogen is greater than that of phosphorus because nitrogen is smaller, but less than that of oxygen because nitrogen has a smaller effective nuclear charge.

15. A sample of a compound known to consist of only carbon, hydrogen, and oxygen is found to have a total mass of 29.05 g. If the mass of the carbon is 18.02 g and the mass of the hydrogen is 3.03 g, what is the empirical formula of the compound?

(A) C_2H_4O
(B) C_3H_6O
(C) $C_2H_6O_3$
(D) $C_3H_8O_2$

Questions 16-18 refer to the following.

A solution of carbonic acid, H_2CO_3, is titrated with sodium hydroxide, NaOH. The following graph is produced:

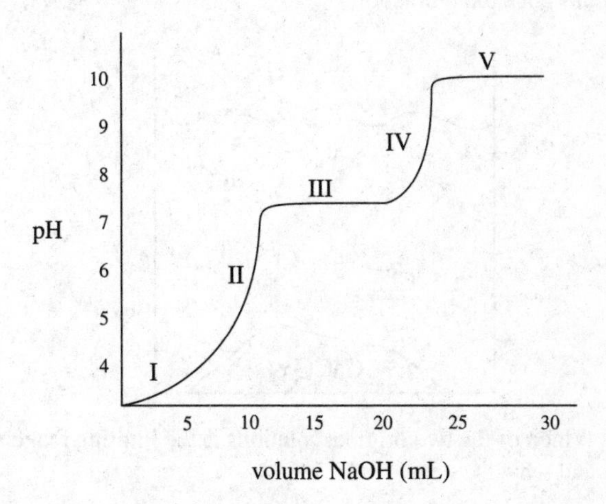

16. In addition to OH^-, what species are present in the solution during section III of the graph?

(A) H_2CO_3, HCO_3^-, and CO_3^{2-}
(B) H_2CO_3 and HCO_3^-
(C) HCO_3^- and CO_3^{2-}
(D) H_2CO_3 and CO_3^{2-}

17. What is the magnitude of the first dissociation constant?

(A) 10^{-2}
(B) 10^{-4}
(C) 10^{-6}
(D) 10^{-8}

GO ON TO THE NEXT PAGE.

18. If the concentration of the sodium hydroxide is increased prior to repeating the titration, what effect, if any, would that have on the graph?

(A) The graph would not change at all.
(B) The pH values at the equivalence points would increase.
(C) The equivalence points would be reached with less volume of NaOH added.
(D) The slope of the equivalence points would decrease.

---

19. Solutions of potassium carbonate and calcium chloride are mixed, and the particulate representation below shows which are present in significant amounts after the reaction has gone to completion.

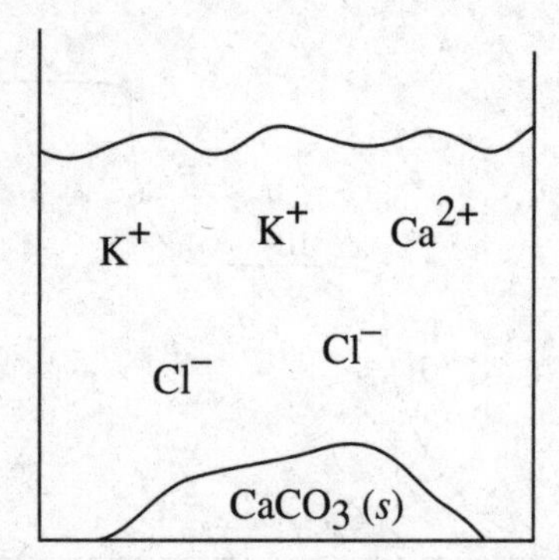

Which of the two original solutions is the limiting reagent and why?

(A) The potassium carbonate, because of the polyatomic anion
(B) The potassium carbonate, because there is no carbonate left after the reaction
(C) The calcium chloride, because there is an excess of calcium ions post-reaction
(D) The calcium chloride, because the component ions are smaller than those in potassium carbonate

20. In which of the following circumstances is the value for K_{eq} always greater than 1?

	ΔH	ΔS
(A)	Positive	Positive
(B)	Positive	Negative
(C)	Negative	Negative
(D)	Negative	Positive

$\Delta G = -RT\ln K$

21. The structure of two oxoacids is shown below:

H — Ö — C̈l: H — Ö — F̈:

Which would be a stronger acid, and why?

(A) HOCl, because the H–O bond is weaker than in HOF as chlorine is larger than fluorine
(B) HOCl, because the H–O bond is stronger than in HOF as chlorine has a higher electronegativity than fluorine
(C) HOF, because the H–O bond is stronger than in HOCl as fluorine has a higher electronegativity than chlorine
(D) HOF, because the H–O bond is weaker than in HOCl as fluorine is smaller than chlorine

22. During a chemical reaction, NO(*g*) gets reduced and no nitrogen-containing compound is oxidized. Which of the following is a possible product of this reaction?

(A) $NO_2(g)$
(B) $N_2(g)$
(C) $NO_3^-(aq)$
(D) $NO_2^-(aq)$

23. Which of the following pairs of substances would make a good buffer solution?

(A) $HC_2H_3O_2(aq)$ and $NaC_2H_3O_2(aq)$
(B) $H_2SO_4(aq)$ and $LiOH(aq)$
(C) $HCl(aq)$ and $KCl(aq)$
(D) $HF(aq)$ and $NH_3(aq)$

Salt of WA/WB how to tell?

Questions 24-27 refer to the following.

Inside a calorimeter, 100.0 mL of 1.0 *M* hydrocyanic acid (HCN), a weak acid, and 100.0 mL of 0.50 *M* sodium hydroxide are mixed. The temperature of the mixture rises from 21.5°C to 28.5°C. The specific heat of the mixture is approximately 4.2 J/g°C, and the density is identical to that of water.

HCN + NaOH → NaC(H)…

24. Identify the correct net ionic equation for the reaction that takes place.

(A) $HCN(aq) + OH^-(aq) \leftrightarrow CN^-(aq) + H_2O(l)$
(B) $HCN(aq) + NaOH(aq) \leftrightarrow NaCN(aq) + H_2O(l)$
(C) $H^+(aq) + OH^-(aq) \leftrightarrow H_2O(l)$
(D) $H^+(aq) + CN^-(aq) + Na^+(aq) + OH^-(aq) \leftrightarrow H_2O(l) + CN^-(aq) + Na^+(aq)$

GO ON TO THE NEXT PAGE.

25. What is the approximate amount of heat released during the reaction?

(A) 1.5 kJ
(B) 2.9 kJ
(C) 5.9 kJ
(D) 11.8 kJ

26. As ΔT increases, what happens to the equilibrium constant and why?

(A) The equilibrium constant increases because more products are created.
(B) The equilibrium constant increases because the rate of the forward reaction increases.
(C) The equilibrium constant decreases because the equilibrium shifts to the left.
(D) The value for the equilibrium constant is unaffected by temperature and will not change.

27. If the experiment is repeated with 200.0 mL of 1.0 *M* HCN and 100. mL of 0.50 *M* NaOH, what would happen to the values for ΔT and ΔH_{rxn}?

	ΔT	ΔH_{rxn}
(A)	Increase	Increase
(B)	Stay the same	Stay the same
(C)	Decrease	Stay the same
(D)	Stay the same	Increase

28. The following diagrams show the Lewis structures of four different molecules. Which molecule would travel the farthest in a paper chromatography experiment using a polar solvent?

```
    H
    |    ..
H - C - O - H
    |    ..
    H
Methanol

  H H H H H
  | | | | |
H-C-C-C-C-C-H
  | | | | |
  H H H H H
Pentane

  H  :O:  H
  |   ||  |
H-C - C - C-H
  |       |
  H       H
Acetone

  H       H
  |   ..  |
H-C - O - C-H
  |   ..  |
  H       H
Ether
```

(A) Methanol
(B) Pentane
(C) Acetone
(D) Ether

29. The first ionization energy for a neutral atom of chlorine is 1.25 MJ/mol and the first ionization energy for a neutral atom of argon is 1.52 MJ/mol. How would the first ionization energy value for a neutral atom of potassium compare to those values?

(A) It would be greater than both because potassium carries a greater nuclear charge then either chlorine or argon.
(B) It would be greater than both because the size of a potassium atom is smaller than an atom of either chlorine or argon.
(C) It would be less than both because there are more electrons in potassium, meaning they repel each other more effectively and less energy is needed to remove one.
(D) It would be less than both because a valence electron of potassium is farther from the nucleus than one of either chlorine or argon.

30. Which net ionic equation below represents a possible reaction that takes place when a strip of magnesium metal is oxidized by a solution of chromium (III) nitrate?

(A) $Mg(s) + Cr(NO_3)_3(aq) \rightarrow Mg^{2+}(aq) + Cr^{3+}(aq) + 3NO_3^-(aq)$
(B) $3Mg(s) + 2Cr^{3+} \rightarrow 3Mg^{2+} + 2Cr(s)$
(C) $Mg(s) + Cr^{3+} \rightarrow Mg^{2+} + Cr(s)$
(D) $3Mg(s) + 2Cr(NO_3)_3(aq) \rightarrow 3Mg^{2+}(aq) + 2Cr(s) + NO_3^-(aq)$

31. $PCl_3(g) + Cl_2(g) \leftrightarrow PCl_5(g) \quad \Delta H = -92.5$ kJ/mol

In which of the following ways could the reaction above be manipulated to create more product?

(A) Decreasing the concentration of PCl_3
(B) Increasing the pressure
(C) Increasing the temperature
(D) None of the above

GO ON TO THE NEXT PAGE.

32. A pure solid substance is heated strongly. It first melts into a liquid, then boils and becomes a gas. Which of the following heating curves correctly shows the relationship between temperature and heat added?

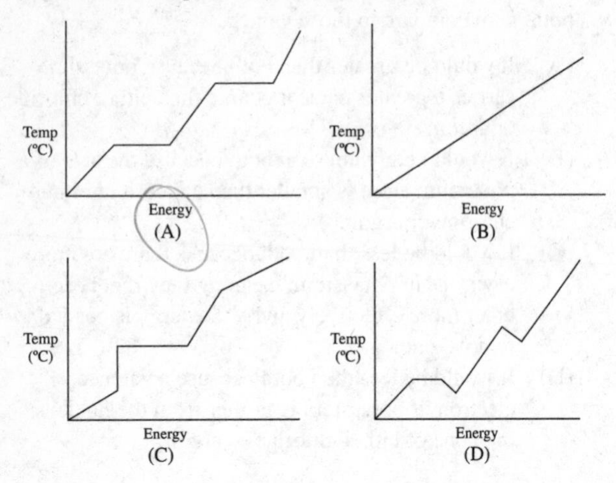

33. Consider the following reaction showing photosynthesis:

$$6CO_2(g) + 6H_2O(l) \rightarrow C_6H_{12}O_6(s) + 6O_2(g)$$
$$\Delta H = +\ 2800 \text{ kJ/mol}$$

Which of the following is true regarding the thermal energy in this system?

(A) It is transferred from the surroundings to the reaction.
(B) It is transferred from the reaction to the surroundings.
(C) It is transferred from the reactants to the products.
(D) It is transferred from the products to the reactants.

34. $$SO_2Cl_2 \rightarrow SO_2(g) + Cl_2(g)$$

At 600 K, SO_2Cl_2 will decompose to form sulfur dioxide and chlorine gas via the above equation. If the reaction is found to be first order overall, which of the following will cause an increase in the half life of SO_2Cl_2?

(A) Increasing the initial concentration of SO_2Cl_2
(B) Increasing the temperature at which the reaction occurs
(C) Decreasing the overall pressure in the container
(D) None of these will increase the half life.

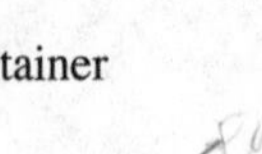

Questions 35-38 refer to the following reaction.

$$2SO_2(g) + O_2(g) \rightarrow 2SO_3(g)$$

4.0 mol of gaseous SO_2 and 6.0 mol of O_2 gas are allowed to react in a sealed container.

35. Which particulate drawing best represents the contents of the flask after the reaction goes to completion?

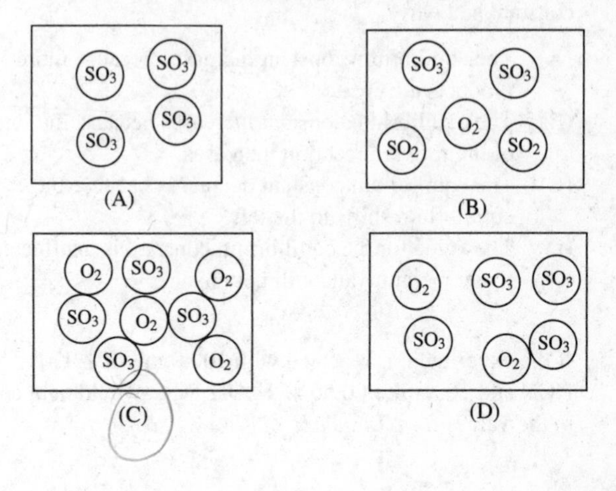

36. If the temperature remains constant, what percentage of the original pressure will the final pressure in the container be equal to?

(A) 67%
(B) 80%
(C) 100%
(D) 133%

37. At a given point in the reaction, all three gases are present at the same temperature. Which gas molecules will have the highest velocity and why?

(A) The O_2 molecules, because they have the least mass
(B) The O_2 molecules, because they are the smallest
(C) The SO_3 molecules, because they are products in the reaction
(D) Molecules of all three gases will have the same speed because they have the same temperature.

GO ON TO THE NEXT PAGE.

38. Under which of the following conditions would the gases in the container most deviate from ideal conditions and why?

(A) Low pressures, because the gas molecules would be spread far apart
(B) High pressures, because the gas molecules will be colliding frequently
(C) Low temperatures, because the intermolecular forces between the gas molecules would increase
(D) High temperatures, because the gas molecules are moving too fast to interact with each other

---

39. Four different acids are added to beakers of water, and the following diagrams represent the species present in each solution at equilibrium. Which acid has the highest pH?

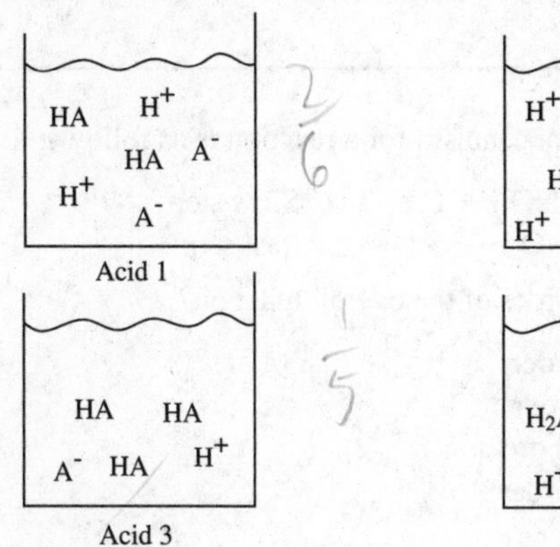

H+ H+
A-
H+ A-
H+ A- A-
Acid 2

H2A H+ A-
H+ A- H2A
Acid 4

(A) Acid 1
(B) Acid 2
(C) Acid 3
(D) Acid 4

40. Which expression below should be used to calculate the mass of copper that can be plated out of a 1.0 *M* $Cu(NO_3)_2$ solution using a current of 0.75 A for 5.0 minutes?

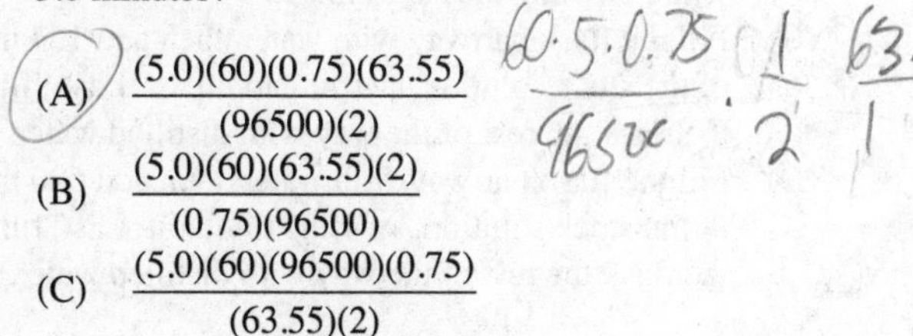

(A) $\frac{(5.0)(60)(0.75)(63.55)}{(96500)(2)}$

(B) $\frac{(5.0)(60)(63.55)(2)}{(0.75)(96500)}$

(C) $\frac{(5.0)(60)(96500)(0.75)}{(63.55)(2)}$

(D) $\frac{(5.0)(60)(96500)(63.55)}{(0.75)(2)}$

41. Lewis diagrams for the nitrate and nitrite ions are shown below. Choose the statement that correctly describes the relationship between the two ions in terms of bond length and bond energy.

Nitrate

Nitrite

(A) Nitrite has longer and stronger bonds than nitrate.
(B) Nitrite has longer and weaker bonds than nitrate.
(C) Nitrite has shorter and stronger bonds than nitrate.
(D) Nitrite has shorter and weaker bonds than nitrate.

42. Examining data obtained from mass spectrometry supports which of the following?

(A) The common oxidation states of elements
(B) Atomic size trends within the periodic table
(C) Ionization energy trends within the periodic table
(D) The existence of isotopes

43. A 2.0 L flask holds 0.40 g of helium gas. If the helium is evacuated into a larger container while the temperature is held constant, what will the effect on the entropy of the helium be?

(A) It will remain constant as the number of helium molecules does not change.
(B) It will decrease as the gas will be more ordered in the larger flask.
(C) It will decrease because the molecules will collide with the sides of the larger flask less often than they did in the smaller flask.
(D) It will increase as the gas molecules will be more dispersed in the larger flask.

GO ON TO THE NEXT PAGE.

Questions 44-48 refer to the following.

NO_2 gas is placed in a sealed, evacuated container and allowed to decompose via the following equation:

$$2NO_2(g) \leftrightarrow 2NO(g) + O_2(g)$$

The graph below indicates the change in concentration for each species over time.

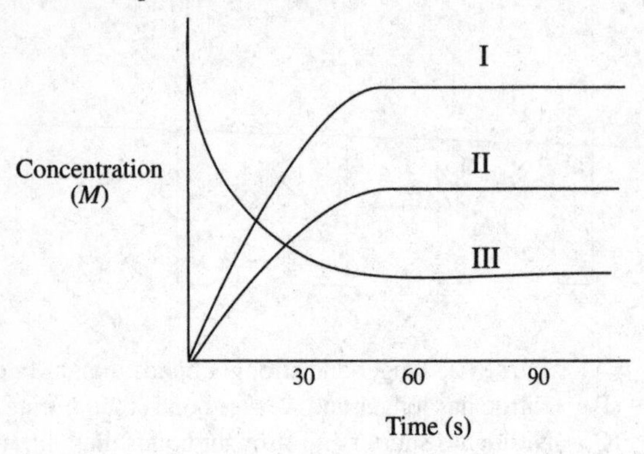

44. Using the numbers of the lines on the graph, identify which line belongs to which species.

	Line I	Line II	Line III
(A)	NO_2	NO	O_2
(B)	O_2	NO_2	NO
(C)	NO	O_2	NO_2
(D)	O_2	NO	NO_2

45. What is happening to the rate of the forward reaction at $t = 60$ s?

(A) It is increasing.
(B) It is decreasing.
(C) It is remaining constant.
(D) It is zero.

46. As the reaction progresses, what happens to the value of the equilibrium constant K_p if the temperature remains constant?

(A) It stays constant.
(B) It increases exponentially.
(C) It increases linearly.
(D) It decreases exponentially.

47. What would happen to the slope of the NO_2 line if additional O_2 were injected into the container?

(A) It would increase, then level off.
(B) It would decrease, then level off.
(C) It would remain constant.
(D) It would increase, then decrease.

48. Using the graph, how could you determine the instantaneous rate of disappearance of NO_2 at $t = 30$ s?

(A) By determining the area under the graph at $t = 30$ s
(B) By taking the slope of a line tangent to the NO_2 curve at $t = 30$ s
(C) By using the values at $t = 30$ s and plugging them into the K_p expression
(D) By measuring the overall gas pressure in the container at $t = 30$ s

---

49. A proposed mechanism for a reaction is as follows:

$NO_2 + F_2 \rightarrow NO_2F + F$ Slow step
$F + NO_2 \rightarrow NO_2F$ Fast step

What is the order of the overall reaction?

(A) Zero order
(B) First order
(C) Second order
(D) Third order

50. Starting with a stock solution of 18.0 M H_2SO_4, what is the proper procedure to create a 1.00 L sample of a 3.0 M solution of H_2SO_4 in a volumetric flask?

(A) Add 167 mL of the stock solution to the flask, then fill the flask the rest of the way with distilled water while swirling the solution.
(B) Add 600 mL of the stock solution to the flask, then fill the flask the rest of the way with distilled water while swirling the solution.
(C) Fill the flask partway with water, then add 167 mL of the stock solution, swirling to mix it. Last, fill the flask the rest of the way with distilled water.
(D) Fill the flask partway with water, then add 600 mL of the stock solution, swirling to mix it. Last, fill the flask the rest of the way with distilled water.

GO ON TO THE NEXT PAGE.

51. The reaction shown in the diagram below is accompanied by a large increase in temperature. If all molecules shown are in their gaseous state, which statement accurately describes the reaction?

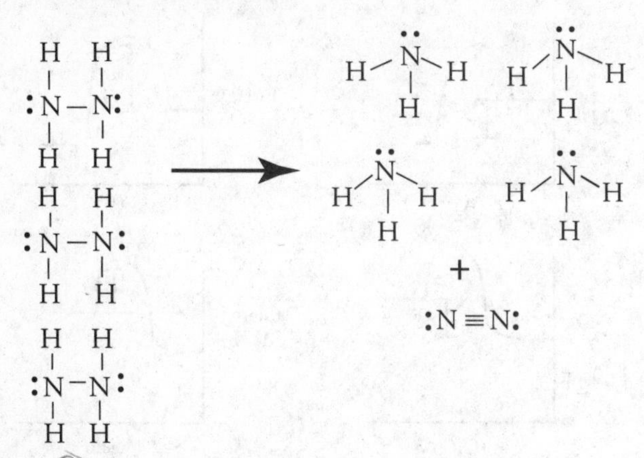

(A) It is an exothermic reaction in which entropy increases.
(B) It is an exothermic reaction in which entropy decreases.
(C) It is an endothermic reaction in which entropy increases.
(D) It is an endothermic reaction in which entropy decreases.

52. A student mixes equimolar amounts of KOH and $Cu(NO_3)_2$ in a beaker. Which of the following particulate diagrams correctly shows all species present after the reaction occurs?

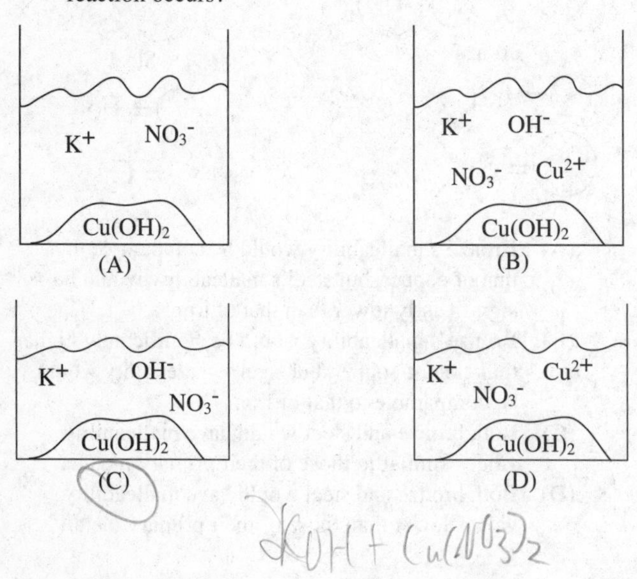

Questions 53-55 refer to the following.

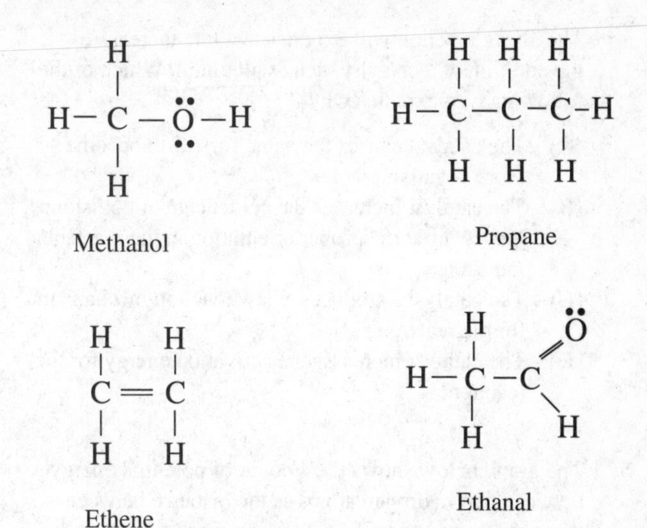

53. Based on the strength of the intermolecular forces in each substance, estimate from greatest to smallest the vapor pressures of each substance in liquid state at the same temperature.

(A) Propane > Ethanal > Ethene > Methanol
(B) Ethene > Propane > Ethanal > Methanol
(C) Ethanal > Methanol > Ethene > Propane
(D) Methanol > Ethanal > Propane > Ethene

54. When in liquid state, which two substances are most likely to be miscible with water?

(A) Propane and ethene
(B) Methanol and propane
(C) Ethene and ethanal
(D) Methanol and ethanal

55. Between propane and ethene, which will likely have the higher boiling point and why?

(A) Propane, because it has a greater molar mass
(B) Propane, because it has a more polarizable electron cloud
(C) Ethene, because of the double bond
(D) Ethene, because it is smaller in size

GO ON TO THE NEXT PAGE.

56. $4NH_3(g) + 5O_2(g) \rightarrow 4NO(g) + 6H_2O(g)$

The above reaction will experience a rate increase by the addition of a cataylst such as platinum. Which of the following best explains why?

(A) The catalyst causes the value for ΔG to become more negative.
(B) The catalyst increases the percentage of collisions that occur at the proper orientation in the reactant molecules.
(C) The catalyst introduces a new reaction mechanism for the reaction.
(D) The catalyst increases the activation energy for the reaction.

57. The graph below shows the amount of potential energy between two hydrogen atoms as the distance between them changes. At which point in the graph would a molecule of H_2 be the most stable?

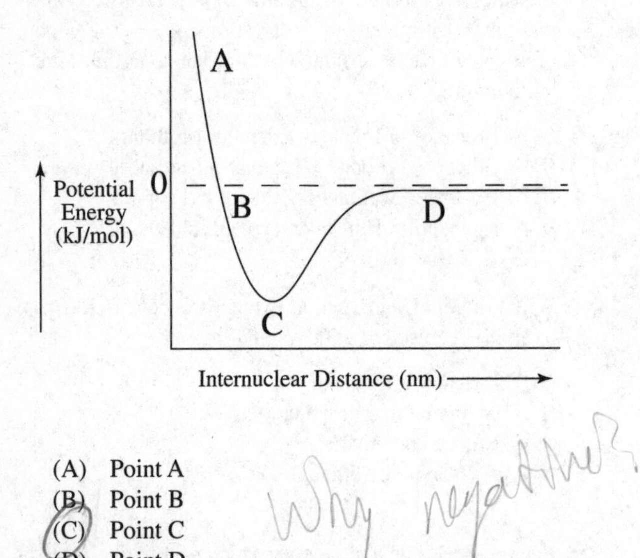

(A) Point A
(B) Point B
(C) Point C
(D) Point D

58.
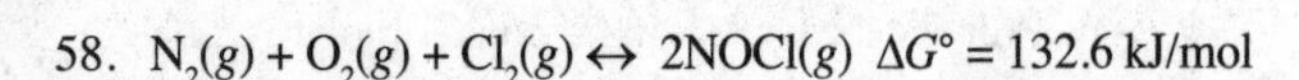
$N_2(g) + O_2(g) + Cl_2(g) \leftrightarrow 2NOCl(g)$ $\Delta G° = 132.6$ kJ/mol

For the equilibrium above, what would happen to the value of $\Delta G°$ if the concentration of N_2 were to increase and why?

(A) It would increase as the reaction would become more thermodynamically favored.
(B) It would increase as the reaction would shift right and create more products.
(C) It would decrease because there are more reactants present.
(D) It would stay the same because the value of K_{eq} would not change.

59. $C(s) + 2S(s) \rightarrow CS_2(l)$ $\Delta H = +92.0$ kJ/mol

Which of the following energy level diagrams gives an accurate representation of the above reaction?

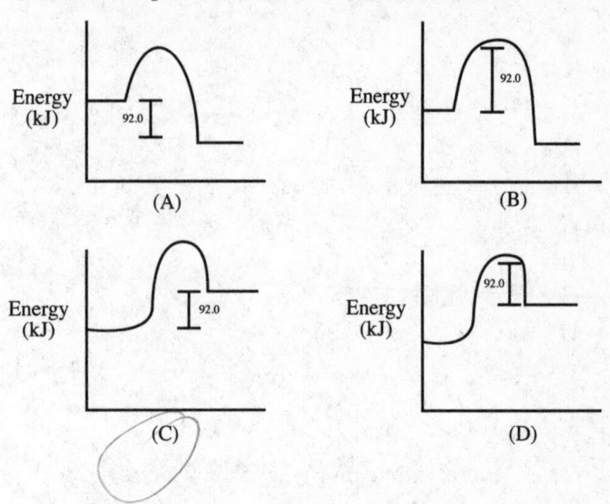

60. Two alloys are shown in the diagrams below—bronze, and steel. Which of the following correctly describes the malleability of both alloys compared to their primary metals?

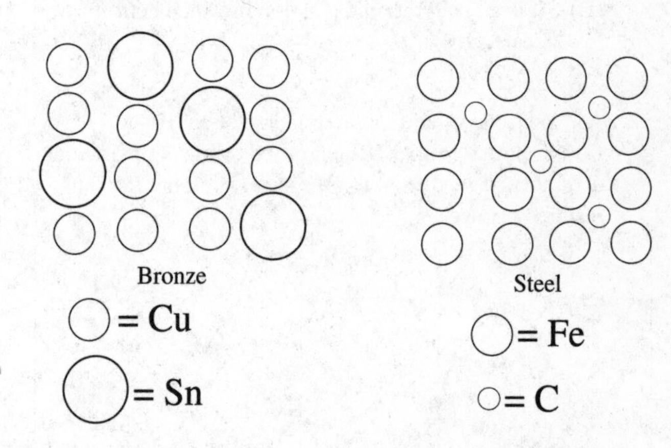

(A) Bronze's malleability would be comparable to that of copper, but steel's malleability would be significantly lower than that of iron.
(B) Bronze's malleability would be significantly higher than that of copper, but steel's malleability would be comparable to that of iron.
(C) Both bronze and steel would have malleability values similar to those of their primary metals.
(D) Both bronze and steel would have malleability values lower than those of their primary metals.

END OF SECTION I

CHEMISTRY
SECTION II
Time—1 hour and 45 minutes

General Instructions

Calculators, including those with programming and graphing capabilities, may be used. However, calculators with typewriter-style (QWERTY) keyboards are NOT permitted.

Pages containing a periodic table and equations commonly used in chemistry will be available for your use.

You may write your answers with either a pen or a pencil. Be sure to write CLEARLY and LEGIBLY. If you make an error, you may save time by crossing it out rather than trying to erase it.

Write all your answers in the essay booklet. Number your answers as the questions are numbered in the examination booklet.

GO ON TO THE NEXT PAGE.

INFORMATION IN THE TABLE BELOW AND ON THE FOLLOWING PAGES MAY BE USEFUL IN ANSWERING THE QUESTIONS IN THIS SECTION OF THE EXAMINATION.

DO NOT DETACH FROM BOOK.

PERIODIC TABLE OF THE ELEMENTS

1 **H** 1.008																	2 **He** 4.00
3 **Li** 6.94	4 **Be** 9.01											5 **B** 10.81	6 **C** 12.01	7 **N** 14.01	8 **O** 16.00	9 **F** 19.00	10 **Ne** 20.18
11 **Na** 22.99	12 **Mg** 24.30											13 **Al** 26.98	14 **Si** 28.09	15 **P** 30.97	16 **S** 32.06	17 **Cl** 35.45	18 **Ar** 39.95
19 **K** 39.10	20 **Ca** 40.08	21 **Sc** 44.96	22 **Ti** 47.90	23 **V** 50.94	24 **Cr** 52.00	25 **Mn** 54.94	26 **Fe** 55.85	27 **Co** 58.93	28 **Ni** 58.69	29 **Cu** 63.55	30 **Zn** 65.39	31 **Ga** 69.72	32 **Ge** 72.59	33 **As** 74.92	34 **Se** 78.96	35 **Br** 79.90	36 **Kr** 83.80
37 **Rb** 85.47	38 **Sr** 87.62	39 **Y** 88.91	40 **Zr** 91.22	41 **Nb** 92.91	42 **Mo** 95.94	43 **Tc** (98)	44 **Ru** 101.1	45 **Rh** 102.91	46 **Pd** 106.42	47 **Ag** 107.87	48 **Cd** 112.41	49 **In** 114.82	50 **Sn** 118.71	51 **Sb** 121.75	52 **Te** 127.60	53 **I** 126.91	54 **Xe** 131.29
55 **Cs** 132.91	56 **Ba** 137.33	57 ***La** 138.91	72 **Hf** 178.49	73 **Ta** 180.95	74 **W** 183.85	75 **Re** 186.21	76 **Os** 190.2	77 **Ir** 192.2	78 **Pt** 195.08	79 **Au** 196.97	80 **Hg** 200.59	81 **Tl** 204.38	82 **Pb** 207.2	83 **Bi** 208.98	84 **Po** (209)	85 **At** (210)	86 **Rn** (222)
87 **Fr** (223)	88 **Ra** 226.02	89 **†Ac** 227.03	104 **Rf** (261)	105 **Db** (262)	106 **Sg** (266)	107 **Bh** (264)	108 **Hs** (277)	109 **Mt** (268)	110 **Ds** (271)	111 **Rg** (272)							

*Lanthanide Series	58 **Ce** 140.12	59 **Pr** 140.91	60 **Nd** 144.24	61 **Pm** (145)	62 **Sm** 150.4	63 **Eu** 151.97	64 **Gd** 157.25	65 **Tb** 158.93	66 **Dy** 162.50	67 **Ho** 164.93	68 **Er** 167.26	69 **Tm** 168.93	70 **Yb** 173.04	71 **Lu** 174.97
†Actinide Series	90 **Th** 232.04	91 **Pa** 231.04	92 **U** 238.03	93 **Np** (237)	94 **Pu** (244)	95 **Am** (243)	96 **Cm** (247)	97 **Bk** (247)	98 **Cf** (251)	99 **Es** (252)	100 **Fm** (257)	101 **Md** (258)	102 **No** (259)	103 **Lr** (262)

GO ON TO THE NEXT PAGE.

ADVANCED PLACEMENT CHEMISTRY EQUATIONS AND CONSTANTS

Throughout the test the following symbols have the definitions specified unless otherwise noted.

L, mL	= liter(s), milliliter(s)	mm Hg	= millimeters of mercury
g	= gram(s)	J, kJ	= joule(s), kilojoule(s)
nm	= nanometer(s)	V	= volt(s)
atm	= atmosphere(s)	mol	= mole(s)

ATOMIC STRUCTURE

$$E = h\nu$$
$$c = \lambda\nu$$

E = energy
ν = frequency
λ = wavelength

Planck's constant, $h = 6.626 \times 10^{-34}$ J s
Speed of light, $c = 2.998 \times 10^{8}$ m s^{-1}
Avogadro's number = 6.022×10^{23} mol^{-1}
Electron charge, $e = -1.602 \times 10^{-19}$ coulombs

EQUILIBRIUM

$$K_c = \frac{[C]^c[D]^d}{[A]^a[B]^b}, \text{ where } a\,A + b\,B \rightleftarrows c\,C + d\,D$$

$$K_p = \frac{(P_C)^c(P_D)^d}{(P_A)^a(P_B)^b}$$

$$K_a = \frac{[H^+][A^-]}{[HA]}$$

$$K_b = \frac{[OH^-][HB^+]}{[B]}$$

$$K_w = [H^+][OH^-] = 1.0 \times 10^{-14} \text{ at } 25°C$$
$$= K_a \times K_b$$

$$pH = -\log[H^+],\ pOH = -\log[OH^-]$$

$$14 = pH + pOH$$

$$pH = pK_a + \log\frac{[A^-]}{[HA]}$$

$$pK_a = -\log K_a,\ pK_b = -\log K_b$$

Equilibrium Constants

K_c (molar concentrations)
K_p (gas pressures)
K_a (weak acid)
K_b (weak base)
K_w (water)

KINETICS

$$\ln[A]_t - \ln[A]_0 = -kt$$

$$\frac{1}{[A]_t} - \frac{1}{[A]_0} = kt$$

$$t_{1/2} = \frac{0.693}{k}$$

k = rate constant
t = time
$t_{1/2}$ = half-life

GO ON TO THE NEXT PAGE.

GASES, LIQUIDS, AND SOLUTIONS

$$PV = nRT$$

$$P_A = P_{total} \times X_A, \text{ where } X_A = \frac{\text{moles A}}{\text{total moles}}$$

$$P_{total} = P_A + P_B + P_C + \ldots$$

$$n = \frac{m}{M}$$

$$K = °C + 273$$

$$D = \frac{m}{V}$$

$$KE \text{ per molecule} = \frac{1}{2}mv^2$$

Molarity, M = moles of solute per liter of solution

$$A = abc$$

P = pressure
V = volume
T = temperature
n = number of moles
m = mass
M = molar mass
D = density
KE = kinetic energy
v = velocity
A = absorbance
a = molar absorptivity
b = path length
c = concentration

Gas constant, $R = 8.314 \text{ J mol}^{-1}\text{K}^{-1}$
$= 0.08206 \text{ L atm mol}^{-1}\text{K}^{-1}$
$= 62.36 \text{ L torr mol}^{-1}\text{K}^{-1}$
1 atm = 760 mm Hg
= 760 torr
STP = 0.00°C and 1.000 atm

THERMOCHEMISTRY/ ELECTROCHEMISTRY

$$q = mc\Delta T$$

$$\Delta S° = \sum S° \text{ products} - \sum S° \text{ reactants}$$

$$\Delta H° = \sum \Delta H_f° \text{ products} - \sum \Delta H_f° \text{ reactants}$$

$$\Delta G° = \sum \Delta G_f° \text{ products} - \sum \Delta G_f° \text{ reactants}$$

$$\Delta G° = \Delta H° - T\Delta S°$$

$$= -RT \ln K$$

$$= -nFE°$$

$$I = \frac{q}{t}$$

q = heat
m = mass
c = specific heat capacity
T = temperature
$S°$ = standard entropy
$H°$ = standard enthalpy
$G°$ = standard free energy
n = number of moles
$E°$ = standard reduction potential
I = current (amperes)
q = charge (coulombs)
t = time (seconds)

Faraday's constant, F = 96,485 coulombs per mole of electrons

$$1 \text{ volt} = \frac{1 \text{ joule}}{1 \text{ coulomb}}$$

GO ON TO THE NEXT PAGE.

CHEMISTRY

Section II

(Total time—105 minutes)

YOU MAY USE YOUR CALCULATOR IN THIS SECTION

THE METHODS USED AND THE STEPS INVOLVED IN ARRIVING AT YOUR ANSWERS MUST BE SHOWN CLEARLY. It is to your advantage to do this since you may obtain partial credit if you do, and you will receive little or no credit if you do not. Attention should be paid to significant figures.

1. A student is tasked with determining the identity of an unknown carbonate compound with a mass of 1.89 g. The compound is first placed in water, where it dissolves completely. The K_{sp} value for several carbonate-containing compounds are given below.

Compound	K_{sp}
Lithium carbonate	8.15×10^{-4}
Nickel (II) carbonate	1.42×10^{-7}
Strontium carbonate	5.60×10^{-10}

(a) In order to precipitate the maximum amount of the carbonate ions from solution, which of the following should be added to the carbonate solution: $LiNO_3$, $Ni(NO_3)_2$, or $Sr(NO_3)_2$? Justify your answer.

(b) For the carbonate compound that contains the cation chosen in part (a), determine the concentration of each ion of that compound in solution at equilibrium.

(c) When mixing the solution, should the student ensure the carbonate solution or the nitrate solution is in excess? Justify your answer.

(d) After titrating sufficient solution to precipitate out all of the carbonate ions, the student filters the solution before placing it in a crucible and heating it to drive off the water. After several heatings, the final mass of the precipitate remains constant and is determined to be 2.02 g.

(i) Determine the number of moles of precipitate.

(ii) Determine the mass of carbonate present in the precipitate.

(e) Determine the percent, by mass, of carbonate in the original sample.

(f) Is the original compound most likely lithium carbonate, sodium carbonate, or potassium carbonate? Justify your answer.

GO ON TO THE NEXT PAGE.

2. The **unbalanced** reaction between potassium permanganate and acidified iron (II) sulfate is a redox reaction that proceeds as follows:

$$H^+(aq) + Fe^{2+}(aq) + MnO_4^-(aq) \rightarrow Mn^{2+}(aq) + Fe^{3+}(aq) + H_2O(l)$$

(a) Provide the equations for both half-reactions that occur below:
(i) Oxidation half-reaction
(ii) Reduction half-reaction

(b) What is the balanced net ionic equation?

A solution of 0.150 *M* potassium permanganate is placed in a buret before being titrated into a flask containing 50.00 mL of iron (II) sulfate solution of unknown concentration. The following data describes the colors of the various ions in solution:

Ion	Color in solution
H^+	Colorless
Fe^{2+}	Pale Green
MnO_4^-	Dark Purple
Mn^{2+}	Colorless
Fe^{3+}	Yellow
K^+	Colorless
SO_4^{2-}	Colorless

(c) Describe the color of the solution in the flask at the following points:
(i) Before titration begins
(ii) During titration prior to the endpoint
(iii) At the endpoint of the titration

(d) (i) If 15.55 mL of permanganate are added to reach the endpoint, what is the initial concentration of the iron (II) sulfate?
(ii) The actual concentration of the $FeSO_4$ is 0.250 *M*. Calculate the percent error.

(e) Could the following errors have led to the experimental result deviating in the direction that it did? You must justify your answers quantitatively.
(i) 55.0 mL of $FeSO_4$ was added to the flask prior to titration instead of 50.0 mL.
(ii) The concentration of the potassium permanganate was actually 0.160 *M* instead of 0.150 *M*.

GO ON TO THE NEXT PAGE.

3. Ammonia gas reacts with dinitrogen monoxide via the following reaction:

$$2NH_3(g) + 3N_2O(g) \rightarrow 4N_2(g) + 3H_2O(g)$$

The absolute entropy values for the varying substances are listed in the table below.

Substance	$S°$ (J/mol·K)
$NH_3(g)$	193
$N_2O(g)$	220
$N_2(g)$	192
$H_2O(g)$	189

(a) Calculate the entropy value for the overall reaction.

Several bond enthalpies are listed in the table below.

Bond	Enthalpy (kJ/mol)	Bond	Enthalpy (kJ/mol)
N–H	388	N=N	409
N–O	210	N≡N	941
N=O	630	O–H	463

(b) Calculate the enthalpy value for the overall reaction.
(c) Is this reaction thermodynamically favored at 25°C? Justify your answer.
(d) If 25.00 g of NH_3 reacts with 25.00 g of N_2O:
(i) Will energy be released or absorbed?
(ii) What is the magnitude of the energy change?
(e) On the reaction coordinates below, draw a line showing the progression of this reaction. Label both ΔH and E_a on the graph.

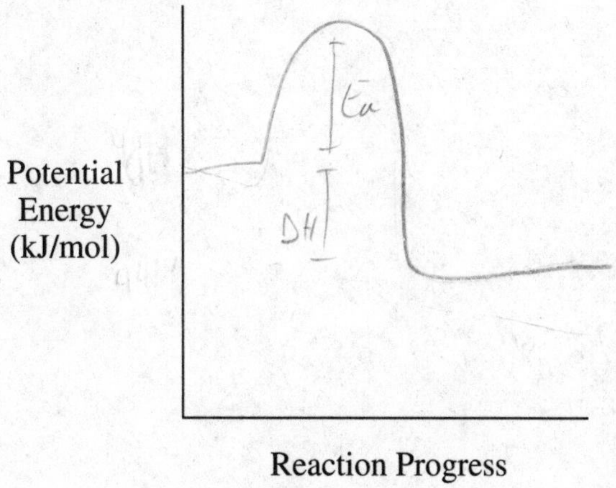

Potential Energy (kJ/mol)

Reaction Progress

GO ON TO THE NEXT PAGE.

4. The acetyl ion has a formula of $C_2H_3O^-$ and two possible Lewis electron-dot diagram representations:

```
H    H                      H
|    |    ..                |      ..   ..
C =  Cx - O :           H - C  -   Cx = O
|         ..                |           ..
H                           H
```

(a) Using formal charge, determine which structure is the most likely correct structure.
(b) For carbon atom "x" in the structure you chose:
  (i) What is the hybridization around the atom?
  (ii) How many sigma and pi bonds has the atom formed?
(c) A hydrogen ion attaches itself to the the acetyl ion, creating C_2H_4O. Draw the Lewis diagram of the new molecule.

5. A student tests the conductivity of three different acid samples, each with a concentration of 0.10 M and a volume of 20.0 mL. The conductivity was recorded in microsiemens per centimeter in the table below:

Sample	Conductivity (μS/cm)
1	26,820
2	8655
3	35,120

(a) The three acids are known to be HCl, H_2SO_4, and H_3PO_4. Identify which sample is which acid. Justify your answer.
(b) The HCl solution is then titrated with a 0.150 M solution of the weak base methylamine, CH_3NH_2. ($K_b = 4.38 \times 10^{-4}$)
  (i) Write out the net ionic equation for this reaction.
  (ii) Determine the pH of the solution after 20.0 mL of methylamine has been added.

GO ON TO THE NEXT PAGE.

6. The photoelectron spectrum of an element is given below:

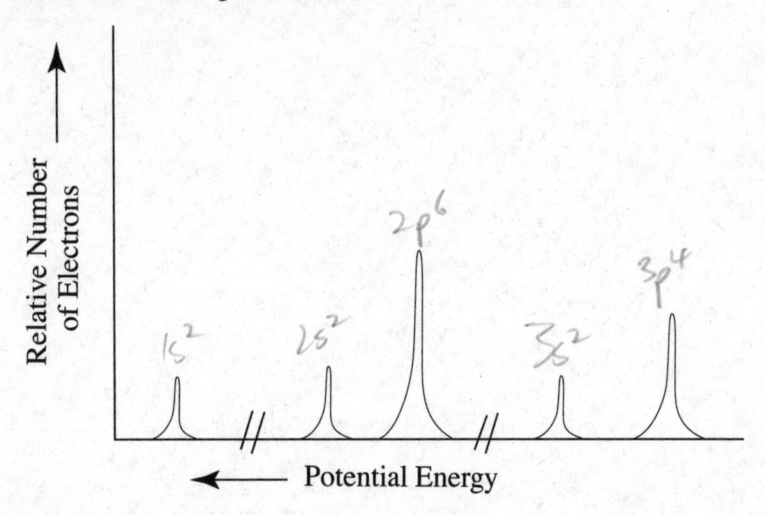

(a) Identify the element this spectra most likely belongs to and write out its full electron configuration.

(b) Using your knowledge of atomic structure, explain the following:

  (i) The reason for the three discrete areas of ionization energies

  (ii) The justification for there being a total of five peaks

  (iii) The relative heights of the peaks when compared to each other

7. Hydrazine (N_2H_4) can be produced commercially via the Raschig process. The following is a proposed mechanism:

Step 1: $NH_3(aq) + OCl^-(aq) \rightarrow NH_2Cl(aq) + OH^-$

Step 2: $NH_2Cl(aq) + NH_3(aq) \rightarrow N_2H_5^+(aq) + Cl^-$

Step 3: $N_2H_5^+(aq) + OH^- \rightarrow N_2H_4(aq) + H_2O(l)$

(a) (i) What is the equation for the overall reaction?

  (ii) Identify any catalysts or intermediates from the reaction mechanism.

(b) The rate law for the reaction is determined to be rate = $k[NH_3][OCl^-]$

  (i) Which elementary step is the slowest one? Justify your answer.

  (ii) If the reaction is measured over the course of several minutes, what would the units of the rate constant be?

STOP

END OF EXAM

Practice Test 1: Answers and Explanations

PRACTICE TEST 1 MULTIPLE-CHOICE ANSWER KEY

1. A
2. D
3. D
4. C
5. D
6. C
7. B
8. B
9. C
10. A
11. D
12. A
13. D
14. D
15. B
16. C
17. B
18. C
19. B
20. D
21. D
22. B
23. A
24. A
25. C
26. C
27. C
28. A
29. D
30. B
31. B
32. A
33. A
34. D
35. C
36. B
37. A
38. C
39. C
40. A
41. C
42. D
43. D
44. C
45. C
46. A
47. A
48. B
49. C
50. C
51. A
52. D
53. B
54. D
55. B
56. C
57. C
58. D
59. C
60. D

Section I—Multiple-Choice Answers and Explanations

1. **A** The reaction is endothermic, meaning energy is a reactant (appears on the left side of the equation). Adding stress to the left side increases the rate of the forward reaction, creating more products.

2. **D** Weak bases do not ionize fully in solution, and most of the methylamine molecules will not deprotonate. The hydroxide and conjugate acid ions are created in a 1:1 ratio and therefore will be equal.

3. **D** Moving across a period, atomic size decreases. Therefore, element Z will be farthest to the right (have the most protons), and thus will have the highest electronegativity value.

4. **C** There are 1 mole of N_2, 1 mole of Ar, and 2 moles of water in the container. The mole fraction of nitrogen is: 1/4 = 0.25.

   $P_{N_2} = (X_{N_2})(P_{total})$

   $P_{N_2} = (0.25)(2.0) = 0.50$ atm

5. **D** When a covalent substance undergoes a phase change, the bonds between the various molecules inside the substance break apart.

6. **C** The oxidation state of copper changes from +1 to 0, meaning it has gained electrons and is being reduced, and reduction occurs at the cathode. Zinc's oxidation state changes from 0 to +2, meaning it has lost electrons and is being oxidized, which occurs at the anode.

7. **B** 127 g is equal to 2 moles of copper, which is what appears on the balanced equation. To change one mole of copper from +1 to 0, 1 mole of electrons is required. Twice as many moles being created means twice as many electrons are needed.

8. **B** $E_{cell} = E_{red} + E_{ox}$

   $1.28\text{ V} = 0.52\text{ V} + E_{ox}$

   $E_{ox} = 0.76\text{ V}$

   $-E_{ox} = E_{red}$

   $E_{red} = -0.76\text{ V}$

9. **C** As the reaction progresses, $[Cu^+]$ will decrease and $[Zn^{2+}]$ will increase. With a lower concentration on the reactants side and a higher concentration on the products side, the reaction will shift left, decreasing the overall potential of the reaction.

10. **A** The electron transfer does not happen across the salt bridge, eliminating (C) and (D). As the reaction progresses and $[Cu^+]$ decreases in the copper half-cell, positively charged sodium ions are transferred in to keep the charge balanced within the half-cell.

11. **D** When $Q > K_c$, the numerator of the equilibrium expression (the product concentration) is too big, and the equation shifts to the left. This is true for both (A) and (B), meaning [NO] would decrease. When $Q < K_c$, the numerator/product concentrations need to increase. This is the case in (C) and (D), but NO(g) is only a product in (D).

12. **A**

:F: — S — F: / :F: / :F: :Cl: — P — Cl: / :Cl / :Cl: / :Cl: :F: — B — :F: / :F: O=C=O

The only tricky bit here is to remember that boron is considered stable with only six electrons in its valence shell.

13. **D** At a higher temperature, the average velocity of the gas molecules would be greater. Additionally, they would have a greater spread of potential velocities, which would lead to a wider curve.

14. **D** Nitrogen only has two shells of electrons while phosphorus has three, making nitrogen smaller and more able to attract additional electrons, meaning a higher electronegativity. Nitrogen and oxygen both have two shells, but oxygen has more protons and an effective nuclear charge of +6 vs. nitrogen's effective nuclear charge of +5. Thus, oxygen has a higher electronegativity.

15. **B** First, the mass of the oxygen must be calculated: 29.05 g – 18.02 g – 3.03 g = 8.00 g.

Converting each of those to moles yields 0.5 moles of oxygen, 1.5 moles of carbon, and 3.0 moles of hydrogen. Thus, for every one oxygen atom there are three carbon atoms and six hydrogen atoms.

16. **C** During sections I and II, the following reaction occurs: $H_2CO_3(aq) + OH^-(aq) \leftrightarrow HCO_3^-(aq) + H_2O(l)$. The endpoint of that is reached when all H_2CO_3 has reacted, meaning that in sections III and IV the following occurs: $HCO_3^-(aq) + OH^-(aq) \leftrightarrow CO_3^{2-}(aq) + H_2O(l)$.

17. **B** $pH = pK_a + [HCO_3^-]/[H_2CO_3]$.

At a point in the graph where half of all the acid has reacted, the last part of the Henderson-Hasselbalch equation cancels out, leaving pH = pK_a. The first equivalence point occurs at a volume of 10 mL, and thus the half-equivalence point is at a volume of 5 mL. The pH at this point is 4, so: $4 = -\log K_a$. K_a is somewhere around 10^{-4}.

18. **C** A more concentrated NaOH solution means more moles of NaOH are added per drop, so a lower volume of NaOH would be needed to add enough moles to reach the equivalence point.

19. **B** If extra Ca^{2+} ions are in solution, that means there were not enough CO_3^{2-} ions present for the Ca^{2+} ions to fully react.

20. D Via $K_{eq} = e^{-\frac{\Delta G}{RT}}$, if ΔG is negative the value for K will be greater than one. Via $\Delta G = \Delta H - T\Delta S$, ΔG is always negative when ΔH is negative and ΔS is positive.

21. D The weaker the O–H bond is in an oxoacid, the stronger the acid will be, because the H^+ ions are more likely to dissociate. The O–F bond in HOF is stronger than the O–Cl bond in HOCl because fluorine is smaller (and thus more electronegative) than chlorine. If the O–F bond is stronger, the O–H bond is correspondingly weaker, making HOF the stronger acid.

22. B The oxidation state of nitrogen in NO is +2. If the nitrogen is reduced, that value must get more negative. The oxidation state of nitrogen in N_2 is 0, so that fits the bill. The oxidation state on the nitrogen in the other choices is greater than +2.

23. A A buffer is made up of either a weak acid and its salt or a weak base and its salt. Choice (B) has a strong acid and strong base, (C) has a strong acid and its salt, and (D) has a weak acid and a weak base.

24. A HCN will lose its proton to the hydroxide, creating a conjugate base and water.

25. C The mass of the mixture is 200.0 g—this is from the volume being 200.0 mL and the density of the mixture being 1.0 g/mL. ΔT is 7.0°C and c is 4.2 J/g°C.

$$q = mc\Delta T$$

$$q = (200.0\text{ g})(4.2\text{ J/g°C})(7.0\text{ °C}) = 5880\text{ J} \approx 5.9\text{ kJ}$$

26. C This is an exothermic reaction; therefore heat is generated as a product. An increase in temperature thus causes a shift to the left, decreasing the numerator and increasing the denominator in the equilibrium expression. This decreases the overall value of K.

27. C The NaOH is limiting (0.050 mol vs. 0.100 mol in the original reaction), and adding even more excess HCN will not change the amount of HCN that acutally reacts with OH^-. So, the value for ΔH_{rxn} stays the same. However, the overall mixture will have a greater mass (300.0 g), which means the temperature change will not be as large.

28. A In a polar solvent, polar molecules will be the most soluble (like dissolves like). Of the four options, methanol and acetone would both have dipoles, but those of methanol would be significantly stronger due to the H-bonding.

29. D Potassium's first valence electron is in the fourth energy level, but both chlorine and argon's first valence electron is in the third energy level.

30. B Cr^{3+} needs to gain three electrons to reduce into Cr(s), while Mg(s) loses 2 electrons to oxidize into Mg^{2+}. To balance the electrons, the reduction half-reaction must be multiplied by two and the oxidation half-reaction must be multiplied by three in order to have six electrons on each side. The nitrate ion is a spectator and would not appear in the net ionic equation.

31. **B** Increasing the pressure in an equilibrium reaction with any gas molecules causes a shift to the side with fewer gas molecules—in this case, the product.

32. **A** During a phase change, all of the energy added goes towards breaking the intermolecular forces holding the molecules together. During this time, the speed of the molecules (and thus the temperature) does not rise.

33. **A** In an endothermic reaction, heat is transferred into the reaction system.

34. **D** None of the options would decrease the rate of reaction, which would be required for the half-life of the reactant to increase.

35. **C** To determine the number of moles of SO_3 created, stoichiometry must be used.

$$SO_2\text{: } 4 \text{ mol } SO_2 \times \frac{2 \text{ mol } SO_3}{2 \text{ mol } SO_2} = 4 \text{ mol } SO_3$$

$$O_2\text{: } 6 \text{ mol } O_2 \times \frac{2 \text{ mol } SO_3}{1 \text{ mol } O_2} = 12 \text{ mol } SO_3$$

The oxygen is in excess, and only 2.0 mol of it will react. (As every 2.0 mol of SO_2 react with 1 mol of O_2.) Thus, 4.0 mol of SO_3 are created and 4.0 mol of O_2 remain.

36. **B** If 4.0 mol of SO_3 are created and 4.0 mol of O_2 remain, there are 8.0 mol of gas present after the reaction. Prior to the reaction there were 10.0 mol of gas. If there are 8/10 = 80% as many moles after the reaction, there is also 80% as much pressure.

37. **A** If all gases are at the same temperature, they have the same amount of kinetic energy. Given that KE = ½ mv^2, if all three gases have the same KE, the gas with the least mass must have the highest velocity in order to compensate.

38. **C** One of the assumptions of kinetic molecular theory is that the amount of intermolecular forces between the gas molecules is negligible. If the molecules are moving very slowly, as happens when the temperature is lowered, the IMFs between them are more likely to cause deviations from ideal behavior.

39. **C** The strength of an acid is dependent on the amount it dissociates in solution. A low dissociation is signified by a low presence of hydrogen ions. The weakest acid is 3, (C).

40. **A** $5.0 \text{ min} \times \frac{60.0 \text{ s}}{1.0 \text{ min}} \times \frac{0.75 \text{ C}}{1.0 \text{ s}} \times \frac{1 \text{ mol } e^-}{96500 \text{ C}} \times \frac{1 \text{ mol Cu}}{2 \text{ mol } e^-} \times \frac{63.55 \text{ g Cu}}{1 \text{ mol Cu}}$

41. C Nitrate has a bond order of $\frac{(1 + 3)}{3} = 1.33$. Nitite has a bond order of $\frac{(1 + 2)}{2} = 1.5$. A higher bond order means shorter and stronger bonds.

42. D Mass spectrometry is used to determine the masses for individual atoms of an element. Through mass spectrometry, it is proven that each element has more than one possible mass.

43. D Entropy is a measure of a system's disorder. In a larger flask, the gas molecules will spread farther apart and become more disordered.

44. C The only species that is present at $t = 0$ is the NO_2, allowing us to identify line III. When identifying line I vs. line II, the NO will be generated twice as quickly as the O_2 due to the coefficients, meaning [NO] will increase about twice as quickly as $[O_2]$.

45. C At equilibrium, the concentrations of all species in the reaction are remaining constant, which shows up as a flat line on the graph. The rate of both the forward and reverse reactions are constant at equilibrium.

46. A The only factor that can affect the value of the equilibrium constant is temperature. If the temperature does not change, neither does the equilibrium constant.

47. A If additional O_2 were injected into the container, the reaction would shift left, increasing the amount of NO_2 present. Eventually, the reaction would reach equilibrium again, meaning the lines would level out.

48. B To determine the change in concentration at a specific time, we would need the slope of the line at that point. As the line is curved, the only way to do that (without calculus) is to draw a line tangent to the curve at that point and measure its slope.

49. C The overall rate law is always equal to the rate law for slowest elementary step, which can be determined using the coefficients of the reactants. In this case, rate = $k[NO_2][F_2]$. To get the overall order, we add the exponents in the rate law: $1 + 1 = 2$.

50. C Use $M_1V_1 = M_2V_2$ to determine the necessary volume of stock solution.

$(18.0\ M)V_1 = (3.0\ M)(1.0\ \text{L})$

$V_1 = 0.167\ \text{L} = 167\ \text{mL}$

When creating solutions with acid, you always add some water first, as the process is extremely exothermic and the water will absorb the generated heat.

51. A The temperature increase is indicative of energy being released, meaning the reaction is exothermic. The entropy (disorder) of the system is increasing as it moves from three gas molecules to five.

52. D The overall reaction (excluding spectator ions) is: $2OH^- + Cu^{2+} \rightarrow Cu(OH)_2(s)$. Both the K^+ and the NO_3^- are spectator ions which are present in solution both before and after the reaction. Additionally, if equimolar amounts of the two reactants are initially present, the OH^- will run out before the Cu^{2+}, meaning that some Cu^{2+} ions will also be present in the final solution.

53. B Vapor pressure is dependent on intermolecular forces. The weaker the IMFs are, the easier it is for molecules to escape from the surface of the liquid. To begin, polar molecules have stronger IMFs than nonpolar molecules. Methanol and ethanal are both polar, but methanol has hydrogen bonding meaning it has stronger IMFs (and thus a lower vapor pressure) than ethanal. Ethene and propane are both nonpolar, but propane is larger meaning it is more polarizable than ethene and thus has stronger IMFs and lower vapor pressure.

54. D Water is polar, and using "like dissolves like," we know that only polar solvents will be able to fully mix with it to create a homogenous solution.

55. B Both are nonpolar, but propane has a lot more electrons and thus is more polarizable than ethene.

56. C Catalysts work by creating a new reaction pathway with a lower activation energy than the original pathway.

57. C The molecule would be the most stable when it has the largest attractive potential energy, which is represented by a negative sign. While the magnitude of the potential energy may be larger at (A), it is repulsive at that point because the nuclei are too close together.

58. D Adding (or removing) any species in an equilibrium reaction does not change the equilibrium constant and also does not change the magnitude of the Gibbs free energy at standard conditions.

59. C A positive *H* means the reaction is endothermic, so the products have more bond energy than the reactants. The difference between the energy levels of the products and reactants is equal to ΔH. (The difference between the energy level of the reactants and the top of the hump is the value for the activation energy.)

60. D The goal of creating an alloy is often to make the base metal stronger. This means that alloys of any type have less malleability than the metals from which they are created.

Section II—Free-Response Answers and Explanations

1. A student is tasked with determining the identity of an unknown carbonate compound with a mass of 1.89 g. The compound is first placed in water, where it dissolves completely. The K_{sp} values for several carbonate-containing compounds are given below.

Compound	K_{sp}
Lithium carbonate	8.15×10^{-4}
Nickel (II) carbonate	1.42×10^{-7}
Strontium carbonate	5.60×10^{-10}

(a) In order to precipitate the maximum amount of the carbonate ions from solution, which of the following should be added to the carbonate solution: $LiNO_3$, $Ni(NO_3)_2$, or $Sr(NO_3)_2$? Justify your answer.

The student should use the strontium nitrate. Using it would create strontium carbonate, which has the lowest K_{sp} value. That means it is the least soluble carbonate compound of the three and will precipitate the most possible carbonate ions out of solution.

(b) For the carbonate compound that contains the cation chosen in part (a), determine the concentration of each ion of that compound in solution at equilibrium.

$SrCO_3(s) \rightleftharpoons Sr^{2+}(aq) + CO_3^{2-}(aq)$

$K_{sp} = [Sr^{2+}][CO_3^{2-}]$

$5.60 \times 10^{-10} = (x)(x)$

$x = 2.37 \times 10^{-5}\ M = [Sr^{2+}] = [CO_3^{2-}]$

(c) When mixing the solution, should the student ensure the carbonate solution or the nitrate solution is in excess? Justify your answer.

The ntirate solution should be in excess. In order to create the maximum amount of precipitate, enough strontium ions need to be added to react with all of the carbonate ions originally in solution. Having excess strontium ions in solution after the precipitate forms will not affect the calculated mass of the carbonate in the original sample.

(d) After titrating sufficient solution to precipitate out all of the carbonate ions, the student filters the solution before placing it in a crucible and heating it to drive off the water. After several heatings, he final mass of the precipitate remains constant and is determined to be 2.02 g.

(i) Determine the number of moles of precipitate.

$$2.02 \text{ g SrCO}_3 \times \frac{1 \text{ mol SrCO}_3}{147.63 \text{ g SrCO}_3} = 1.37 \times 10^{-2} \text{ mol SrCO}_3$$

(ii) Determine the mass of carbonate present in the precipitate.

$$1.37 \times 10^{-2} \text{ mol SrCO}_3 \times \frac{1 \text{ mol CO}_3^{2-}}{1 \text{ mol SrCO}_3} \times \frac{60.01 \text{ g CO}_3^{2-}}{1 \text{ mol CO}_3^{2-}} = 0.822\text{g CO}_3^{2-}$$

(e) Determine the percent, by mass, of carbonate in the original sample.

$$\frac{0.82 \text{ g}}{1.89 \text{ g}} \times 100 = 43.4\% \text{ CO}_3^{2-}$$

(f) Is the original compound most likely lithium carbonate, sodium carbonate, or potassium carbonate? Justify your answer.

The mass percent of carbonate in each compound must be compared to the experimentally determined mass percent of carbonate in the sample.

$$\text{Li}_2\text{CO}_3\text{: } \frac{60.01}{73.89} \times 100 = 81.2\%$$

$$\text{Na}_2\text{CO}_3\text{: } \frac{60.01}{105.99} \times 100 = 56.6\%$$

$$\text{K}_2\text{CO}_3\text{: } \frac{60.01}{138.21} \times 100 = 45.4\%$$

The compound is most likely potassium carbonate.

2. The unbalanced reaction between potassium permanganate and acidified iron (II) sulfate is a redox reaction that proceeds as follows:

$$H^+(aq) + Fe^{2+}(aq) + MnO_4^-(aq) \rightarrow Mn^{2+}(aq) + Fe^{3+}(aq) + H_2O(l)$$

(a) Provide the equations for both half-reactions that occur below:

(i) Oxidation half-reaction:

Oxidation: $Fe^{2+}(aq) \rightarrow Fe^{3+}(aq) + e^-$

(ii) Reduction half-reaction:

Reduction: $5e^- + 8H^+(aq) + MnO_4^-(aq) \rightarrow Mn^{2+}(aq) + 4H_2O(l)$

(b) What is the balanced net ionic equation?

The oxidation reaction must be multiplied by a factor of 5 in order for the electrons to balance out. So:

$$5Fe^{2+}(aq) + 8H^{+}(aq) + MnO_4^{-}(aq) \rightarrow 5Fe^{3+}(aq) + Mn^{2+}(aq) + 4H_2O(l)$$

A solution of 0.150 *M* potassium permanganate, a dark purple solution, is placed in a buret before being titrated into a flask containing 50.00 mL of iron (II) sulfate solution of unknown concentration. The following data describes the colors of the various ions in solution:

Ion	Color in solution
H^+	Colorless
Fe^{2+}	Pale Green
MnO_4^-	Dark Purple
Mn^{2+}	Colorless
Fe^{3+}	Yellow
K^+	Colorless
SO_4^{2-}	Colorless

(c) Describe the color of the solution in the flask at the following points:

(i) Before titration begins

The only ions present in the flask are Fe^{2+}, SO_4^{2-}, and H^+. The latter two are colorless, so the solution would be pale green.

(ii) During titration prior to the endpoint

The MnO_4^- is reduced to Mn^{2+} upon entering the flask, and the Fe^{2+} ions are oxidized into Fe^{3+} ions. The solution would become less green as more yellow as more Fe^{3+} ions are formed, as all other ions present are colorless.

(iii) At the endpoint of the titration

After the Fe^{2+} ions have all been oxidized, there is nothing left to donate electrons to the MnO_4^- ions. Therefore, they will no longer be reduced upon entering the flask, and the solution will take on a light purplish/yellow hue due to the mixture of MnO_4^- and Fe^{3+} ions.

(d) (i) If 15.55 mL of permanganate are added to reach the endpoint, what is the initial concentration of the iron (II) sulfate?

First the moles of permanganate added must be calculated:

$$0.150\ M = \frac{n}{0.0155\ \text{L}} \qquad n = 2.33 \times 10^{-3}\ \text{mol}\ MnO_4^-$$

Then the moles of iron (II) can be determined via stoichiometry:

$$2.33 \times 10^{-3}\ \text{mol}\ MnO_4^- \times \frac{5\ \text{mol}\ Fe^{2+}}{1\ \text{mol}\ MnO_4} = 0.0117\ \text{mol}\ Fe^{2+}$$

Finally the concentration of the $FeSO_4$ can be determined:

$$\text{Molarity} = \frac{0.0117\ \text{mol}}{0.050\ \text{L}} = 0.234\ M$$

(ii) The actual concentration of the $FeSO_4$ is 0.250 M. Calculate the percent error.

$$\text{Percent error} = \frac{\text{actual value} - \text{experimental value}}{\text{actual value}} \times 100$$

$$\text{Percent error} = \frac{0.250 - 0.234}{0.250} \times 100 = 6.40\%\ \text{error}$$

(e) Could the following errors have led to the experimental result deviating in the direction that it did? You must justify your answers quantitatively.

(i) 55.0 mL of $FeSO_4$ was added to the flask prior to titration instead of 50.0 mL.

If the volume of $FeSO_4$ was artificially low in the calculations, that would lead to the experimental value for the concentration of $FeSO_4$ being artificially high. As the calculated value for the concentration of $FeSO_4$ was too low, this error source is not supported by data.

(ii) The concentration of the potassium permanganate was actually 0.160 M instead of 0.150 M.

If the molarity of the permanganate was artificially low in the calculations, the moles of permanganate, and by extension, the moles of Fe^{2+} would also be artificially low. This would lead to an artificially low value for the concentration of $FeSO_4$. This matches with the experimental results and is thus supported by the data.

3. Ammonia gas reacts with dinitrogen monoxide via the following reaction:

$$2NH_3(g) + 3N_2O(g) \rightarrow 4N_2(g) + 3H_2O(g)$$

The standard entropy of formation values for the varying substances are listed in the table below.

Substance	$S°$ (J/mol·K)
$NH_3(g)$	193
$N_2O(g)$	220
$N_2(g)$	192
$H_2O(g)$	189

(a) Calculate the entropy value for the overall reaction.

$\Delta S = \Delta S_{products} - \Delta S_{reactants}$

$\Delta S = [3(H_2O) + 4(N_2)] - [2(NH_3) + 3(N_2O)]$

$\Delta S = [3(189) + 4(192)] - [2(193) + 3(220)]$

$\Delta S = 1335 - 1046$

$\Delta S = 289$ J/mol·K

Several bond enthalpies are listed in the table below.

Bond	Enthalpy (kJ/mol)	Bond	Enthalpy (kJ/mol)
N–H	388	N=N	409
N–O	210	N≡N	941
N=O	630	O–H	463

(b) Calculate the enthalpy value for the overall reaction.

To determine the amount of enthalpy change, Lewis structures of all of the species should be drawn.

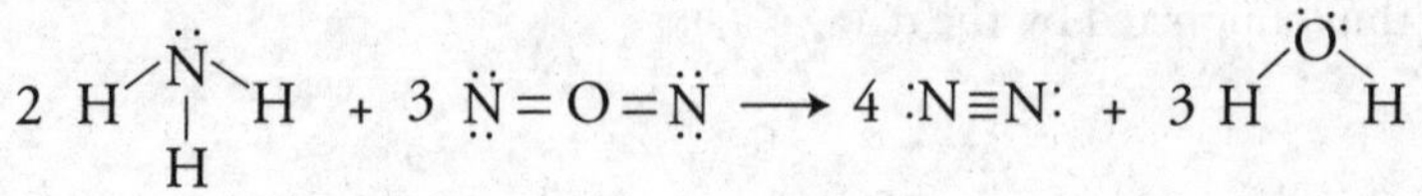

$$2\ H{-}\ddot{N}(H){-}H + 3\ \ddot{N}{=}O{=}\ddot{N} \longrightarrow 4\ :N{\equiv}N: + 3\ H{-}\ddot{O}{-}H$$

You have to account for not only how many bonds are within each molecule, but also how many of those molecules there are. For instance, there are three N–H bonds in NH_3, so with two NH_3 molecules there will be six total N–H bonds broken. The bond energy for the other three molecules must be calculated the same way.

ΔH = Bonds broken (reactants) – bonds formed (products)

$\Delta H = [6(\text{N–H}) + 6(\text{N=O})] - [4(\text{N}\equiv\text{N}) + 6(\text{O–H})]$

$\Delta H = [6(388) + 6(630)] - [4(941) + 6(463)]$

$\Delta H = 6108 - 6542$

$\Delta H = -434$ kJ/mol

(c) Is this reaction thermodynamically favored at 25°C? Justify your answer.

We will use the Gibbs free energy equation here. Keep in mind the units have to match. In this solution, the units for entropy have been converted to kJ.

$\Delta G = \Delta H - T\Delta S$

$\Delta G = -434 \text{ kJ/mol} - (298 \text{ K})(0.289 \text{ kJ/mol·K})$

$\Delta G = -434 \text{ J/mol} - 86.1 \text{ kJ/mol}$

$\Delta G = -520$ kJ/mol

The value for ΔG is negative, therefore the reaction is thermodynamically favored at 25°C.

(d) If 25.00 g of NH_3 reacts with 25.00 g of N_2O:

(i) Will energy be released or absorbed?

The value for ΔH is negative; therefore it is an exothermic reaction and energy will be released.

(ii) What is the magnitude of the energy change?

To determine how much energy is released, the limiting reagent must be determined.

$$25.00 \text{ g NH}_3 \times \frac{1 \text{ mol NH}_3}{17.04 \text{ g NH}_3} \times \frac{1 \text{ mol}_{rxn}}{2 \text{ mol NH}_3} \times \frac{-434 \text{ kJ}}{1 \text{ mol}_{rxn}} = -318.4 \text{ kJ}$$

$$25.00 \text{ g N}_2\text{O} \times \frac{1 \text{ mol N}_2\text{O}}{44.02 \text{ g N}_2\text{O}} \times \frac{1 \text{ mol}_{rxn}}{3 \text{ mol N}_2\text{O}} \times \frac{-434 \text{ kJ}}{1 \text{ mol}_{rxn}} = -82.16 \text{ kJ}$$

As the N_2O would produce less energy, it would run out first and is thus limiting. The answer is thus –82.16 kJ.

(e) On the reaction coordinates below, draw a line showing the progression of this reaction. Label both ΔH and E_a on the graph.

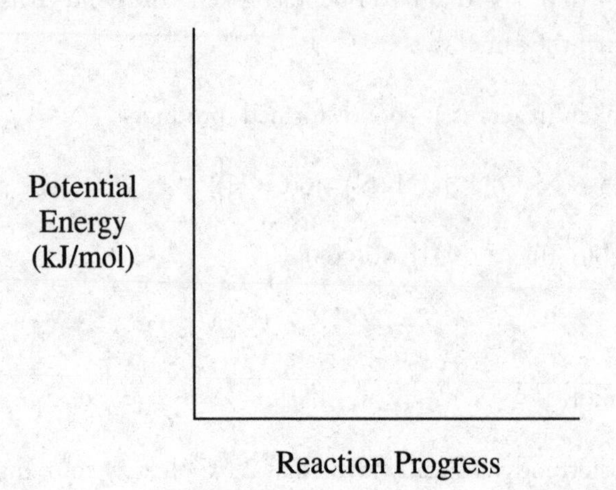

The activation energy describes the distance between the bond energy of the reactants and the energy of the activated state. In terms of enthalpy, the value for ΔH is negative, so that means that the bond energy of the products will be lower than that of the reactants.

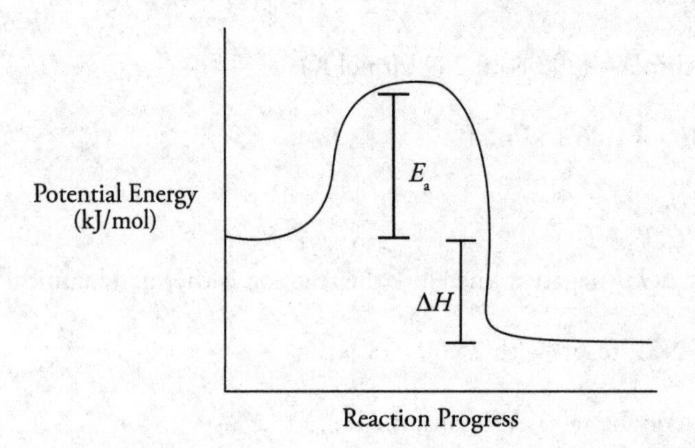

4. The acetyl ion has a formula of $C_2H_3O^-$ and two possible Lewis electron-dot diagram representations:

$$\begin{array}{ccccc} H & & H & & \\ | & & | & & \\ C & = & C_X & - & \ddot{\underset{\cdot\cdot}{O}}: \\ | & & & & \\ H & & & & \end{array} \qquad \begin{array}{ccccc} & & H & & & & \\ & & | & & & & \\ H & - & C & - & \ddot{C}_X & = & \ddot{\underset{\cdot\cdot}{O}} \\ & & | & & & & \\ & & H & & & & \end{array}$$

(a) Using formal charge, determine which structure is the most likely correct and circle it.

For this formal charge calculation, the H atoms are left out as they are identically bonded/drawn in both structures

	C	C_x	O	C	C_x	O
Valence	4	4	6	4	4	6
Assigned	−4	4	7	−4	5	6
Formal Charge	0	0	−1	0	−1	0

As oxygen is more electronegative than carbon, an oxygen atom is more likely to have the negative formal charge than a carbon atom. The left-hand structure is most likely correct.

(b) For carbon atom "x" in the structure you chose:

(i) What is the hybridization around the atom?

There are three charge groups around the carbon atom, so the hybridization is sp^2.

(ii) How many sigma and pi bonds has the atom formed?

Single bonds consist of sigma bonds, and double bonds consist of one sigma and one pi bond.

There are a total of three sigma bonds and one pi bound around the carbon atom.

(c) A hydrogen ion attaches itself to the the acetyl ion, creating C_2H_4O. Draw the Lewis diagram of the new molecule.

The hydrogen ion will attach to the negatively-charged oxygen.

$$\begin{array}{ccccccc} H & & H & & & & \\ | & & | & & & & \\ C & = & C & - & \ddot{\underset{\cdot\cdot}{O}} & - & H \\ | & & & & & & \\ H & & & & & & \end{array}$$

5. A student tests the conductivity of three different acid samples, each with a concentration of 0.10 M and a volume of 20.0 mL. The conductivity was recorded in microsiemens per centimeter in the table below:

Sample	Conductivity (μS/cm)
1	26,820
2	8655
3	35,120

(a) The three acids are known to be HCl, H_2SO_4, and H_3PO_4. Identify which sample is which acid. Justify your answer.

H_3PO_4 is the only weak acid, meaning it will not dissociate completely in solution. Therefore, it is acid 2. H_2SO_4 and HCl are both strong acids that will dissociate completely, but H_2SO_4 is diprotic, meaning there will be more H^+ ions in solution and thus a higher conductivity. Therefore, H_2SO_4 is acid 3, and HCl is acid 1.

(b) The HCl solution is then titrated with a 0.150 M solution of the weak base methylamine, CH_3NH_2. ($K_b = 4.38 \times 10^{-4}$)

(i) Write out the net ionic equation for this reaction.

$H^+ + CH_3NH_2 \leftrightarrow CH_3NH_3^+$

(ii) Determine the pH of the solution after 20.0 mL of methylamine has been added.

First, the number of moles of hydrogen ions and methylamine need to be determined.

H^+: $0.10\ M = \dfrac{n}{0.020\text{ L}}$ $\quad n = 0.0020$ mol

CH_3NH_2: $0.150\ M = \dfrac{n}{0.020\text{ L}}$ $\quad n = 0.0030$ mol

	H^+	+	CH_3NH_2	$\leftrightarrow$	$CH_3NH_3^+$
I	0.0020		0.0030		0
C	−0.0020		−0.0020		+0.0020
E	0		0.0010		0.0020

The new volume of the solution is 40.0 mL, which we used to calculated the concentrations of the ions in solution at equilibrium:

$$[CH_3NH_2] = \frac{0.0010 \text{ mol}}{0.040 \text{ L}} = 0.025\ M$$

$$[CH_3NH_3^+] = \frac{0.0020 \text{ mol}}{0.040 \text{ L}} = 0.050\ M$$

Finally, use the Henderson-Hasselbalch equation.

$$\text{pOH} = \text{p}K_b + \log\frac{[CH_3NH_3^+]}{[CH_3NH_2]}$$

$$\text{pOH} = -\log(4.38 \times 10^{-4}) + \log\frac{0.050}{0.025}$$

$$\text{pOH} = 3.36 + 0.30 = 3.66$$

$$\text{pH} + \text{pOH} = 14$$

$$\text{pH} + 3.66 = 14$$

$$\text{pH} = 10.34$$

6. The photoelectron spectrum of an element is given below:

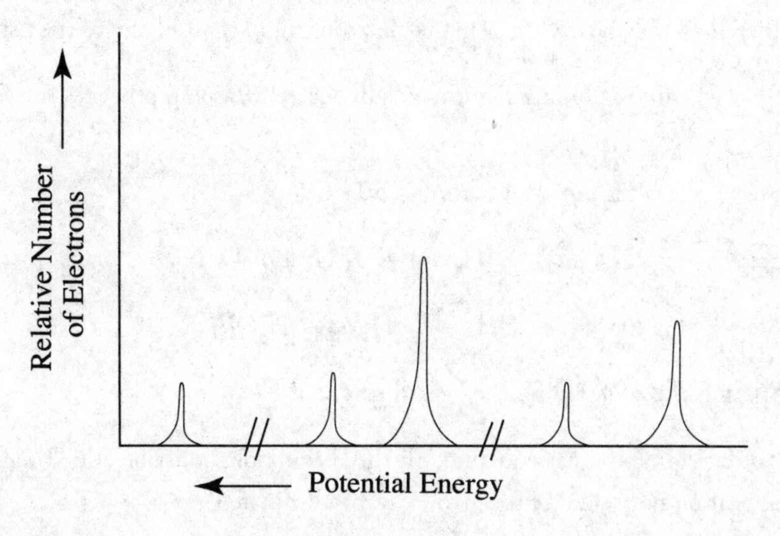

(a) Identify the element this spectra most likely belongs to and write out its full electron configuration.

This PES belongs to sulfur: $1s^2 2s^2 2p^6 3s^2 3p^4$.

(b) Using your knowledge of atomic structure, explain the following:

(i) The reason for the three discrete areas of ionization energies:

Each discrete area of ionization energy represents a different energy level of the electrons. The closer the electrons are to the nucleus, the more ionization energy will be required to remove the electrons. Sulfur has electrons present at three different energy levels, thus, there are three different areas for the peaks.

(ii) The justification for there being a total of five peaks:

Within each energy level (except for the first), there are subshells which are not the exact same distance from the nucleus. Both energy levels 2 and 3 have *s* and *p* subshells, and while the electrons in those shells will have similar ionization energy values, they will not be identical. Thus, the five peaks represent 1*s*, 2*s*, 2*p*, 3*s*, and 3*p*.

(iii) The relative heights of the peaks when compared to each other:

The heights of the peaks represents the ratio of electrons present in each of them. All three *s* peaks are exactly one third the height of the 2*p* peak, meaning there are three times more electrons in 2*p* than in any of the *s* subshells. The 3*p* peak is only twice as high as the *s* peaks, therefore, there are twice as many electrons in 3*p* than in any of the *s* subshells.

7. Hydrazine (N_2H_4) can be produced commercially via the Raschig process. The following is a proposed mechanism:

Step 1: $NH_3(aq) + OCl^-(aq) \rightarrow NH_2Cl(aq) + OH^-$

Step 2: $NH_2Cl(aq) + NH_3(aq) \rightarrow N_2H_5^+(aq) + Cl^-$

Step 3: $N_2H_5^+(aq) + OH^- \rightarrow N_2H_4(aq) + H_2O(l)$

(a) (i) What is the equation for the overall reaction?

To determine the net equation, all three equations must be added together, and species that appear on both sides of the arrow can be eliminated.

$NH_3(aq) + OCl^-(aq) + \cancel{NH_2Cl(aq)} + NH_3(aq) + \cancel{N_2H_5^+(aq)} + \cancel{OH^-} \rightarrow \cancel{NH_2Cl(aq)} + \cancel{OH^-}$
$+ \cancel{N_2H_5^+(aq)} + Cl^- + N_2H_4(aq) + H_2O(l)$

$2NH_3(aq) + OCl^-(aq) \rightarrow N_2H_4(aq) + H_2O(l) + Cl^-$

(ii) Identify any catalysts or intermediates from the reaction mechanism.

There are no catalysts present, but NH_2Cl, $N_2H_5^+$, and OH^- are all intermediates in the process.

(b) The rate law for the reaction is determined to be rate = $k[NH_3][OCl^-]$.

(i) Which elementary step is the slowest one? Justify your answer.

The overall rate law will match the rate law of the slowest step. The rate law of an elementary step can be determined by the reactants present, and in this case, the rate law for Step 1 matches the overall rate law. Therefore, Step 1 is the slowest step.

(ii) If the reaction is measured over the course of several minutes, what would the units of the rate constant be?

Using unit analysis:

rate = $k[NH_3][OCl^-]$
$M/\text{min} = k(M)(M)$
$k = M^{-1}\text{min}^{-1}$

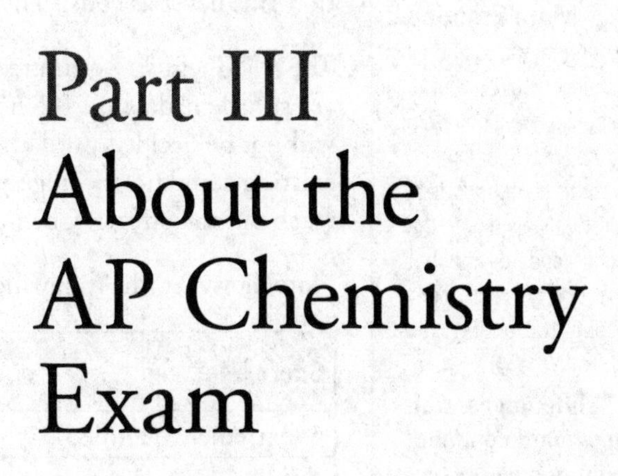

Part III About the AP Chemistry Exam

THE STRUCTURE OF THE AP CHEMISTRY EXAM

The AP Chemistry Exam is a three-hour-long, two-section test that attempts to cover the material you would learn in a college first-year chemistry course. The first section is a 90-minute, 60-question multiple-choice section. As of 2015, the second section is 105 minutes and consists of three long-form free-response questions and four short-form free-response questions.

More Practice
If you're eager for additional review, you can turn to free-response questions from older exams. Bear in mind that this content doesn't perfectly align with the current standards and that the question types are likely to be formatted and written differently on the 2017 exam:
http://apcentral.collegeboard.com/apc/public/exam/exam_information/221837.html

The multiple-choice section is scored by a computer, and the free-response questions are scored by a committee of high school and college teachers. The free-response questions are graded according to a standard set at the beginning of the grading period by the chief faculty consultants. Inevitably, the grading of Section II is never as consistent or accurate as the grading of Section I.

The AP Chemistry exam recently changed in May 2014, and as a result there isn't yet sufficient data on the necessary raw score to get a 3, 4, or 5. The breakdown will not be decided until after more exams have been scored. In the absence of an existing standard to judge yourself against, just do your best! Don't be afraid to reach out to your AP Chemistry teacher to see what he or she thinks of your work.

Note, however, the following data on how students scored on the 2015 test:

Score (Meaning)	Percentage of Test Takers	Equivalent Grade in a first-year college course
5 (extremely qualified)	9.2%	A
4 (well qualified)	16.1%	A–, B+, B
3 (qualified)	28.1%	B–, C+, C
2 (possibly qualified)	24.9%	C–
1 (no recommendation)	21.7%	D

OVERVIEW OF CONTENT TOPICS

The concepts of the AP Chemistry exam are broken down into six major themes defined by the College Board as the Big Ideas. Rather than learning multiple disparate topics (as has been done in the past), these Big Ideas interconnect principles within these topics that describe fundamental chemical phenomena. The six Big Ideas are as follows:

- Big Idea #1: The chemical elements are fundamental building materials of matter, and all matter can be understood in terms of arrangements of atoms. These atoms retain their identity in chemical reactions.
- Big Idea #2: Chemical and physical properties of materials can be explained by the structure and the arrangement of atoms, ions, or molecules and the forces between them.
- Big Idea #3: Changes in matter involve the rearrangement and/or reorganization of atoms and/or the transfer of electrons.
- Big Idea #4: Rates of chemical reactions are determined by details of the molecular collisions.

- Big Idea #5: The laws of thermodynamics describe the essential role of energy and explain and predict the direction of changes in matter.
- Big Idea #6: Any bond or intermolecular attraction that can be formed can be broken. These two processes are in a dynamic competition, sensitive to initial conditions and external perturbations.

This book has been arranged to teach chemistry topics grouped to each of these Big Ideas.

HOW AP EXAMS ARE USED

Different colleges use AP Exam scores in different ways, so it is important that you go to a particular college's web site to determine how it uses AP Exam scores. The three items below represent the main ways in which AP Exam scores can be used.

- **College Credit**. Some colleges will give you college credit if you score well on an AP Exam. These credits count towards your graduation requirements, meaning that you can take fewer courses while in college. Given the cost of college, this could be quite a benefit, indeed.
- **Satisfy Requirements.** Some colleges will allow you to "place out" of certain requirements if you do well on an AP Exam, even if they do not give you actual college credits. For example, you might not need to take an introductory-level course, or perhaps you might not need to take a class in a certain discipline at all.
- **Admissions Plus**. Even if your AP Exam will not result in college credit or even allow you to place out of certain courses, most colleges will respect your decision to push yourself by taking an AP Course or even an AP Exam outside of a course. A high score on an AP Exam shows mastery of more difficult content than is taught in many high school courses, and colleges may take that into account during the admissions process.

OTHER RESOURCES

There are many resources available to help you improve your score on the AP Chemistry Exam, not the least of which are your teachers. If you are taking an AP class, you may be able to get extra attention from your teacher, such as obtaining feedback on your free-response questions. If you are not in an AP course, reach out to a teacher who teaches chemistry, and ask if the teacher will review your free-response questions or otherwise help you with content.

Another wonderful resource is **AP Students,** the official site of the AP Exams. The scope of the information at this site is quite broad and includes:

- Course Description, which includes details on what content is covered and sample multiple-choice and free-response questions
- Exam practice tips
- Free-response question prompts and scoring guidelines from the 2015 and 2014 exams

The AP Students home page address is: **https://apstudent.collegeboard.org.**

Finally, The Princeton Review offers tutoring and small group instruction. Our expert instructors can help you refine your strategic approach and add to your content knowledge. For more information, call 1-800-2REVIEW.

DESIGNING YOUR STUDY PLAN

As part of the Introduction, you identified some areas of potential improvement. Let's now delve further into your performance on Test 1, with the goal of developing a study plan appropriate to your needs and time commitment.

Read the answers and explanations associated with the multiple-choice questions (starting at page 36). After you have done so, respond to the following:

- Review the Overview of Content Topics at page 56. Next to each topic, indicate your rank of the topic as follows: "1" means "I need a lot of work on this," "2" means "I need to beef up my knowledge," and "3" means "I know this topic well."
- How many days/weeks/months away is your exam?
- What time of day is your best, most focused study time?
- How much time per day/week/month will you devote to preparing for your exam?
- When will you do this preparation? (Be as specific as possible: Mondays & Wednesdays from 3:00 P.M.–4:00 P.M., for example)
- Based on the answers above, will you focus on strategy (Part IV) or content (Part V) or both?
- What are your overall goals in using this book?

Part IV
Test-Taking Strategies for the AP Chemistry Exam

Chapter 1
How to Approach Multiple-Choice Questions

THE BASICS

Section I of the test is composed of 60 multiple-choice questions, for which you are allotted 90 minutes. This part is worth 50 percent of your total score.

For this section, you will be given a periodic table of the elements along with a sheet that lists common chemistry formulas, and you may NOT use a calculator. The College Board says that this is because the new scientific calculators not only program and graph but also store information—and they are afraid you'll use this function to cheat!

On the multiple-choice section, you receive 1 point for a correct answer. There is no penalty for leaving a question blank or getting a question wrong.

PACING

According to the College Board, the multiple-choice section of the AP Chemistry Exam covers more material than any individual student is expected to know. Nobody is expected to get a perfect or even near perfect score. What does that mean to you?

Use the Two-Pass System

Go through the multiple-choice section twice. The first time, do all the questions that you can get answers to immediately. That is, do the questions with little or no math and questions on chemistry topics in which you are well versed. Skip questions on topics that make you uncomfortable. Also, you want to skip the ones that look like number crunchers (even without a calculator, you may still be expected to crunch a few numbers). Circle the questions that you skip in your test booklet so you can find them easily during the second pass. Once you've done all the questions that come easily to you, go back and pick out the tough ones that you have the best shot at.

In general, the questions near the end of the section are tougher than the questions near the beginning. You should keep that in mind, but be aware that each person's experience will be different. If you can do acid-base questions in your sleep, but you'd rather have your teeth drilled than draw a Lewis diagram, you may find questions near the end of the section easier than questions near the beginning.

That's why the Two-Pass System is so handy. By using it, you make sure you get to see all the questions you can get right, instead of running out of time because you got bogged down on questions you couldn't do earlier in the test.

Don't Turn a Question into a Crusade!

Most people don't run out of time on standardized tests because they work too slowly. Instead, they run out of time because they spend half the test wrestling with two or three particular questions.

You should never spend more than a minute or two on any question. If a question doesn't involve calculation, then either you know the answer, you can make an educated guess, or you don't know the answer. Figure out where you stand on a question, make a decision, and move on.

Any question that requires more than two minutes worth of calculations probably isn't worth doing. Remember: Skipping a question early in the section is a good thing if it means that you'll have time to get two correct answers later on.

GUESSING

You get one point for every correct answer on the multiple-choice section. Guessing randomly neither helps you nor hurts you. Educated guessing, however, will help you.

Use Process of Elimination (POE) to Find Wrong Answers

There is a fundamental weakness to a multiple-choice test. The test makers must show you the right answer, along with three wrong answers. Sometimes seeing the right answer is all you need. Other times you may not know the right answer, but you may be able to identify one or two of the answers that are clearly wrong. Here is where you should use process of elimination (POE) to take an educated guess.

Look at this hypothetical question.

1. Which of the following compounds will produce a purple solution when added to water?

    (A) Brobogdium rabelide
    (B) Diblythium perjuvenide
    (C) Sodium chloride
    (D) Carbon dioxide

You should have no idea what the correct answer is because two of these compounds are made up, but you do know something about the obviously wrong answers. You know that sodium chloride, (C), and carbon dioxide, (D), do not turn water purple. So, using POE, you have a 50 percent chance at guessing the correct answer. Now the odds are in your favor. Now you should guess.

Guess and Move On

Remember that you're guessing. Pondering the possible differences between brobogdium rabelide and diblythium perjuvenide is a waste of time. Once you've taken POE as far as it will go, pick your favorite letter and move on.

The multiple-choice section is the exact opposite of the free-response section. It's scored by a machine. There's no partial credit. The computer doesn't know, or care if you know, why an answer is correct. All the computer cares about is whether you blackened in the correct oval on your score sheet. You get the same number of points for picking (B) because you know that (A) is wrong and that (B) is a nicer letter than (C) or (D) as you would get for picking (B) because you fully understood the subtleties of an electrochemical process.

ABOUT CALCULATORS

You will NOT be allowed to use a calculator on this section. That shouldn't worry you. All it means is that there won't be any questions in the section that you'll need a calculator to solve.

Most of the calculation problems will have fairly user-friendly numbers—that is, numbers with only a couple of significant digits, or things like "11.2 liters of gas at STP" or "160 grams of oxygen" or "a temperature increase from 27°C to 127°C." Sometimes these user-friendly numbers will actually point you toward the proper steps to take in your calculations.

Don't be afraid to make rough estimates as you do your calculations. Sometimes knowing that an answer is closer to 50 than to 500 will enable you to pick the correct answer on a multiple-choice test (if the answer choices are far enough apart). Once again, the rule against calculators works in your favor because the College Board will not expect you to do very precise calculations by hand.

REFLECT

Respond to the following questions:

- How long will you work on multiple-choice questions?
- How will you change your approach to multiple-choice questions?
- What is your multiple-choice guessing strategy?

Chapter 2
How to Approach Free-Response Questions

OVERVIEW OF THE FREE-RESPONSE SECTION

Section II is composed of seven free-response questions. You will be given 105 minutes to complete this section, which is worth 50 percent of your total score.

The first three free-response questions are longer, and are worth 10 points each. The College Board recommends you take about 22 minutes per question on these. The last four questions are much shorter, and only worth 4 points each. For these, the College Board recommends you take about 9 minutes each.

The suggested times are just that—suggestions. Your results may vary based on how comfortable you are with the topic being tested within each question. They do give you a good guideline, though. You probably shouldn't be spending more than 20 minutes on any of the first three questions. Don't be afraid to cut yourself off and come back to questions later if time allows.

CRACKING THE MATH PROBLEMS

You want to show the graders that you understand the math behind these chemistry problems, so here are some suggestions.

Show Every Step of Your Calculations on Paper

This section is the opposite of multiple choice. You don't just get full credit for writing the correct answer. You get most of your points on this section for showing the process that got you to the answer. The graders give you partial credit when you show them that you know what you're doing. So even if you can do a calculation in your head, you should set it up and show it on the page.

By showing every step, or explaining what you're doing in words, you ensure that you'll get all the partial credit possible, even if you screw up a calculation.

Include Units in All Your Calculations

Scientists like units in calculations. Units make scientists feel secure. You'll get points for including them and you may lose points for leaving them out.

Remember Significant Figures

You can lose 1 point per question if your answer is off by more than one significant figure. Without getting too bent out of shape about it, try to remember that a calculation is only as accurate as the least accurate number in it.

(For a detailed explanation of significant figures, see Chapter 9.)

The Graders Will Follow Your Reasoning, Even if You've Made a Mistake

Often, you are asked to use the result of a previous part of a problem in a later part. If you got the wrong answer in part (a) and used it in part (c), you can still get full credit for part (c), as long as your work is correct based on the number that you used. That's important, because it means that botching the first part of a question doesn't necessarily sink the whole question.

Remember the Mean!

So let's say that you could complete only parts (a) and (b) on the required equilibrium problem. That's 4 or 5 points out of 10, tops. Are you doomed? Of course not. You're above average. If this test is hard on you, it's probably just as hard on everybody else. Remember: You don't need anywhere near a perfect score to get a 5, and you can leave half the test blank and still get a 3!

STRATEGIES FOR CRACKING THE FREE-RESPONSE SECTION

This section is here to test whether you can translate chemistry into English. Most of the questions can be answered in two or three simple sentences, or with a simple diagram or two. Here are some tips for answering the seven free-response questions on Part II.

Show That You Understand the Terms Used in the Question

If you are asked why sodium and potassium have differing first ionization energies, the first thing you should do is tell them what ionization energy is. That's probably worth the first point of partial credit. Then you should tell them how the differing structures of the atoms make for differing ionization energies. That leads to the next tip.

Take a Step-by-Step Approach

Grading these tests is hard work. Breaking a question into parts in this way makes it easier on the grader, who must match your response to a set of guidelines he or she has been given that describe how to assign partial and full credit.

Each grader scores each test based on these rough guidelines that are established at the beginning of the grading period. For instance, if a grader has 3 points for the question about ionization energies, the points might be distributed the following way:

- one point for understanding ionization energy
- one point for explaining the structural difference between sodium and potassium
- one point for showing how this difference affects the ionization energy

You can get all three points for this question if the grader thinks that all three concepts are addressed implicitly in your answer, but by taking a step-by-step approach, you improve your chances of explicitly addressing the things that a grader has been instructed to look for. Once again, grading these tests is hard work; graders won't know for sure if you understand something unless you tell them.

Write Neatly

This simple concept cannot be stressed enough: write neatly, even if that means working at half-speed. You can't get points for answers if the graders can't understand them. Of course, this applies to the rest of the free-response section as well.

Graders Will Follow Your Reasoning, Even If You Make A Mistake

Just like in the multiple-choice section, you might be asked to use the result of a previous part of a problem in a later part. If you decide (incorrectly) that an endothermic reaction in part (a) is exothermic, you can still get full credit in part (c) for your wrong answer about the reaction's favorability, as long as your answer in (c) is correct based on an exothermic reaction.

REFLECT

Respond to the following questions:

- How much time will you spend on the short free-response questions? What about the long free-response questions?
- What will you do before you begin writing your free-response answers?
- Will you seek further help, outside of this book (such as a teacher, tutor, or AP Students), on how to approach the questions that you will see on the AP Chemistry Exam?

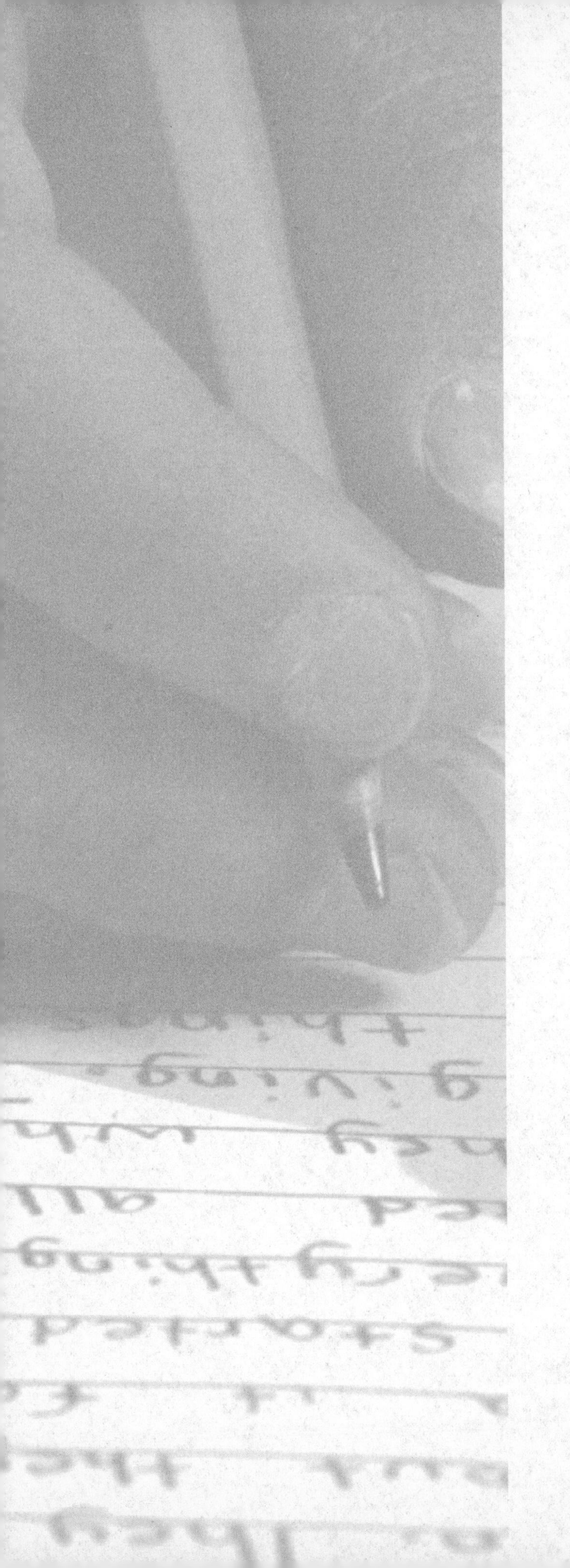

Part V
Content Review for the AP Chemistry Exam

Chapter 3
Big Idea #1: Atoms, Elements, and the Building Blocks of Matter

The chemical elements are fundamental building materials of matter, and all matter can be understood in terms of arrangements of atoms. These atoms retain their identity in chemical reactions.

THE PERIODIC TABLE

The most important tool you will use on this test is the Periodic Table of the Elements.

PERIODIC TABLE OF THE ELEMENTS

1 IA	2 IIA	3 IIIB	4 IVB	5 VB	6 VIB	7 VIIB	8	9 VIIIB	10	11 IB	12 IIB	13 IIIA	14 IVA	15 VA	16 VIA	17 VIIA	18 VIIIA
1 H 1.008																	2 He 4.00
3 Li 6.94	4 Be 9.01											5 B 10.81	6 C 12.01	7 N 14.01	8 O 16.00	9 F 19.00	10 Ne 20.18
11 Na 22.99	12 Mg 24.30											13 Al 26.98	14 Si 28.09	15 P 30.97	16 S 32.06	17 Cl 35.45	18 Ar 39.95
19 K 39.10	20 Ca 40.08	21 Sc 44.96	22 Ti 47.90	23 V 50.94	24 Cr 52.00	25 Mn 54.94	26 Fe 55.85	27 Co 58.93	28 Ni 58.69	29 Cu 63.55	30 Zn 65.39	31 Ga 69.72	32 Ge 72.59	33 As 74.92	34 Se 78.96	35 Br 79.90	36 Kr 83.80
37 Rb 85.47	38 Sr 87.62	39 Y 88.91	40 Zr 91.22	41 Nb 92.91	42 Mo 95.94	43 Tc (98)	44 Ru 101.1	45 Rh 102.91	46 Pd 106.42	47 Ag 107.87	48 Cd 112.41	49 In 114.82	50 Sn 118.71	51 Sb 121.75	52 Te 127.60	53 I 126.91	54 Xe 131.29
55 Cs 132.91	56 Ba 137.33	57 *La 138.91	72 Hf 178.49	73 Ta 180.95	74 W 183.85	75 Re 186.21	76 Os 190.2	77 Ir 192.2	78 Pt 195.08	79 Au 196.97	80 Hg 200.59	81 Tl 204.38	82 Pb 207.2	83 Bi 208.98	84 Po (209)	85 At (210)	86 Rn (222)
87 Fr (223)	88 Ra 226.02	89 †Ac 227.03	104 Rf (261)	105 Db (262)	106 Sg (266)	107 Bh (264)	108 Hs (277)	109 Mt (268)	110 Ds (271)	111 Rg (272)							

*Lanthanide Series	58 Ce 140.12	59 Pr 140.91	60 Nd 144.24	61 Pm (145)	62 Sm 150.4	63 Eu 151.97	64 Gd 157.25	65 Tb 158.93	66 Dy 162.50	67 Ho 164.93	68 Er 167.26	69 Tm 169.93	70 Yb 173.04	71 Lu 174.97
†Actinide Series	90 Th 232.04	91 Pa 231.04	92 U 238.03	93 X (237)	94 Pu (244)	95 Am (243)	96 Cm (247)	97 Bk (247)	98 Cf (251)	99 Es (252)	100 Fm (257)	101 Md (258)	102 No (259)	103 Lr (262)

The Periodic Table gives you very basic but very important information about each element.

Bad Joke Alert
To prepare for the AP Chemistry Exam, familiarize yourself with the periodic table of elements by looking at it periodically.

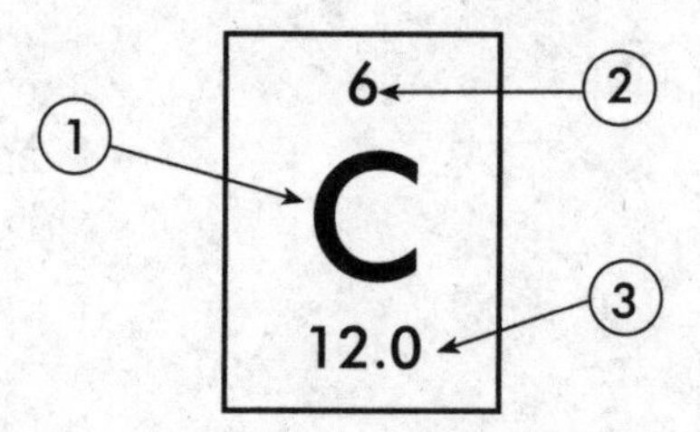

1. This is the **symbol** for the element; carbon, in this case. On the test, the symbol for an element is used interchangeably with the name of the element.
2. This is the **atomic number** of the element. The atomic number is the same as the number of protons in the nucleus of an element; it is also the same as the number of electrons surrounding the nucleus of an element when it is neutrally charged.
3. This number represents the average atomic mass of a single atom of carbon, measured in atomic mass units (amus). It also represents the average mass for a mole (see p. 76) of carbon atoms, measured in grams. Thus, one mole of carbon atoms has a mass of 12.01 g. This is called the **molar mass** of the element.

The horizontal rows of the periodic table are called **periods.**

The vertical columns of the periodic table are called **groups.**

Groups can be numbered in two ways. The old system used Roman numerals to indicate groups. The new system simply numbers the groups from 1 to 18. While it is not important to know the specific group numbers, it is important to know the names of some groups.

Group IA/1 – Alkali Metals
Group IIA/2 – Alkaline Earth Metals
Group B/3-12 – Transition Metals
Group VIIA/17 – Halogens
Group VIIIA/18 – Noble Gases

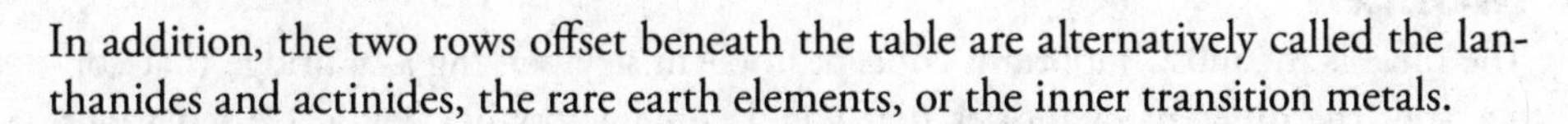

In addition, the two rows offset beneath the table are alternatively called the lanthanides and actinides, the rare earth elements, or the inner transition metals.

The identity of an atom is determined by the number of protons contained in its nucleus. The nucleus of an atom also contains neutrons. The mass number of an atom is the sum of its neutrons and protons. Electrons have significantly less mass than protons or neutrons and do not contribute to an element's mass.

Atoms of an element with different numbers of neutrons are called **isotopes**; for instance, carbon-12, which contains 6 protons and 6 neutrons, and carbon-14, which contains 6 protons and 8 neutrons, are isotopes of carbon. The molar mass given on the periodic table is the average of the mass numbers of all known isotopes weighted by their percent abundance.

The mass of various isotopes of an element can be determined by a technique called mass spectrometry. A mass spectrum of selenium looks like the following:

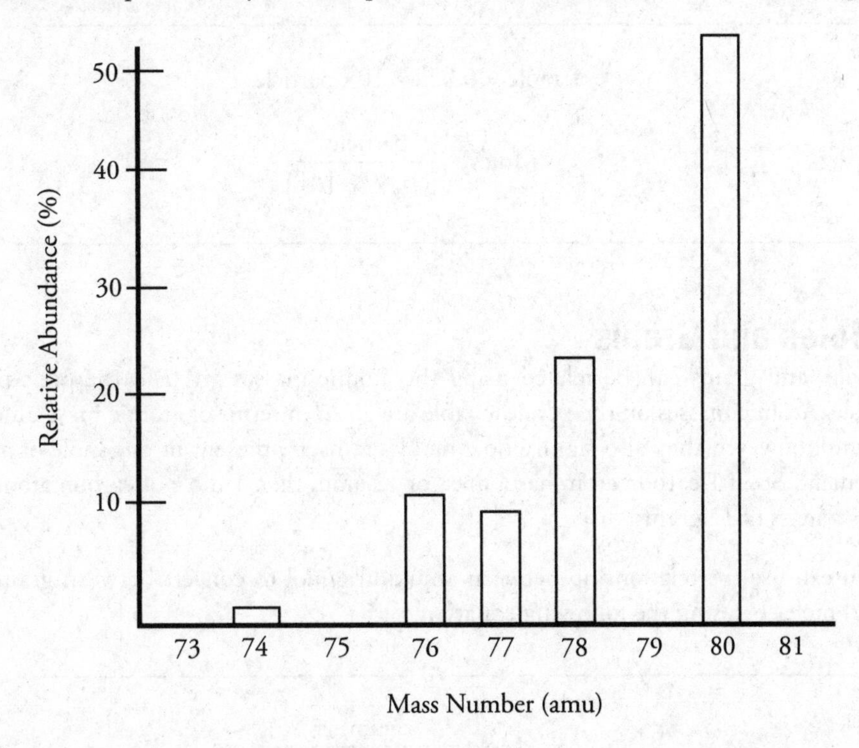

As you can see, the most abundant isotope of selenium has a mass of 80, but there are four other naturally occurring isotopes of selenium. The average atomic mass is the weighted average of all five isotopes of selenium shown on this spectra.

The molar mass of an element will give you a pretty good idea of the most common isotope of that element. For instance, the molar mass of carbon is 12.01 and about 99 percent of the carbon in existence is carbon-12.

MOLES

The mole is the most important concept in chemistry, serving as a bridge that connects all the different quantities that you'll come across in chemical calculations. The coefficients in chemical reactions tell you about the reactants and products in terms of moles, so most of the stoichiometry questions you'll see on the test will be exercises in converting between moles and grams, liters, molarities, and other units.

Moles and Molecules

Avogadro's number describes how many atoms are in a single mole of any given element. Much like a dozen is always 12, Avogadro's number is always 6.02×10^{23}. While it technically can be used to count anything, due to its extremely large value it is usually only used to count extraordinarily small things. Within the confines of this book, it will be used to count atoms, molecules, electrons, or ions, depending on the problem.

$$1 \text{ mole} = 6.022 \times 10^{23} \text{ particles}$$

$$\text{Moles} = \frac{\text{particles}}{(6.022 \times 10^{23})}$$

Moles and Grams

Moles and grams can be related using the atomic masses given in the periodic table. Atomic masses on the periodic table are given in terms of atomic mass units (amu); however, they also signify how many grams are present in one mole of an element. So, if 1 carbon atom has a mass of 12 amu, then 1 mole of carbon atoms has a mass of 12 grams.

You can use the relationship between amu and g/mol to convert between grams and moles by using the following equation:

$$\text{Moles} = \frac{\text{grams}}{\text{molar mass}}$$

Moles and Gases

We'll talk more about the ideal gas equation in Chapter 4, but for now, you should know that you can use it to calculate the number of moles of a gas if you know some of the gas's physical properties. All you need to remember at this point is that in the equation $PV = nRT$, n stands for moles of gas.

$$\text{Moles} = \frac{PV}{RT}$$

P = pressure (atm)
V = volume (L)
T = temperature (K)
R = the gas constant, 0.0821 L·atm/mol·K

The equation above gives the general rule for finding the number of moles of a gas. Many gas problems will take place at STP, or standard temperature and pressure, where $P = 1$ atmosphere and $T = 273$ K. At STP, the situation is much simpler and you can convert directly between the volume of a gas and the number of moles. That's because at STP, one mole of gas always occupies 22.4 liters.

$$\text{Moles} = \frac{\text{liters}}{(22.4\,\text{L/mol})}$$

Moles and Solutions

We'll talk more about molarity in Chapter 4, but for now you should realize that you can use the equations that define these common measures of concentration to find the number of moles of solute in a solution. Just rearrange the equations to isolate moles of solute.

$$\text{Moles} = (\text{molarity})(\text{liters of solution})$$

Percent Composition

Percent composition is the percent by mass of each element that makes up a compound. It is calculated by dividing the mass of each element or component in a compound by the total molar mass for the substance.

Calculate the percent composition of each element in calcium nitrate, $Ca(NO_3)_2$.

To do this, you need to first separate each element and count how many atoms are present. Subscripts outside of parentheses apply to all atoms inside of those parentheses.

Calcium: 1
Nitrogen: 2
Oxygen: 6

Then, multiply the number of atoms by the atomic mass of each element.

Ca: $40.08 \times 1 = 40.08$
N: $14.01 \times 2 = 28.02$
O: $16.00 \times 6 = 96.00$

Adding up the masses of the individual elements will give you the atomic mass of that compound. Divide each individual mass by the total molar mass to get your percent composition.

$$40.08 + 28.02 + 96.00 = 164.10 \text{ g/mol}$$

Ca: $40.08/164.10 \times 100\% = 24.42\%$
N: $28.02/164.10 \times 100\% = 17.07\%$
O: $96.00/164.10 \times 100\% = 58.50\%$

You can check your work at the end by making sure your percents add up to 100% (taking rounding into consideration).

$24.42\% + 17.07\% + 58.50\% = 99.99\%$. Close enough!

Empirical and Molecular Formulas

You will also need to know how to determine the empirical and molecular formulas of a compound given masses or mass percents of the components of that compound. Remember that the empirical formula represents the simplest ratio of one element to another in a compound (e.g., CH_2O), while the molecular formula represents the actual formula for the substance (e.g., $C_6H_{12}O_6$).

Let's take a look at the following example:

A compound is found to contain 56.5% carbon, 7.11% hydrogen, and 36.4% phosphorus.

a) Determine the empirical formula for the compound

We start by assuming a 100 gram sample; this allows us to convert those percentages to grams. After we have that done, each element needs to be converted to moles.

$$\text{C: } 56.5 \text{ g C} \times \frac{1 \text{ mol C}}{12.01 \text{ g C}} = 4.71 \text{ mol C}$$

$$H: 7.11\text{ g Cl} \times \frac{1\text{ mol H}}{1.01\text{ g H}} = 7.04\text{ mol H}$$

$$P: 36.4\text{ g P} \times \frac{1\text{ mol P}}{30.97\text{ g P}} = 1.18\text{ mol P}$$

We then divide each mole value by the lowest of the values. In this example, that would be the phosphorus. It is acceptable to round your answers if they are close (within 0.1) to a whole number.

$$C: \frac{4.71\text{ mol}}{1.18\text{ mol}} = 4$$

$$H: \frac{7.04\text{ mol}}{1.18\text{ mol}} = 6$$

$$P: \frac{1.18\text{ mol}}{1.18\text{ mol}} = 1$$

Those values become subscripts, so the empirical formula for the compound is C_4H_6P.

b) If the compound has a molar mass of 170.14 g/mol, what is its molecular formula?

First, we determine the molar mass of the empirical formula.

$$(12.01 \times 4) + (1.01 \times 6) + 30.97 = 85.07\text{ g/mol}$$

Then, we divide that mass into the molar mass.

$$\frac{170.14}{85.07} = 2$$

Finally, multiply all subscripts in the empirical formula by that value. So, the molecular formula is $C_8H_{12}P_2$.

Electron Configurations and the Periodic Table

The positively-charged nucleus is always pulling at the negatively-charged electrons around it, and the electrons have potential energy that increases with their distance from the nucleus. It works the same way that the gravitational potential energy of a brick on the third floor of a building is greater than the gravitational potential energy of a brick nearer to ground level.

The energy of electrons, however, is **quantized**. That's important. It means that electrons can exist only at specific energy levels, separated by specific intervals. It's kind of like if the brick in the building could be placed only on the first, second, or third floor of the building, but not in-between.

The Aufbau Principle

The **Aufbau principle** states that when building up the electron configuration of an atom, electrons are placed in orbitals, subshells, and shells in order of increasing energy.

The Pauli Exclusion Principle

The **Pauli Exclusion Principle** states that the two electrons which share an orbital cannot have the same spin. One electron must spin clockwise, and the other must spin counterclockwise.

Hund's Rule

Hund's rule says that when an electron is added to a subshell, it will always occupy an empty orbital if one is available. Electrons always occupy orbitals singly if possible and pair up only if no empty orbitals are available.

Watch how the 2*p* subshell fills as we go from boron to neon.

	1*s*	2*s*	2*p*		
Boron	⥮	⥮	↿		
Carbon	⥮	⥮	↿	↿	
Nitrogen	⥮	⥮	↿	↿	↿
Oxygen	⥮	⥮	⥮	↿	↿
Fluorine	⥮	⥮	⥮	⥮	↿
Neon	⥮	⥮	⥮	⥮	⥮

COULOMB'S LAW

The amount of energy that an electron has depends on its distance from the nucleus of the atom. This can be calculated using Coulomb's law:

$$E = \frac{k(+q)(-q)}{r}$$

E = energy

k = Coulomb's constant

$+q$ = magnitude of the positive charge (nucleus)

$-q$ = magnitude of the negative charge (electron)

r = distance between the charges

While on the exam, you will not be required to mathematically calculate the amount of energy a given electron has, you should be able to qualitatively apply Coulomb's law. Essentially, the greater the charge of the nucleus, the more energy an electron will have (as all electrons have the same amount of charge). Coulombic potential energy is considered to be 0 at a distance of infinity. The Coulombic potential energy for a 1*s* electron is lower (more negative) than that of say, a 3*s* electron. The amount of energy required to remove a 1*s* electron, thereby bringing its Coulombic potential energy to zero, will thus be greater than the amount needed to remove a 3*s* electron. This removal energy is called the binding energy of the electron and is always a positive value.

Quantum Theory

Max Planck figured out that electromagnetic energy is quantized. That is, for a given frequency of radiation (or light), all possible energies are multiples of a certain unit of energy, called a quantum (mathematically, that's $E = h\nu$). So, energy changes do not occur smoothly but rather in small but specific steps.

The Bohr Model

Neils Bohr took the quantum theory and used it to predict that electrons orbit the nucleus at specific, fixed radii, like planets orbiting the Sun.

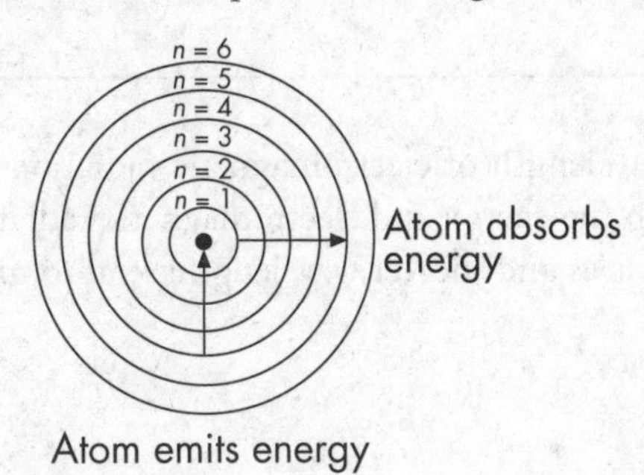

Each energy level is represented by a row on the periodic table. There are currently seven known energy levels, which correspond with $n = 1$ to $n = 7$. The closer an energy level is to an atom, the less energy electrons on that level have. While the Bohr model is not a perfect model of the atom, it serves as an excellent basis to understand atomic structure.

When atoms absorb energy in the form of **electromagnetic radiation,** electrons jump to higher energy levels. When electrons drop from higher to lower energy levels, atoms give off energy in the form of electromagnetic radiation.

The relationship between the change in energy level of an electron and the electromagnetic radiation absorbed or emitted is given below.

Energy and Electromagnetic Radiation

$$\Delta E = h\nu = \frac{hc}{\lambda}$$

ΔE = energy change
h = Planck's constant, 6.63×10^{-34} joule·sec
ν = frequency of the radiation
λ = wavelength of the radiation
c = the speed of light, 3.00×10^{8} m/sec

For a particular atom, the energy level changes of the electrons are always the same, so atoms can be identified by their emission and absorption spectra.

Frequency and Wavelength

$$c = \lambda\nu$$

c = speed of light in a vacuum (2.998×10^{8} ms^{-1})
λ = wavelength of the radiation
ν = frequency of the radiation

The frequency and wavelength of electromagnetic radiation are inversely proportional. Combined with the energy and electromagnetic radiation equation, we can see that higher frequencies and shorter wavelengths lead to more energy.

PHOTOELECTRON SPECTROSCOPY

If an atom is exposed to electromagnetic radiation at an energy level that exceeds the various binding energies of the electrons of that atom, the electrons can be ejected. The amount of energy necessary to do that is called the **ionization energy** for that electron. For the purposes of this exam, ionization energy and binding energy can be considered to be synonymous terms. When examining the spectra for electrons from a single atom or a small number of atoms, this energy is usually measured in electronvolts, eV (1 eV = 1.60 × 10–19 Joules). If moles of atoms are studied, the unit for binding energy is usually either kJ/mol or MJ/mol.

All energy of the incoming radiation must be conserved and any of that energy that does not go into breaking the electron free from the nucleus will be converted into **kinetic energy** (the energy of motion) for the ejected electron. So:

Incoming Radiation Energy =
Binding Energy + Kinetic Energy (of the ejected electron)

The faster an ejected electron is going, the more kinetic energy it has. Electrons that were originally further away from the nucleus require less energy to eject, and thus will be moving faster. So, by examining the speed of the ejected electrons, we can determine how far they were from the nucleus of the atom in the first place. Usually, it takes electromagnetic radiation in either the visible or ultraviolet range to cause electron emission, while radiation in the infrared range is often used to study chemical bonds. Radiation in the microwave region is used to study the shape of molecules.

Spectra

If the amount of ionization energy for all electrons ejected from a nucleus is charted, you get what is called a **photoelectron spectra** (PES) that looks like the following:

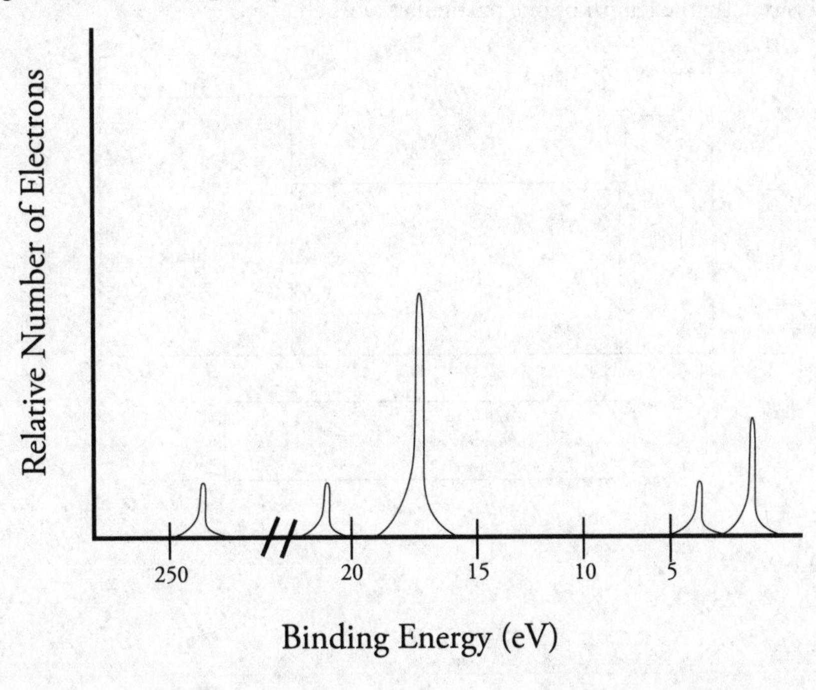

The *y*-axis describes the relative number of electrons that are ejected from a given energy level, and the *x*-axis shows the binding energy of those electrons. Unlike most graphs, binding energy (ionization energy) decreases going from left to right in a PES. The spectra on the previous page is for sulfur.

Each section of peaks in the PES represents a different energy level. The number of peaks in a section shows us that not all electrons at $n = 2+$ are located the same distance from the nucleus. With each energy level, there are **subshells**, which describe the shape of the space the electron can be found in (remember, we are in three dimensional space here). The Bohr model is limited to two dimensions and does not represent the true positions of electrons due to that reason. Electrons do not orbit the nucleus as planets orbit the Sun. Instead, they are found moving about in a certain area of space (the subshell) a given distance (the energy level) away from the nucleus.

In all energy levels, the first subshell is called the *s*-subshell and can hold a maximum of two electrons. The second subshell is called the *p*-subshell and can hold a maximum of six electrons. In the spectra on the previous page, we can see the peak for the *p*-subshell in energy level 2 is three times higher than that of the *s*-subshell. The relative height of the peaks helps determine the number of electrons in that subshell.

In the area for the third energy level, the *p*-subshell peak is only twice as tall as the *s*-subshell. This indicates there are only four electrons in the *p*-subshell of this particular atom.

Electron Configuration

Studying the PES of elements allows scientists to understand more about the structure of the atom. In addition to the *s* and *p* subshells, two others exist: *d* (10 electrons max) and *f* (14 electrons max). The periodic table is designed so that each area is exactly the length of one particular subshell.

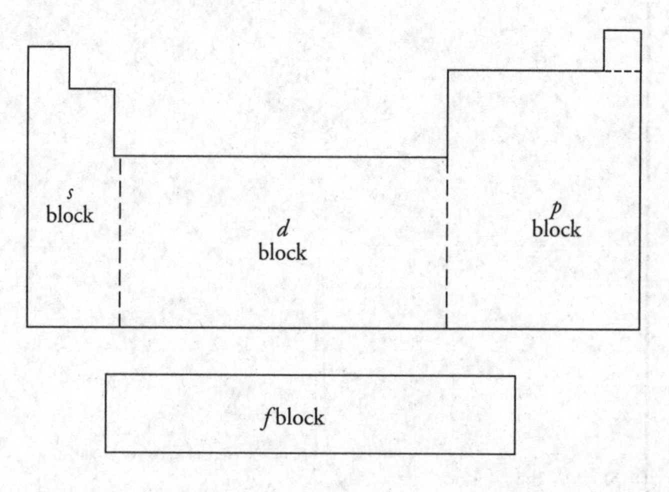

As you can see, the first two groups (plus helium) are in what is called the *s*-block. The groups on the right of the table are in the *p*-block, while transition metals make up the *d*-block. The inner transition metals below the table inhabit the *f*-block. The complete description of the energy level and subshell that each electron on an element inhabits is called its **electron configuration.**

Example 1: Determine the electron configuration for sulfur.

Sulfur has 16 electrons. The first two go into energy level 1 subshell *s*; this is represented by $1s^2$. The next two go into energy level 2 subshell *s*—$2s^2$. Six more fill energy level *p* ($2p^6$) then two more go into 3s ($3s^2$) and the final four enter into $3p$ ($3p^4$). So the final configuration is $1s^22s^22p^63s^23p^4$.

Using the periodic table as a reference can allow for the determination of any electron configuration. One thing to watch out for is that when entering the *d*-block (transition metals), the energy level drops by 1. The why behind that isn't important right now, but you should still be able to apply that rule.

Example 2: Determine the electron configuration for nickel.

$$1s^22s^22p^63s^23p^63d^84s^2$$

Electron configurations can also be written "shorthand" by replacing parts of them with the symbol for the noble gas at the end of the highest energy level which has been filled.

For example, the shorthand notation for Si is $[Ne]3s^23p^3$ and the shorthand for nickel would be $[Ar]4s^23d^8$.

PREDICTING IONIC CHARGES

One of the great rules of chemistry is that the most stable configurations from an energy standpoint are those in which the outermost energy level is full. For anything in the *s* or *p* blocks, that means achieving a state in which there are eight electrons in the outermost shell (2 in the s subshell and 6 in the p subshell).

Elements that are close to a full energy level, such as the halogens or those in the oxygen group, tend to gain electrons to achieve a stable configuration. An **ion** is an atom which has either gained or lost electrons, while the number of protons and neutrons remains constant. Halogens need only gain one electron to achieve a stable configuration, and as such, typically form ions with a charge of negative one. F^-, Cl^-, etc. Any particle with more electrons than protons is called an **anion** (negatively charged ion). Those elements in the oxygen group need two electrons for stability; thus, they have a charge of negative 2.

It's tempting to say that fluorine atoms "want" one electron or are "happy" with a full valence shell, but in reality, atoms don't have feelings and don't want anything! Instead, say fluorine atoms need only attract one electron to have a stable energy level. Assigning feelings and desires to inanimate particles won't do you any favors on the exam

On the other end of the table, the alkali metals can most easily achieve a full valence shell by losing a single electron, rather than by gaining seven. So, they will form positively charged ions (**cations**), which have more protons than electrons. Alkali metals typically have a charge of +1, alkaline earth metals of +2, and so forth. The table below gives a good overview of the common ionic charges for various groups:

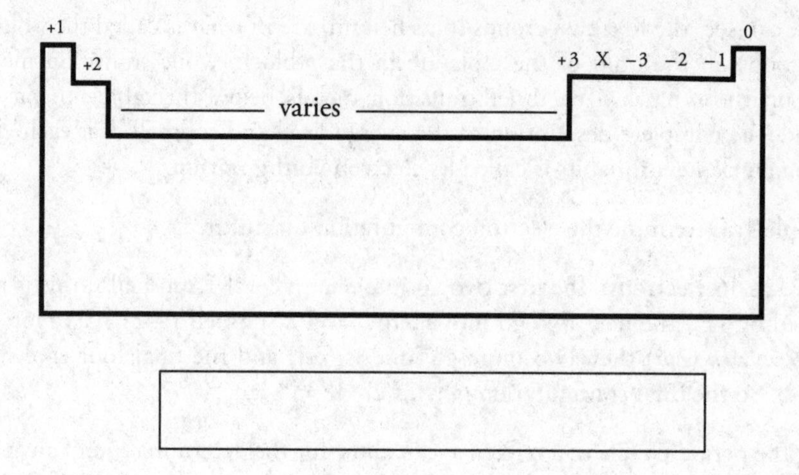

Note that the transition metals tend to have varying charges. They do all form cations, but most can form cations with multiple charges, depending on the situation.

There's also a slightly different rule as to how transition metals lose electrons when forming those cations. Transition metals, when losing electrons, will lose their higher-level *s* electrons before losing any of the lower-level *d* electrons. Thus, while an iron atom has a configuration of $[Ar]4s^2 3d^6$, an iron ion that has lost two electrons (Fe^{2+}) would have a configuration of $[Ar]3d^6$. If it were to lose further electrons, those would then come from the *d*-orbital.

Names and Theories

Dalton's Elements

In the early 1800s, John Dalton presented some basic ideas about atoms that we still use today. He was the first to say that there are many different kinds of atoms, which he called elements. He said that these elements combine to form compounds and that these compounds always contain the same ratios of elements. Water (H_2O), for instance, always has two hydrogen atoms for every oxygen atom. He also said that atoms are never created or destroyed in chemical reactions.

Development of the Periodic Table

In 1869, Dmitri Mendeleev and Lothar Meyer independently proposed arranging the elements into early versions of the periodic table, based on the trends of the known elements.

Thomson's Experiment

In the late 1800s, J. J. Thomson watched the deflection of charges in a cathode ray tube and put forth the idea that atoms are composed of positive and negative charges. The negative charges were called electrons, and Thomson guessed that they were sprinkled throughout the positively charged atom like chocolate chips sprinkled throughout a blob of cookie dough.

The Plum Pudding Model of an Atom

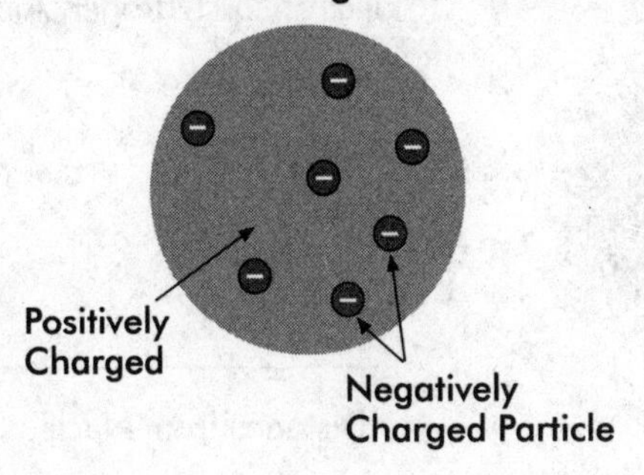

Millikan's Experiment

Robert Millikan was able to calculate the charge on an electron by examining the behavior of charged oil drops in an electric field.

Rutherford's Experiment

In the early 1900s, Ernest Rutherford fired alpha particles at gold foil and observed how they were scattered. This experiment led him to conclude that all of the positive charge in an atom was concentrated in the center and that an atom is mostly empty space. This led to the idea that an atom has a positively charged nucleus, which contains most of the atom's mass, and that the tiny, negatively charged electrons travel around this nucleus.

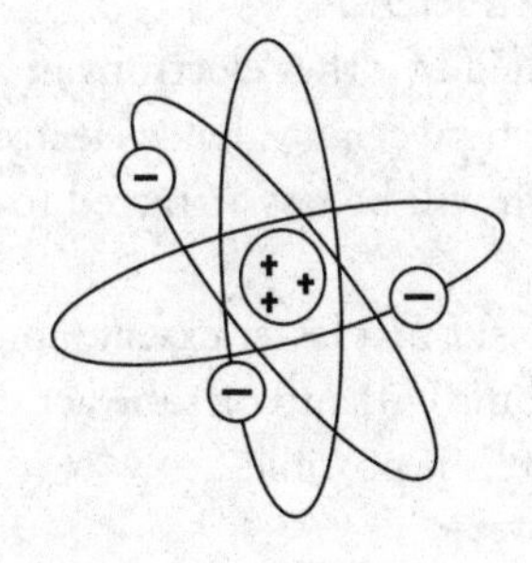

The Heisenberg Uncertainty Principle

Werner Heisenberg said that it is impossible to know both the position and momentum of an electron at a particular instant. In terms of atomic structure, this means that electron orbitals do not represent specific orbits like those of planets. Instead, an electron orbital is a probability function describing the possibility that an electron will be found in a region of space.

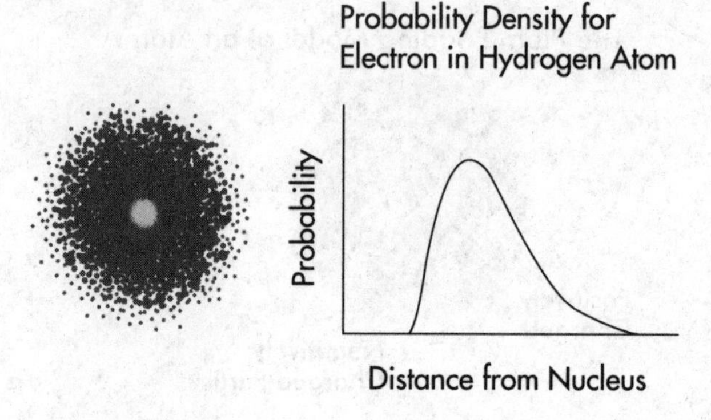

PERIODIC TRENDS

You can make predictions about certain behavior patterns of an atom and its electrons based on the position of the atom in the periodic table. All the periodic trends can be understood in terms of three basic rules.

1. Electrons are attracted to the protons in the nucleus of an atom.
    a. The closer an electron is to the nucleus, the more strongly it is attracted.
    b. The more protons in a nucleus, the more strongly an electron is attracted.
2. Electrons are repelled by other electrons in an atom. So, if other electrons are between a valence electron and the nucleus, the valence electron will be less attracted to the nucleus. That's called shielding.
3. Completed shells (and to a lesser extent, completed subshells) are very stable. Atoms will add or subtract valence electrons to create complete shells if possible.

The atoms in the left-hand side of the periodic table are called metals. Metals give up electrons when forming bonds. Most of the elements in the table are metals. The elements in the upper right-hand portion of the table are called nonmetals. Nonmetals generally gain electrons when forming bonds. The metallic character of the elements decreases as you move from left to right across the periodic table. The elements in the borderline between metal and nonmetal, such as silicon and arsenic, are called metalloids.

Atomic Radius

The atomic radius is the approximate distance from the nucleus of an atom to its valence electrons.

Moving from Left to Right Across a Period (Li to Ne, for Instance), Atomic Radius Decreases

Moving from left to right across a period, protons are added to the nucleus, so the valence electrons are more strongly attracted to the nucleus; this decreases the atomic radius. Electrons are also being added, but they are all in the same shell at about the same distance from the nucleus, so there is not much of a shielding effect.

Moving Down a Group (Li to Cs, for Instance), Atomic Radius Increases

Moving down a group, shells of electrons are added to the nucleus. Each shell shields the more distant shells from the nucleus and the valence electrons get farther away from the nucleus. Protons are also being added, but the shielding effect of the negatively charged electron shells cancels out the added positive charge.

Cations (Positively Charged Ions) Are Smaller than Atoms

Generally, when electrons are removed from an atom to form a cation, the outer shell is lost, making the cation smaller than the atom. Also, when electrons are removed, electron–electron repulsions are reduced, allowing all of the remaining valence electrons to move closer to the nucleus.

Anions (Negatively Charged Ions) Are Larger than Atoms

When an electron is added to an atom, forming an anion, electron–electron repulsions increase, causing the valence electrons to move farther apart, which increases the radius.

Ionization Energy

Electrons are attracted to the nucleus of an atom, so it takes energy to remove an electron. The energy required to remove an electron from an atom is called the first ionization energy. Once an electron has been removed, the atom becomes a positively charged ion. The energy required to remove the next electron from the ion is called the second ionization energy, and so on.

Moving from Left to Right Across a Period, Ionization Energy Increases

Moving from left to right across a period, protons are added to the nucleus, which increases its positive charge. For this reason, the negatively charged valence electrons are more strongly attracted to the nucleus, which increases the energy required to remove them.

Moving Down a Group, Ionization Energy Decreases

Moving down a group, shells of electrons are added to the nucleus. Each inner shell shields the more distant shells from the nucleus, reducing the pull of the nucleus on the valence electrons and making them easier to remove. Protons are also being added, but the shielding effect of the negatively charged electron shells cancels out the added positive charge.

The Second Ionization Energy Is Greater than the First Ionization Energy

When an electron has been removed from an atom, electron–electron repulsion decreases and the remaining valence electrons move closer to the nucleus. This increases the attractive force between the electrons and the nucleus, increasing the ionization energy.

As Electrons Are Removed, Ionization Energy Increases Gradually Until a Shell Is Empty, Then Makes a Big Jump

- For each element, when the valence shell is empty, the next electron must come from a shell that is much closer to the nucleus, making the ionization energy for that electron much larger than for the previous ones.
- For Na, the second ionization energy is much larger than the first.
- For Mg, the first and second ionization energies are comparable, but the third is much larger than the second.
- For Al, the first three ionization energies are comparable, but the fourth is much larger than the third.

Electronegativity

Electronegativity refers to how strongly the nucleus of an atom attracts the electrons of other atoms in a bond. Electronegativity is affected by two factors. The smaller an atom is, the more effectively its nuclear charge will be felt past its outermost energy level and the higher its electronegativity will be. Second, the closer an element is to having a full energy level, the more likely it is to attract the necessary electrons to complete that level. In general

- Moving from left to right across a period, electronegativity increases.
- Moving down a group, electronegativity decreases.

The various periodic trends are summarized in the diagram below. The primary exception to these trends are the electronegativity values for the three smallest noble gases. As helium, neon, and argon do not form bonds, they have zero electronegativity. The larger noble gases, however, can form bonds under certain conditions and do follow the general trends as outlined in this section. This will be discussed in more detail in the next chapter.

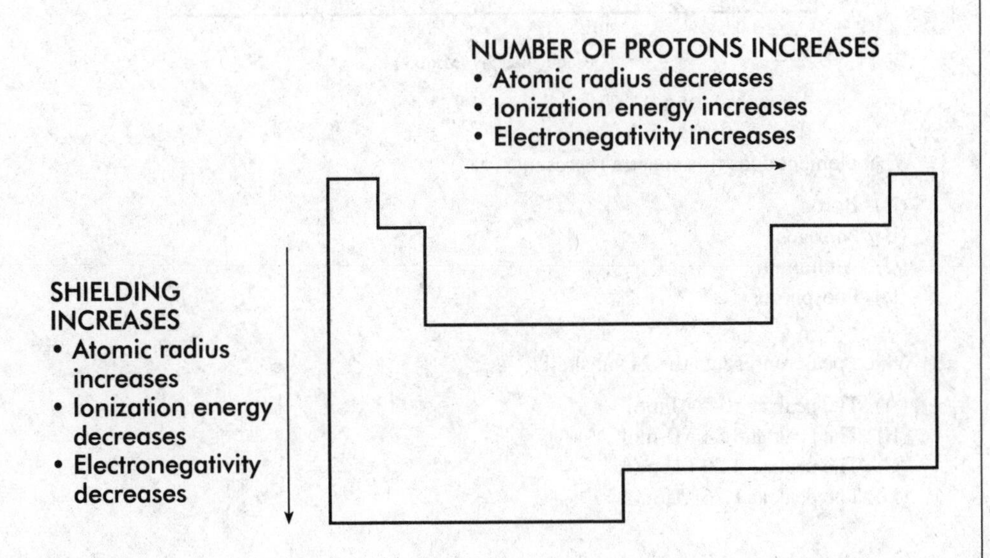

CHAPTER 3 QUESTIONS

Multiple-Choice Questions

Use the PES spectra below to answer questions 1-4.

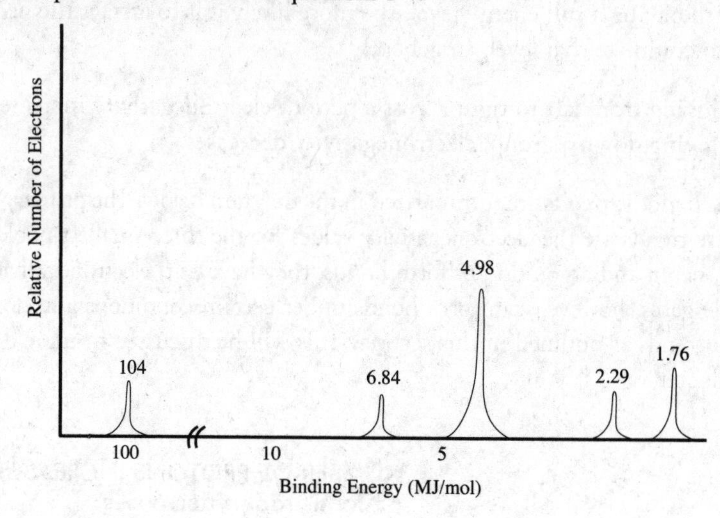

1. What element does this spectra represent?
   (A) Boron
   (B) Nitrogen
   (C) Aluminum
   (D) Phosphorus

2. Which peak represents the 2*s* subshell?
   (A) The peak at 104 MJ/mol
   (B) The peak at 6.84 MJ/mol
   (C) The peak at 2.29 MJ/mol
   (D) The peak at 1.76 MJ/mol

3. An electron from which peak would have the greatest velocity after ejection?
   (A) The peak at 104 MJ/mol
   (B) The peak at 6.84 MJ/mol
   (C) The peak at 4.98 MJ/mol
   (D) The peak at 1.76 MJ/mol

4. How many valence electrons does this atom have?
   (A) 2
   (B) 3
   (C) 4
   (D) 5

5. Why does an ion of phosphorus, P^{3-}, have a larger radius than a neutral atom of phosphorus?

   (A) There is a greater Coulombic attraction between the nucleus and the electrons in P^{3-}.
   (B) The core electrons in P^{3-} exert a weaker shielding force than those of a neutral atom.
   (C) The nuclear charge is weaker in P^{3-} than it is in P.
   (D) The electrons in P^{3-} have a greater Coulombic repulsion than those in the neutral atom.

6. Which neutral atom of the following elements would have the most unpaired electrons?

   (A) Titanium
   (B) Manganese
   (C) Nickel
   (D) Zinc

7. Which element will have a higher electronegativity value: chlorine or bromine? Why?

   (A) Chlorine, because it has less Coulombic repulsion among its electrons
   (B) Bromine, because it has more protons
   (C) Chlorine, because it is smaller
   (D) Bromine, because it is larger

8. Which of the following elements has its highest energy subshell completely full?

   (A) Sodium
   (B) Aluminum
   (C) Chlorine
   (D) Zinc

9. Which of the following isoelectric species has the smallest radius?

   (A) S^{2-}
   (B) Cl^-
   (C) Ar
   (D) K^+

10. What is the most likely electron configuration for a sodium ion?

    (A) $1s^22s^22p^5$
    (B) $1s^22s^22p^6$
    (C) $1s^22s^22p^63s^1$
    (D) $1s^22s^22p^53s^2$

11. Which of the following statements is true regarding sodium and chlorine?

    (A) Sodium has greater electronegativity and a larger first ionization energy.
    (B) Sodium has a larger first ionization energy and a larger atomic radius.
    (C) Chlorine has a larger atomic radius and a greater electronegativity.
    (D) Chlorine has greater electronegativity and a larger first ionization energy.

12. An atom of silicon in its ground state is subjected to a frequency of light that is high enough to cause electron ejection. An electron from which subshell of silicon would have the highest kinetic energy after ejection?

(A) 1*s*
(B) 2*p*
(C) 3*p*
(D) 4*s*

13. The wavelength range for infrared radiation is 10^{-5} m, while that of ultraviolet radiation is 10^{-8} m. Which type of radiation has more energy, and why?

(A) Ultraviolet has more energy because it has a higher frequency.
(B) Ultraviolet has more energy because it has a longer wavelength.
(C) Infrared has more energy because it has a lower frequency.
(D) Infrared has more energy because it has a shorter wavelength.

14. Which of the following nuclei has 3 more neutrons than protons? (Remember: The number before the symbol indicates atomic mass.)

(A) ^{11}B
(B) ^{37}Cl
(C) ^{24}Mg
(D) ^{70}Ga

15. Which of the following is true of the halogens in periods two and three when comparing them to other elements in the same period?

(A) Halogens have larger atomic radii than other elements within their period.
(B) Halogens have less ionization energy than other elements within their period.
(C) Halogens have fewer peaks on a PES than other elements within their period.
(D) The electronegativity of halogens is higher than other elements within their period.

16. In general, do metals or nonmetals from the same period have higher ionization energies? Why?

(A) Metals have higher ionization energies because they usually have more protons than nonmetals.
(B) Nonmetals have higher ionization energies because they are larger than metals and harder to ionize.
(C) Metals have higher ionization energies because there is less electron shielding than there is in nonmetals.
(D) Nonmetals have higher ionization energies because they are closer to having filled a complete energy level.

17. The ionization energies for an element are listed in the table below.

First	Second	Third	Fourth	Fifth
8 eV	15 eV	80 eV	109 eV	141 eV

Based on the ionization energy table, the element is most likely to be

(A) sodium
(B) magnesium
(C) aluminum
(D) silicon

Use the following information to answer questions 18-20.

The outermost electron of an atom has a binding energy of 2.5 eV. The atom is exposed to light of a high enough frequency to cause exactly one electron to be ejected. The ejected electron is found to have a KE of 2.0 eV.

18. How much energy did photons of the incoming light contain?

(A) 0.50 eV
(B) 0.80 eV
(C) 4.5 eV
(D) 5.0 eV

19. If the wavelength of the light were to be shortened, how would that effect the kinetic energy of the ejected electron?

(A) A shorter wavelength would increase the kinetic energy.
(B) A shorter wavelength would decrease the kinetic energy.
(C) A shorter wavelength would stop all electron emissions completely.
(D) A shorter wavelength would have no effect on the kinetic energy of the ejected electrons.

20. If the intensity of the light were to be decreased (that is, if the light is made dimmer), how would that affect the kinetic energy of the ejected electron?

(A) The decreased intensity would increase the kinetic energy.
(B) The decreased intensity would decrease the kinetic energy.
(C) The decreased intensity would stop all electron emissions completely.
(D) The decreased intensity would have no effect.

21. Which type of radiation would be most useful in examining the dimensionality of molecules?

(A) Ultraviolet
(B) Visible
(C) Infrared
(D) Microwave

22. Which of the following ions would have the most unpaired electrons?

(A) Mn^{2+}
(B) Ni^{3+}
(C) Ti^{2+}
(D) Cr^{6+}

23. Which of the following statements best explains why phosphorus ions tend to have a charge of –3?

(A) Phosphorus atoms want to gain three electrons in order to fill their p-subshell.
(B) Phosphorus is only three electrons away from having the same number of electrons as a noble gas.
(C) Phosphorus will be the most stable with a full third energy level.
(D) Phosphorus is in period three of the periodic table, and thus needs three additional electrons.

Free-Response Questions

1. Explain each of the following in terms of atomic and molecular structures and/or forces.

   (a) The first ionization energy for magnesium is greater than the first ionization energy for calcium.

   (b) The first and second ionization energies for calcium are comparable, but the third ionization energy is much greater.

   (c) There are three peaks of equal height in the PES of carbon, but on the PES of oxygen the last peak has a height twice as high as all the others.

   (d) The first ionization energy for aluminum is lower than the first ionization energy for magnesium.

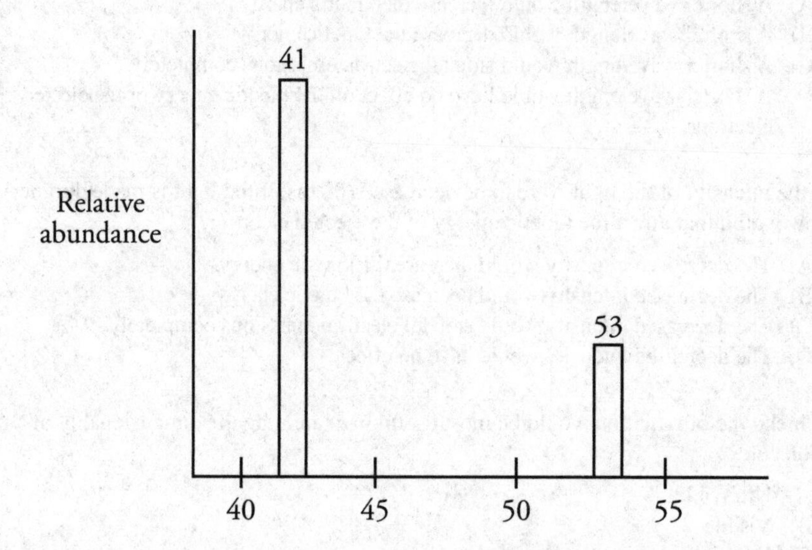

2. The above mess spectra is for the hypochlorite ion, ClO^-. Oxygen has only one stable isotope, which has a mass of 16 amu.

   (a) How many neutrons does the most common isotope of chlorine have?

   (b) Using the spectra, calculate the average mass of a hypochlorite ion.

   (c) Does the negative charge on the ion affect the spectra? Justify your answer.

   (d) The negative charge in the ion is located around the oxygen atom. Speculate as to why.

3. The table below gives data on four different elements, in no particular order:

Carbon, Oxygen, Phosphorus, and Chlorine

	Atomic radius (pm)	First Ionization Energy (kJ/mol^{-1})
Element 1	170	1086.5
Element 2	180	1011.8
Element 3	175	1251.2
Element 4	152	1313.9

(a) Which element is number 3? Justify your answer using both properties.
(b) What is the outermost energy level that has electrons in element 2? How many valence electrons does element 2 have?
(c) Which element would you expect to have the highest electronegativity? Why?
(d) How many peaks would the PES for element 4 have and what would the relative heights of those peaks be to each other?

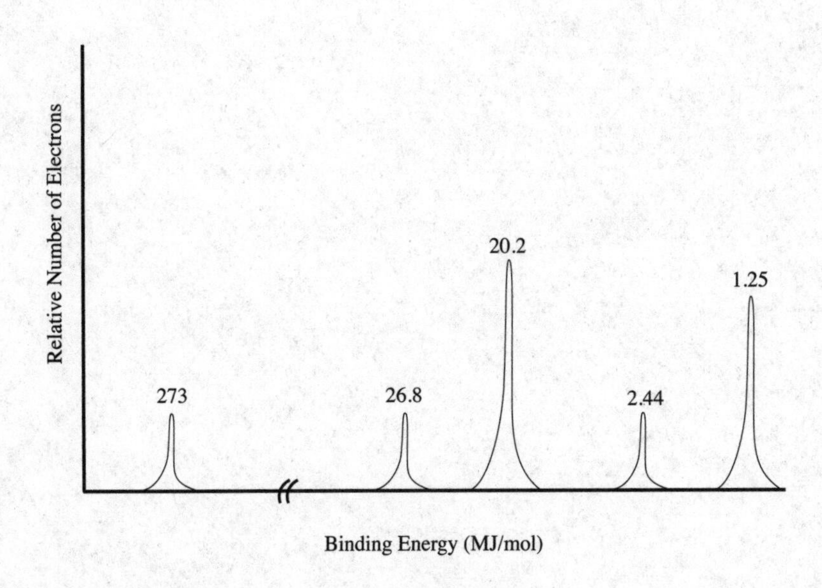

4. The above PES belongs to a neutral chlorine atom.
(a) What wavelength of light would be required to eject a 3*s* electron from chlorine?
(b) For the PES of a chloride ion, how would the following variables compare to the peaks on the PES above? Justify your answers.
(i) Number of peaks
(ii) Height of the peaks

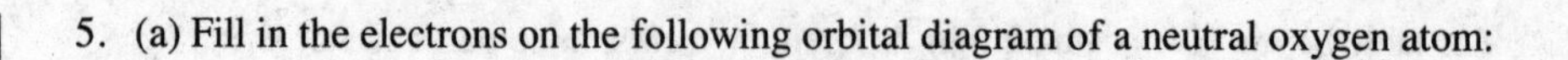

5. (a) Fill in the electrons on the following orbital diagram of a neutral oxygen atom:

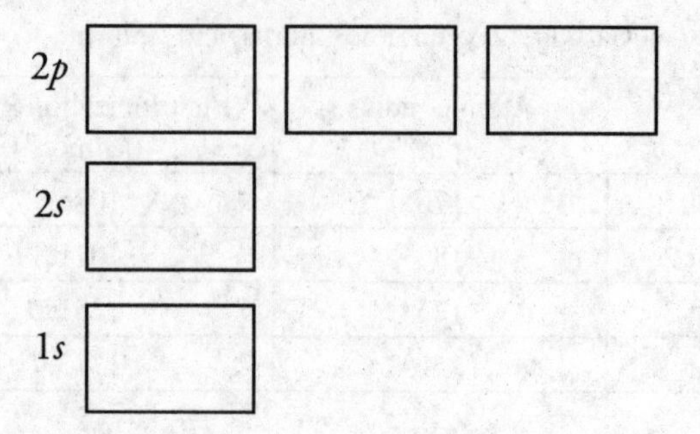

(b) Using your diagram above, explain why the most common charge on oxygen ions is –2.

(c) Would oxygen atoms be deflected if they were shot through a magnetic field? Why or why not?

CHAPTER 3 ANSWERS AND EXPLANATIONS

Multiple-Choice

1. **D** This element has five peaks, meaning a total of five subshells. The final peak, which would be located in the 3*p* subshell, is slightly higher than the 3*s* peak to the left of it. A full 3*s* peak has two electrons, therefore there must be at least three electrons in the 3*p* subshell. The element that best fits this is phosphorus.

2. **B** The peaks, in order, represent 1*s*, 2*s*, 2*p*, 3*s*, and 3*p*.

3. **D** The less ionization energy that is required to remove an electron, the more kinetic energy that electron will have after ejection.

4. **D** Valence electrons are those in the outermost energy level. In this case, that is the third level, which has five valence electrons in it (two in 3*s* and three in 3*p*).

5. **D** The ion has three more electrons than the neutral atom, meaning the overall repulsion will be greater. The electrons will "push" each other away more effectively, creating a bigger radius.

6. **B** In all of these transition metals, the 3*d* subshell is the last one to receive electrons (Aufbau's Principle). A 3*d* subshell has 5 orbitals which can hold two electrons each (Pauli's Exclusion Principle). The electrons will enter one by one into each orbital before any pair up (Hund's Rule), meaning manganese will have all five of its 3*d* electrons unpaired.

7. **C** The smaller an atom is, the more the "pull" of the nucleus can be felt and the easier it will be for that element to attract more electrons; thus the best answer is chlorine. Choice (B) is a tempting answer; however, the additional protons in bromine are much farther away from the outside of the atom, as it has one full energy level greater than chlorine.

8. **D** Zinc has 10 electrons in the 3*d* subshell, filling it completely.

9. **D** Potassium has the highest number of protons out of all the options, therefore it exerts a higher nuclear charge on the electrons, pulling them in closer and creating a smaller atomic radius.

10. **B** Neutral sodium in its ground state has the electron configuration shown in (C). Sodium forms a bond by giving up its one valence electron ($3s^1$) and becoming a positively charged ion with the same electron configuration as neon, which is (B).

11. **D** As we move from left to right across the periodic table within a single period (from sodium to chlorine), we add protons to the nuclei, which progressively increases the pull of each nucleus on its electrons. So chlorine will have a larger first ionization energy, greater electronegativity, and a smaller atomic radius.

12. **C** Electrons that are further away from the nucleus require less ionization energy to remove, and thus will have a greater velocity after being ejected. An electron from $3p$ would be further away from the nucleus than electrons in either $1s$ or $2p$. A $4s$ electron would be faster; however, silicon has no electrons in the $4s$ subshell while in its ground state.

13. **A** Via $c = \nu\lambda$, we can see that there is an inverse relationship between wavelength and frequency (as one goes up, the other goes down). So, ultraviolet radiation has a higher frequency, and via $E = h\nu$, also has more energy.

14. **B** The atomic mass is the sum of the neutrons and protons in any atom's nucleus. Since atomic number, which indicates the number of protons, is unique to each element we can subtract this from the weight to find the number of neutrons. The atomic number of B is 5; hence, the ^{11}B has 6 neutrons, only 1 in excess. The atomic number of Cl is 17, so ^{37}Cl has 20 neutrons, 3 in excess. The atomic number of Mg is 12, so ^{24}Mg has the same number of neutrons as protons. The element Ga has an atomic number of 31, meaning there are 39 neutrons in the given nucleus.

15. **D** Electronegativity describes how easy it is for an element to attract additional electrons. Because halogens are smaller than other elements in their period (save for the noble gases, which have full energy levels and are not likely to gain additional electrons), they tend to have the highest electronegativity values within their period.

16. **D** Nonmetals appear on the right side of the periodic table, and so tend to be smaller than the other elements in their period. For this reason, it is easier for them to attract additional electrons, which means they have higher electronegativity values.

17. **B** The ionization energy will show a large jump when an electron is removed from a full shell. In this case, the jump occurs between the second and third electrons removed, so the element is stable after two electrons are removed. Magnesium (Mg) is the only element on the list with exactly two valence electrons.

18. **C** Radiation Energy = Binding Energy + Kinetic Energy

Radiation energy (photon) = 2.5 eV + 2.0 eV

Radiation energy (photon) = 4.5 eV

19. **A** If the wavelength is decreased, that increases the frequency via $c = \lambda v$ (remember, the speed of light is constant). An increased frequency increases the amount of photon energy via $\Delta E = hv$. As the binding energy of the electron would not change, the excess radiation energy would turn into kinetic energy, increasing the velocity of the electron.

20. **D** The intensity of the light is independent of the amount of energy that the light has. Energy is entirely based on frequency and wavelength, and the brightness of the light would not change the amount of radiation energy. Thus, the amount of kinetic energy of the ejected electrons would not change either.

21. **D** Microwave radiation is used to determine the shape of full molecules, including whether or not they are fully planar or three-dimensional.

22. **A** Remember, transition metals lose their *s*-electrons first when forming an ion. A manganese atom is initially $[Ar]4s^2 3d^5$, but the ion becomes $[Ar]3d^5$, which has a total of five unpaired electrons (as Hund's rule states the electrons will remain unpaired as long as there are empty orbitals for them to enter).

23. **C** Ions form in order to create a stable energy configuration, and the most stable energy configuration is one in which the outermost energy level is completely full. In (A), the problem is the word "want"—atoms don't have feelings. And while (B) may be true, that doesn't explain the reason for the ionic charge; it just gives a handy way to predict it!

Free-Response

1. (a) Ionization energy is the energy required to remove an electron from an atom. The outermost electron in Ca is at the 4*s* energy level. The outermost electron in Mg is at the 3*s* level. The outermost electron in Ca is at a higher energy level and is more shielded from the nucleus, making it easier to remove.

   (b) Calcium has two electrons in its outer shell. The second ionization energy will be larger than the first but still comparable because both electrons are being removed from the same energy level. The third electron is much more difficult to remove because it is being removed from a lower energy level, so it will have a much higher ionization energy than the other two.

   (c) The height of the peaks on a PES represent the relative number of electrons in each subshell. In carbon, all three subshells hold two electrons ($1s^2 2s^2 2p^2$), and thus all peaks are the same height. In oxygen, the 2*p* subshell has four electrons, meaning its peak will be twice as high as the other two.

   (d) The valence electron to be removed from magnesium is located in the completed 3*s* subshell, while the electron to be removed from aluminum is the lone electron in the 3*p* subshell. It is easier to remove the electron from the higher-energy 3*p* subshell than from the lower energy (completed) 3*s* subshell, so the first ionization energy is lower for aluminum.

2. (a) The most common mass of a ClO^- ion is 51 amu. 51 amu-16 amu = 35 amu, which must be the mass of the most common isotope of chlorine. As mass number is equal to protons + neutrons, and chlorine has 17 protons (its atomic number), 35 – 17 = 18 neutrons.

   (b) 51(0.75) + 53(0.25) = 51.5 amu

   (c) No. The only subatomic particles that contribute to the mass of any atom are neutrons and protons. Changing the number of electrons does not change the mass significantly.

(d) An oxygen atom is smaller than a chlorine atom, and as such is more electronegative. The electrons in the bond are thus more attracted to oxygen than chlorine.

3. (a) Element 3 is chlorine. Chlorine and phosphorus would have the largest atomic radii as they both have three energy levels with electrons present. However, chlorine would be smaller than phosphorus because it has more protons (a higher effective nuclear charge). Additionally, chlorine would have a higher ionization energy than phosphorus due to its smaller size and greater number of protons.

   (b) Element 2 is phosphorus, and therefore the outermost energy level would be $n = 3$. Phosphorus has two electrons in 3*s* and three electrons in 3*p* for a total of five valence electrons.

   (c) Electronegativity increases as atomic radius decreases, so it is expected that element 4 (oxygen) would have the highest electronegativity value. Alternatively, electronegativity increases as an energy level comes close to being full, so it is possible that element 3 (chlorine) may have the highest electronegativity as it is only one electron away from filling its outermost energy level. (Either answer is acceptable with the proper justification.)

   (d) Element 4 is oxygen, so it would be expected to have three peaks in a PES, one for each subshell. The first two peaks would be the same height because there are two electrons each in the 1*s* and 2*s* subshells. The final peak would be twice the height of the others, as there are four electrons in oxygen's 2*p* subshell.

4. (a) The 3*s* electron belongs to the peak located at 2.44 MJ/mol. First, you need to calculate the amount of energy needed to remove a single 3*s* electron (rather than a mole of them):

$$\frac{2.44 \text{ MJ}}{1 \text{ mol}} \times \frac{1 \text{ mol}}{6.02 \times 10^{23} \text{ electrons}} = 4.05 \times 10^{-24} \text{ MJ} = 4.05 \times 10^{-18} \text{ J}$$

Then, you can use $E = hc/\lambda$ to calculate the wavelength of light that would have sufficient energy:

$$4.05 \times 10^{-18}\text{ J} = \frac{(6.63 \times 10^{-34}\text{ Js})(3.0 \times 10^{8}\text{ m/s})}{\lambda}$$

$$\lambda = 4.9 \times 10^{-8}\text{ m}$$

(b) (i) The chloride ion would have one more electron, which would enter the $3p$ energy level. This is represented by the rightmost peak, and so the PES of chloride would have the same number of peaks as that of the chlorine atom.

(ii) The first four peaks would have the same height, but the last peak would have an additional electron. That would give it six total electrons, making it the same height as the $2p$ (middle) peak shown in the PES of the neutral chlorine atom.

5. (a)

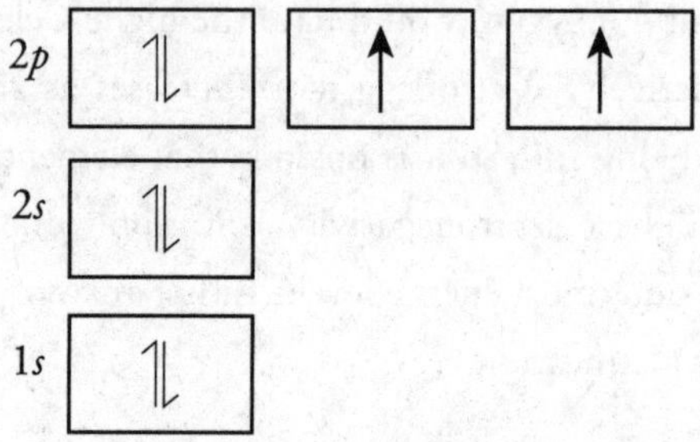

$2p$	↑↓	↑	↑
$2s$	↑↓		
$1s$	↑↓		

Via Hund's rule, we know the electrons in the $2p$ energy level will fill each empty orbital before pairing up.

(b) The oxygen atom will be most stable when it has full energy levels. The second energy level is only two electrons away from being full, thus gaining two electrons causes an oxygen ion to be in its most stable state.

(c) Yes. Atoms are paramagnetic when they have unpaired electrons, as the oxygen atom does. This causes them to be deflected if exposed to a magnetic field. Only atoms with no unpaired electrons (such as the alkaline earth metals or noble gases) would be unaffected by a magnetic field.

Chapter 4
Big Idea #2: Bonding and Phases

Chemical and physical properties of materials can be explained by the structure and the arrangement of atoms, ions, or molecules and the forces between them.

BONDS OVERVIEW

Atoms engage in chemical reactions in order to reach a more stable, lower-energy state. This requires the transfer or sharing of electrons, a process that is called **bonding**. Atoms of elements are usually at their most stable when they have eight electrons in their valence shells. As a result, atoms with too many or too few electrons in their valence shells will find one another and pass the electrons around until all the atoms in the molecule have stable outer shells. Sometimes an atom will give up electrons completely to another atom, forming an ionic bond. Sometimes atoms share electrons, forming covalent bonds.

IONIC BONDS

An ionic solid is held together by the electrostatic attractions between ions that are next to one another in a lattice structure. They often occur between metals and nonmetals. In an ionic bond, electrons are not shared. Instead, the cation gives up an electron (or electrons) to the anion.

The two ions in an ionic bond are held together by electrostatic forces. In the diagram below, a sodium atom has given up its single valence electron to a chlorine atom, which has seven valence electrons and uses the electron to complete its outer shell (with eight). The two atoms are then held together by the positive and negative charges on the ions.

$$[Na]^+[:\ddot{\underset{..}{Cl}}:]^-$$

The electrostatic attractions that hold together the ions in the NaCl lattice are very strong and any substance held together by ionic bonds will usually be a solid at room temperature and have very high melting and boiling points.

Two factors affect the melting points of ionic substances. The primary factor is the charge on the ions. According to Coulomb's law, a greater charge leads to a greater bond energy (often called lattice energy in ionic bonds), so a compound composed of ions with charges of +2 and –2 (such as MgO) will have a higher melting point than a compound composed of ions with charges of +1 and –1 (such as NaCl). If both compounds are made up of ions with equal charges, then the size of the ions must be considered. Smaller ions will have greater Coulombic attraction (remember, size is inversely proportional to bond energy), so a substance like LiF would have a greater melting point than KBr.

In an ionic solid, each electron is localized around a particular atom, so electrons do not move around the lattice; this makes ionic solids poor conductors of electricity. Ionic liquids, however, do conduct electricity because the ions themselves are free to move about in the liquid phase, although the electrons are still localized around particular atoms. Salts are held together by ionic bonds.

METALLIC BONDS

When examining metals, the sea of electrons model can be used. The positively-charged core of a metal, consisting of its nucleus and core electrons, is generally stationary, while the valence electrons on each atom do not belong to a specific atom and are very mobile. These mobile electrons explain why metals are such good conductors of electricity. The delocalized structure of a metal also explains why metals are both malleable and ductile, as deforming the metal does not change the environment immediately surrounding the metal cores.

Metals can also bond with each other to form alloys. This typically occurs when two metals are melted into their liquid phases, and are then poured together before cooling and creating the alloy. In an **interstitial alloy**, metal atoms with two vastly different radii combine. Steel is one such example—the much smaller carbon atoms occupy the interstices of the iron atoms. A **substitutional alloy** forms between atoms of similar radii. Brass is a good example, atoms of zinc are substituted for some copper atoms to create the alloy.

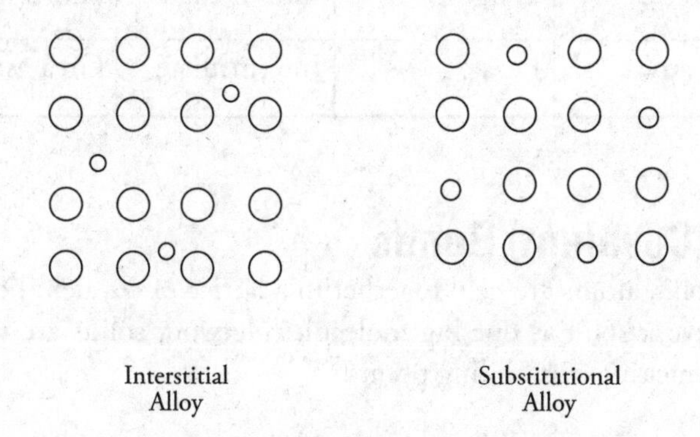

COVALENT BONDS

In a covalent bond, two atoms share electrons. Each atom counts the shared electrons as part of its valence shell. In this way, both atoms achieve complete outer shells.

In the diagram below, two fluorine atoms, each of which has seven valence electrons and needs one electron to complete its valence shell, form a covalent bond. Each atom donates an electron to the bond, which is considered to be part of the valence shell of both atoms.

$$:\ddot{\underset{..}{F}}\cdot \; + \; \cdot\ddot{\underset{..}{F}}: \;\Rightarrow\; :\ddot{\underset{..}{F}}:\ddot{\underset{..}{F}}:$$

The number of covalent bonds an atom can form is the same as the number of unpaired electrons in its valence shell.

Single bonds have one sigma (σ) bond and a bond order of one. The single bond has the longest bond length and the least bond energy.

The first covalent bond formed between two atoms is called a sigma (σ) bond. All single bonds are sigma bonds. If additional bonds between the two atoms are formed, they are called pi (π) bonds. The second bond in a double bond is a pi bond and the second and third bonds in a triple bond are also pi bonds. Double and triple bonds are stronger and shorter than single bonds, but they are not twice or triple the strength.

Summary of Multiple Bonds

Bond type:	Single	Double	Triple
Bond designation:	One sigma (σ)	One sigma (σ) and one pi (π)	One sigma (σ) and two pi (π)
Bond order:	One	Two	Three
Bond length:	Longest	Intermediate	Shortest
Bond energy:	Least	Intermediate	Greatest

Network (Covalent) Bonds

In a network solid, atoms are held together in a lattice of covalent bonds. You can visualize a network solid as one big molecule. Network solids are very hard and have very high melting and boiling points.

The electrons in a network solid are localized in covalent bonds between particular atoms, so they are not free to move about the lattice. This makes network solids poor conductors of electricity.

The most commonly seen network solids are compounds of carbon (such as diamond or graphite) and silicon (SiO_2—quartz). This is because both carbon and silicon have four valence electrons, meaning they are able to form a large number of covalent bonds.

Silicon also serves as a semiconductor when it is doped with other elements. **Doping** is a process in which an impurity is added to an existing lattice. In a normal silicon lattice, each individual silicon atom is bonded to four other silicon atoms. When some silicon atoms are replaced with elements that have only three valence electrons (such as boron or aluminum), the neighboring silicon atoms will lack one bond apiece.

This missing bond (or "hole") creates a positive charge in the lattice, and the hole attracts other electrons to it, increasing conductivity. Those electrons leave behind holes when they move, creating a chain reaction in which the conductivity of the silicon increases. This type of doping is called **p-doping** for the positively charged holes.

If an element with five valence electrons (such as phosphorus or arsenic) is used to add impurities to a silicon lattice, there is an extra valence electron that is free to move around the lattice, causing an overall negative charge that increases the conductivity of the silicon. This type of doping is called **n-doping** due to the free-moving negatively charged electrons.

Polarity

In the F_2 molecule shown on page 107, the two fluorine atoms share the electrons equally, but that's not usually the case in molecules. Usually, one of the atoms (the more electronegative one) will exert a stronger pull on the electrons in the bond—not enough to make the bond ionic, but enough to keep the electrons on one side of the molecule more than on the other side. This gives the molecule a dipole. That is, the side of the molecule where the electrons spend more time will be negative and the side of the molecule where the electrons spend less time will be positive.

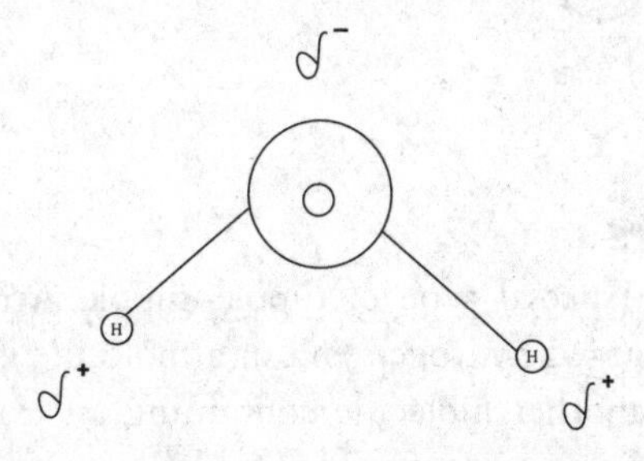

In the water molecule above, oxygen has a higher electronegativity than hydrogen and thus will have the electrons closer to it more often. This gives the oxygen a negative dipole and each hydrogen a positive dipole.

Dipole Moment

The polarity of a molecule is measured by the dipole moment. The more polar the molecule is, the larger the dipole moment is. You will not need to calculate the strength of a dipole, but you should be familiar with the unit with which that strength is quantified. That unit is called the debye (D).

INTERMOLECULAR FORCES

Intermolecular forces (IMFs) are the forces that exist between molecules in a covalently bonded substance. These forces are what need to be broken apart in order for covalent substances to change phases. Note that when ionic substances change phase, bonds between the individual ions are actually broken. When covalent substances change phase, the bonds between the individual atoms remain in place, it is just the forces that hold the molecules to other molecules that break apart.

Dipole–Dipole Forces

Dipole–dipole forces occur between polar molecules: The positive end of one polar molecule is attracted to the negative end of another polar molecule.

Molecules with greater polarity will have greater dipole–dipole attraction, so molecules with larger dipole moments tend to have higher melting and boiling points. Dipole–dipole attractions are relatively weak, however, and these substances melt and boil at very low temperatures. Most substances held together by dipole–dipole attraction are gases or liquids at room temperature.

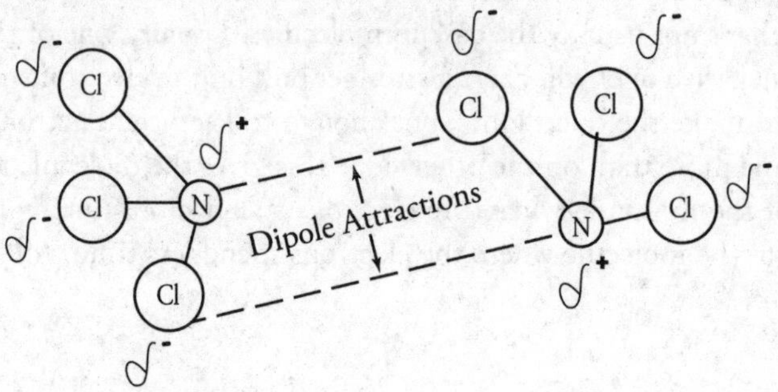

Hydrogen Bonds

Hydrogen bonds are a special type of dipole–dipole attraction. In a hydrogen bond, the positively charged hydrogen end of a molecule is attracted to the negatively charged end of another molecule containing an extremely electronegative element (fluorine, oxygen, or nitrogen—F, O, N).

Hydrogen bonds are much stronger than normal dipole–dipole forces because when a hydrogen atom gives up its lone electron to a bond, its positively charged nucleus is left virtually unshielded. Substances that have hydrogen bonds, such as water and ammonia, have higher melting and boiling points than substances that are held together only by other types of intermolecular forces.

Water is less dense as a solid than as a liquid because its hydrogen bonds force the molecules in ice to form a crystal structure, which keeps them farther apart than they are in the liquid form.

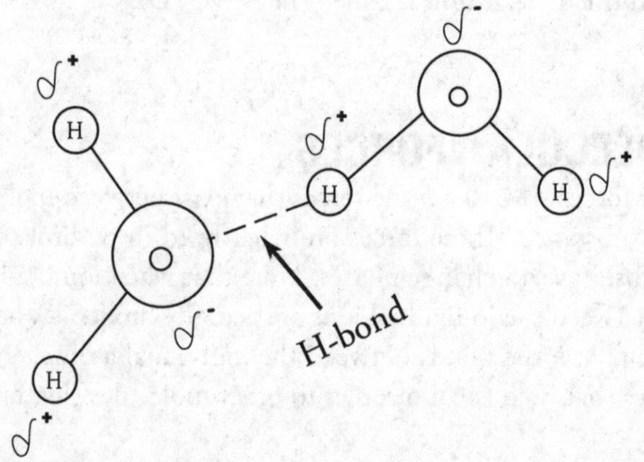

London Dispersion Forces

London dispersion forces occur between all molecules. These very weak attractions occur because of the random motions of electrons on atoms within molecules. At a given moment, a nonpolar molecule might have more electrons on one side than on the other, giving it an instantaneous polarity. For that fleeting instant, the molecule will act as a very weak dipole.

Since London dispersion forces depend on the random motions of electrons, molecules with more electrons will experience greater London dispersion forces. So among substances that experience only London dispersion forces, the one with more electrons will generally have higher melting and boiling points. London dispersion forces are even weaker than dipole–dipole forces, so substances that experience only London dispersion forces melt and boil at extremely low temperatures and tend to be gases at room temperature.

As molecules gain more electrons, the London dispersion forces between them start to become much more significant. Comparing the boiling point of a nonpolar substance with a large number of electrons vs. a polar substance with less electrons is difficult, and there is no simple rule to follow. For instance, water has hydrogen bonds and a boiling point of 100°C. Butane (C_4H_8) and octane (C_8H_{18}) are both completely nonpolar molecules, and while butane's boiling point is 34°C, octane's is 125°C. Even though octane has no permanent dipoles, it has so many electrons that its London dispersion forces are significant enough that they create greater intermolecular attractions than even the hydrogen bonds in water.

The role of London dispersion forces is often determined by comparing the molar mass of molecules. However, it is not the mass itself which affects the strength of the IMFs. Rather, it is simply that as mass (based on protons and neutrons) increases, so too do the number of electrons, as the molecule must remain electrically neutral.

IMF Questions: Not So Impossible

The AP Exam will most likely focus questions regarding IMFs on molecules with similar masses, as it is much easier to compare the strengths of their IMFs. However, understanding that London dispersion forces can have an impact on IMF-related trends when considering molecules with a large difference in their masses is worth keeping in mind.

Bond Strength

Ionic substances are generally solids at room temperature, and turning them into liquids (melting them), requires the bonds holding the lattice together to be broken. The amount of energy needed for that is based on the Coulombic attraction between the molecules.

Covalent substances, which are liquid at room temperature, will boil when the intermolecular forces between them are broken. For molecules with similar sizes, the following IMF ranking can help you determine the relative strength of the IMFs within the molecules.

a. Hydrogen bonds
b. Non-hydrogen bond permanent dipoles
c. London dispersion forces (temporary dipoles)
i. Larger molecules are more polarizable and have stronger London dispersion forces because they have more electrons.

The melting and boiling points of covalent substances are almost always lower than the melting and boiling point of ionic ones.

Metallic bonding, which often only involves one type of atom, tends to be very strong and thus metals (particularly the transition metals) tend to have high melting points. Network covalent bonding is the strongest type of bonding there is, and it is very difficult to cause substances that exhibit network covalent bonding to melt.

Bonding and Phases

The phase of a substance is directly related to the strength of its intermolecular forces. Solids have highly ordered structures where the atoms are packed tightly together, while gases have atoms spread so far apart that most of the volume is free space.

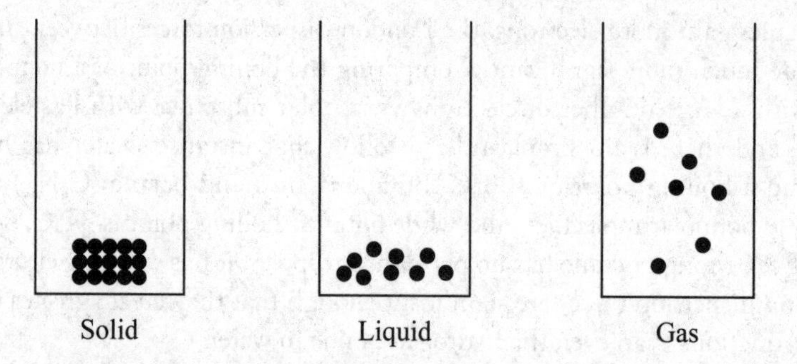

In other words, substances that exhibit weak intermolecular forces (such as London Dispersion forces) tend to be gases at room temperature. Nitrogen (N_2) is an example of this. Substances that exhibit strong intermolecular forces (such as hydrogen bonds) tend to be liquids at room temperature. A good example is water.

Ionic substances do not experience intermolecular forces. Instead, their phase is determined by the ionic bond holding the ions together in the lattice. Because ionic bonds are generally significantly stronger than intermolecular forces in covalent molecules, ionic substances are usually solid at room temperatures.

VAPOR PRESSURE

Beyond helping to determine the melting point and boiling point of covalent substances, the relative strength of the intermolecular forces in a substance can also predict several other properties of that substance. The most important of these is vapor pressure. Vapor pressure arises from the fact that the molecules inside a liquid are in constant motion. If those molecules hit the surface of the liquid with enough kinetic energy, they can escape the intermolecular forces holding them to the other molecules and transition into the gas phase.

This process is called vaporization. It is not to be confused with a liquid boiling. When a liquid boils, energy (in the form of heat) is added, increasing the kinetic energy of all of the molecules in the liquid until all of the intermolecular forces are broken. For vaporization to occur, no outside energy needs to be added. Note that there is a direct relationship between temperature and vapor pressure.

The higher the temperature of a liquid, the faster the molecules are moving and the more likely they are to break free of the other molecules. So, temperature and vapor pressure are directly proportional.

If two liquids are at the same temperature, the vapor pressure is dependent primarily on the strength of the intermolecular forces within that liquid. The stronger those intermolecular forces are, the less likely it is that molecules will be able to escape the liquid, and the lower the vapor pressure for that liquid will be.

LEWIS DOT STRUCTURES

Drawing Lewis Dot Structures

At some point on the test, you'll be asked to draw the Lewis structure for a molecule or polyatomic ion. Here's how to do it.

1. Count the valence electrons in the molecule or polyatomic ion; refer to page 74 for the periodic table.
2. If a polyatomic ion has a negative charge, add electrons equal to the charge of the total in (1). If a polyatomic ion has a positive charge, subtract electrons equal to the charge of the electrons from the total in (1).
3. Draw the skeletal structure of the molecule and place two electrons (or a single bond) between each pair of bonded atoms. If the molecule contains three or more atoms, the least electronegative atom will usually occupy the central position.
4. Add electrons to the surrounding atoms until each has a complete outer shell.
5. Add the remaining electrons to the central atom.
6. Look at the central atom.
   (a) If the central atom has fewer than eight electrons, remove an electron pair from an outer atom and add another bond between that outer atom and the central atom. Do this until the central atom has a complete octet.
   (b) If the central atom has a complete octet, you are finished.
   (c) If the central atom has more than eight electrons, that's okay, too, as long as the total does not exceed twelve.

Let's find the Lewis dot structure for the CO_3^{2-} ion.

1. Carbon has 4 valence electrons; oxygen has 6.
   $4 + 6 + 6 + 6 = 22$
2. The ion has a charge of –2, so add 2 electrons.
   $22 + 2 = 24$
3. Carbon is the central atom.

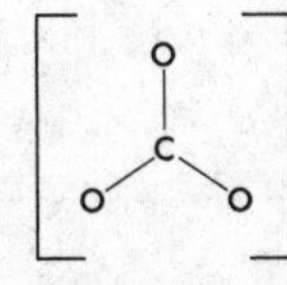

4. Add electrons to the oxygen atoms.

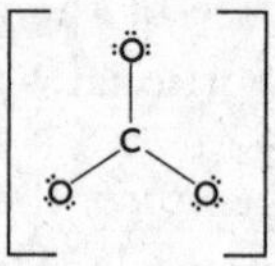

5. We've added all 24 electrons, so there's nothing left to put on the carbon atom.
6. (a) We need to give carbon a complete octet, so we take an electron pair away from one of the oxygens and make a double bond instead. Place a bracket around the model and add a charge of negative two.

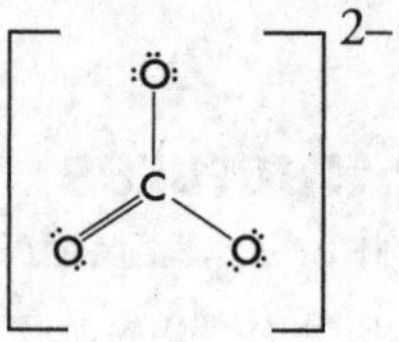

Resonance Forms

When we put a double bond into the CO_3^{2-} ion, we place it on any one of the oxygen atoms, as shown below.

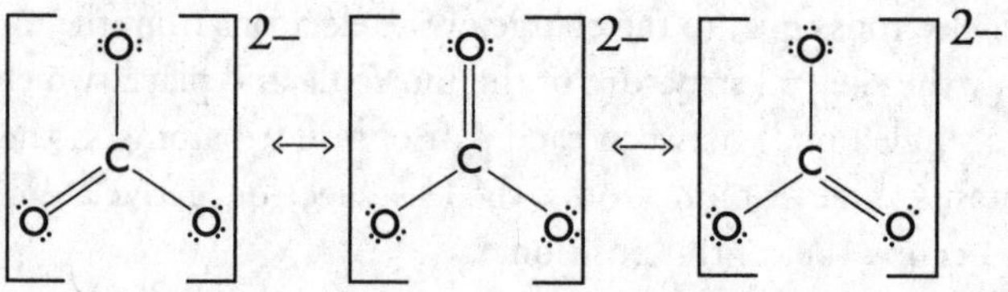

Don't Be Fooled By Appearances
Despite the conventional way the Lewis diagrams have been drawn, remember that the bonds of a compound that displays resonance are all identical. The only reason it looks like a mix of single and double bonds is because any attempt to illustrate a bond somewhere between these two strengths and lengths would be awkward.

All three resonance forms are considered to exist simultaneously, and the strength and lengths of all three bonds are the same: somewhere between the strength and length of a single bond and a double bond.

To determine the relative length and strength of a bond in a resonance structure, a bond order calculation can be used. A single bond has a bond order of 1, and a double bond has an order of 2. When resonance occurs, pick one of the bonds in the resonance structure and add up the total bond order across the resonance forms, then divide that sum by the number of resonance forms.

For example, in the carbonate ion above, the top C–O bond would have a bond order of $\frac{1+2+1}{3}$, or 1.33. Bond order can be used to compare the length and strength of resonance bonds with pure bonds as well as other resonance bonds.

Incomplete Octets

Some atoms are stable with less than eight electrons in their outer shell. Hydrogen only requires two electrons, as does helium (although helium never forms bonds). Boron is considered to be stable with six electrons, as in the BF_3 diagram below. All other atoms involved in covalent bonding require a minimum of eight electrons to be considered stable.

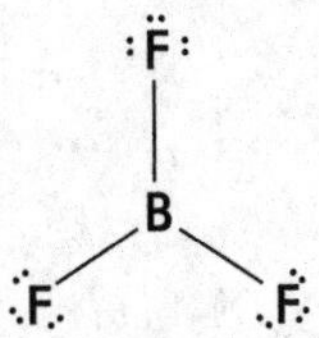

Expanded Octets

In molecules that have *d* subshells available, the central atom can have more than eight valence electrons, but never more than twelve. This means any atom of an element from $n = 3$ or greater can have expanded octets, but NEVER elements in $n = 2$ (C, N, O, etc.). Expanded octets also explains why some noble gases can actually form bonds; the extra electrons go into the empty *d*-orbital.

Here are some examples.

PCl_5

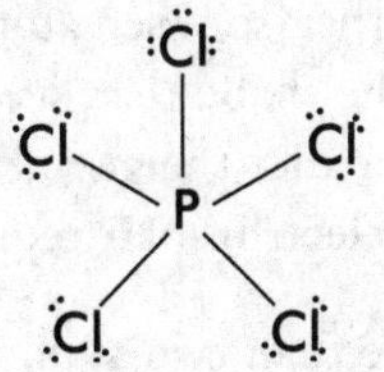

SF_4

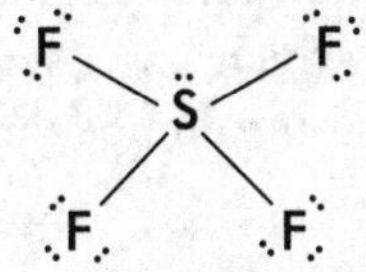

XeF_4

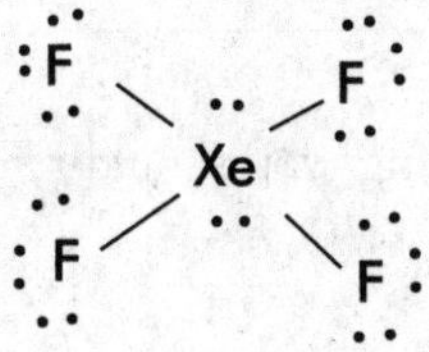

Formal Charge

Sometimes, there is more than one valid Lewis structure for a molecule. Take CO_2; it has two valid structures as shown below. To determine the more likely structure, formal charge is used. To calculate the formal charge on atoms in a molecule, take the number of valence electrons for that atom and subtract the number of assigned electrons in the Lewis structure. When counting assigned electrons, lone pairs count as two and bonds count as one.

O	C	O		O	C	O
$\ddot{\underset{\cdot\cdot}{O}}=C=\ddot{\underset{\cdot\cdot}{O}}$				$:\ddot{\underset{\cdot\cdot}{O}}-C\equiv O:$		
6	4	6	valence e^-	6	4	6
− 6	4	6	assigned e^-	− 7	4	5
0	0	0	formal charge	−1	0	+1

The total formal charge for a neutral molecule should be zero, which it is on both diagrams. Additionally, the fewer number of atoms there are with an actual formal charge, the more likely the structure will be—so the left structure is the more likely one for CO_2. For polyatomic ions, the sum of the formal charges on each atom should equal the overall charge on the ion.

Molecular Geometry

Electrons repel one another, so when atoms come together to form a molecule, the molecule will assume the shape that keeps its different electron pairs as far apart as possible. When we predict the geometries of molecules using this idea, we are using the **valence shell electron-pair repulsion (VSEPR) model.**

In a molecule with more than two atoms, the shape of the molecule is determined by the number of electron pairs on the central atom. The central atom forms **hybrid orbitals,** each of which has a standard shape. Variations on the standard shape occur depending on the number of bonding pairs and lone pairs of electrons on the central atom.

Here are some things you should remember when dealing with the VSEPR model.

- Double and triple bonds are treated in the same way as single bonds in terms of predicting overall geometry for a molecule; however, multiple bonds have slightly more repulsive strength and will therefore occupy a little more space than single bonds.
- Lone electron pairs have a little more repulsive strength than bonding pairs, so molecules with lone pairs will have slightly reduced bond angles between terminal atoms.

The following pages show the different hybridizations and geometries that you might see on the test.

If the central atom has 2 electron pairs, then it has *sp* hybridization and its basic shape is **linear.**

Number of lone pairs	Geometry	Examples
0	B—A—B linear	$BeCl_2$ CO_2

If the central atom has 3 electron pairs, then it has sp^2 hybridization and its basic shape is **trigonal planar**; its bond angles are about 120°.

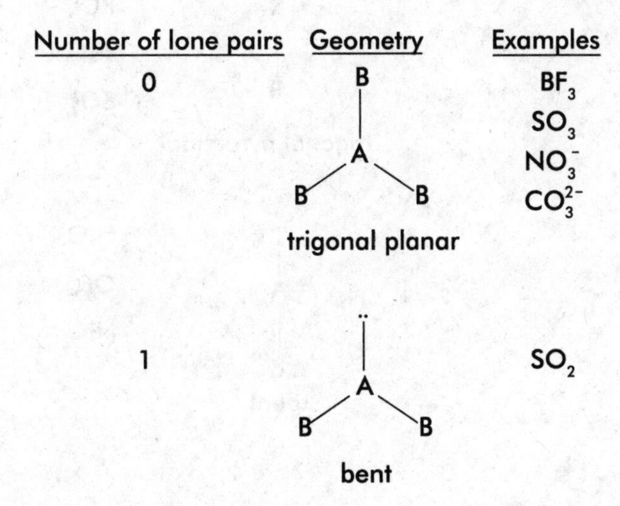

The angle between the terminal atoms in the bent shape is slightly less than 120° because of the extra lone pair repulsion.

If the central atom has 4 electron pairs, then it has sp^3 hybridization and its basic shape is **tetrahedral**; its bond angles are about 109.5°.

The angle between the terminal atoms in the trigonal pyramidal and bent shapes is slightly less than 109.5° because of the extra lone pair repulsion.

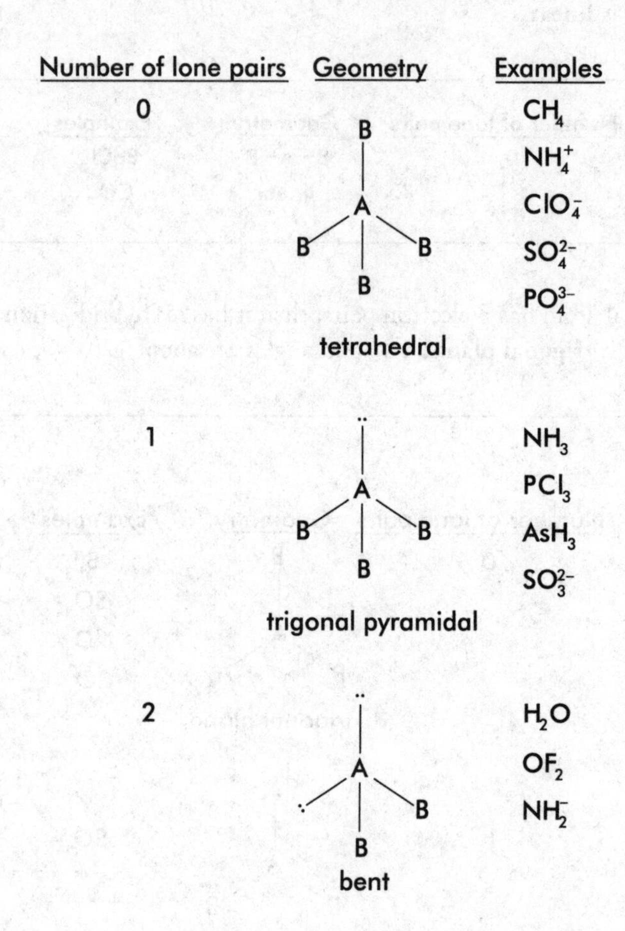

If the central atom has 5 electron pairs, its basic shape is **trigonal bipyramidal.**

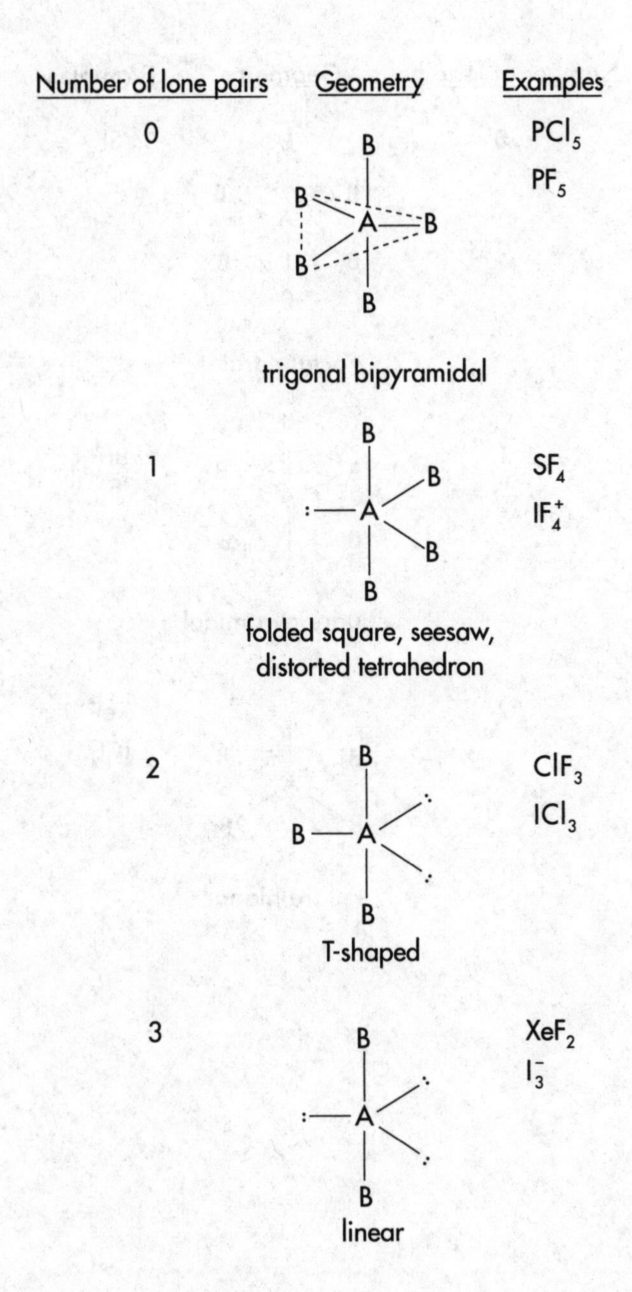

In trigonal bipyrimidal shapes, place the lone pairs in axial position first. In octahedral shapes, place lone pairs in equatorial position first.

If the central atom has 6 electron pairs, its basic shape is **octahedral.**

Number of lone pairs	Geometry	Examples
0	octahedral	SF_6
1	square pyramidal	BrF_5 IF_5
2	square planar	XeF_4 ICl_4^-

KINETIC MOLECULAR THEORY

For ideal gases, the following assumptions can be made:

- The kinetic energy of an ideal gas is directly proportional to its absolute temperature: The greater the temperature, the greater the average kinetic energy of the gas molecules.

The Average Kinetic Energy of a Single Gas Molecule

$$KE = \frac{1}{2}mv^2$$

m = mass of the molecule (kg)
v = speed of the molecule (meters/sec)
KE is measured in joules

- If several different gases are present in a sample at a given temperature, all the gases will have the same average kinetic energy. That is, the average kinetic energy of a gas depends only on the absolute temperature, not on the identity of the gas.
- The volume of an ideal gas particle is insignificant when compared with the volume in which the gas is contained.
- There are no forces of attraction between the gas molecules in an ideal gas.
- Gas molecules are in constant motion, colliding with one another and with the walls of their container without losing any energy.

MAXWELL-BOLTZMANN DIAGRAMS

A Maxwell-Boltzmann diagram shows the range of velocities for molecules of a gas. Molecules at a given temperature are not all moving at the same velocity. When determining the temperature, we take the average velocity of all the molecules and use that in the relevant equation to calculate temperature. You do not need to know that equation (unless you are taking AP Physics!). All you need to know here is that temperature is directly proportional to kinetic energy.

The first type of Maxwell-Boltzmann diagram involves plotting the velocity distributions for the molecules of one particular gas at multiple temperatures. In the diagram on the next page, there are three curves representing a sample of nitrogen gas at 100 K, 300 K, and 500 K.

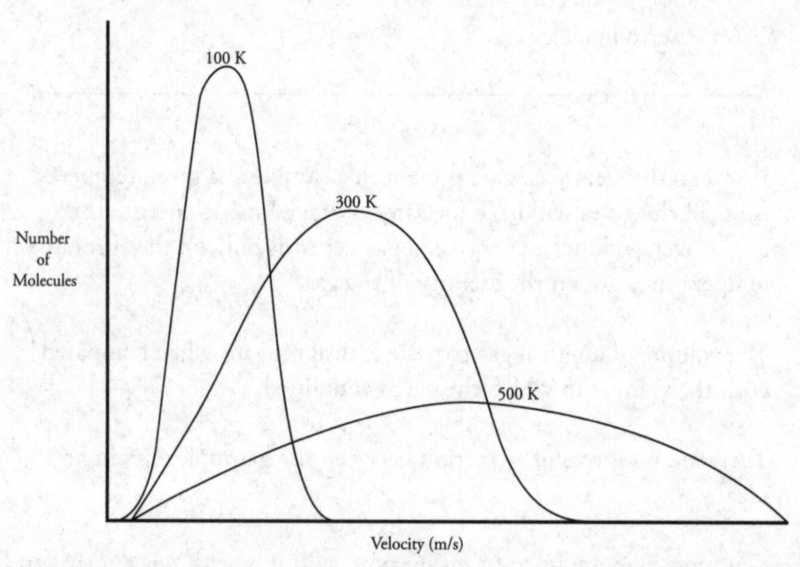

As you can see, the higher the temperature of the gas, the larger the range is for the velocities of the individual molecules. Gases at higher temperatures have greater kinetic energy (KE), and as all the molecules in this example have the same mass, the increased KE is due to the increased velocity of the gas molecules.

Maxwell-Boltzmann diagrams are also used to show a number of different gases at the same temperature. The diagram on the next page shows helium, argon, and xenon gas, all at 300 K:

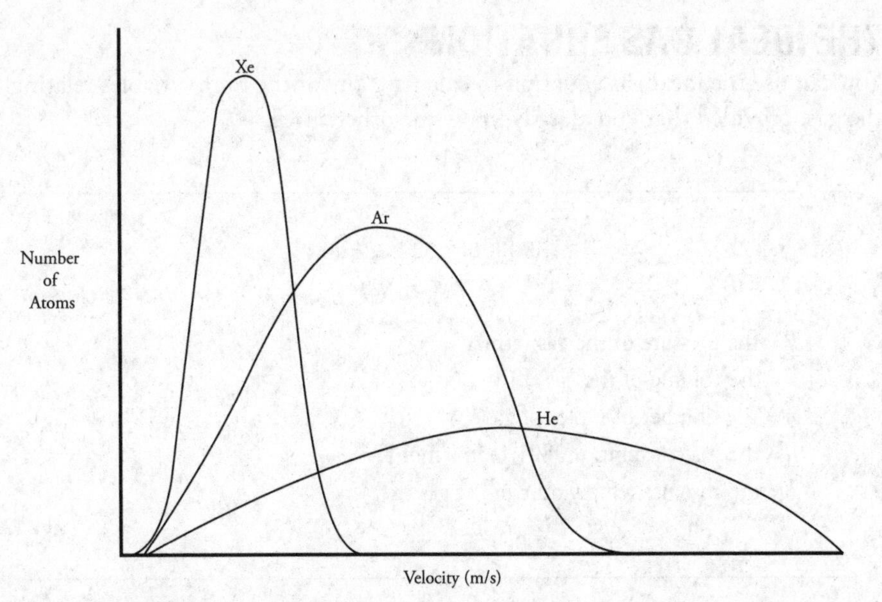

In this case, all of the gases have the same amount of total kinetic energy because they have identical temperatures. However, not all of the atoms have the same mass. If the atoms have smaller masses, they must have greater velocities in order to have a kinetic energy identical to that of atoms with greater mass. Because helium atoms have the least mass, they have the highest average velocity. Xenon atoms, which have a much greater mass, have a correspondingly lower velocity.

EFFUSION

Effusion is the rate at which a gas will escape from a container through microscopic holes in the surface of the container. For instance, even though the rubber or latex that makes up a balloon may seem solid, after the balloon is filled with a gas, it will gradually shrink over time. This is due to the fact that there are tiny holes in the surface of the balloon, through which the even tinier gas molecules can escape.

The rate at which a gas effuses from a container is dependent on the speed of the gas particles. The faster the particles are moving, the more often they hit the sides of the container, and the more likely they are to hit a hole and escape. The rate of effusion thus increases with temperature, but also, if examining gases at the same temperature, the gas with the lower molar mass will effuse first.

It is likely you have experienced this; a balloon filled with helium will deflate more rapidly than one filled with air (which is composed primarily of nitrogen and oxygen) or one filled with carbon dioxide. There is a formula that quantifies the rate at which a gas will effuse, but that is beyond the scope of the exam. As long as you understand the basic principles behind effusion, you should be able to answer any questions on this topic that may come up.

THE IDEAL GAS EQUATION

You can use the ideal gas equation to calculate any of the four variables relating to the gas, provided that you already know the other three.

The Ideal Gas Equation

$$PV = nRT$$

P = the pressure of the gas (atm)
V = the volume of the gas (L)
n = the number of moles of gas
R = the gas constant, 0.0821 L·atm/mol·K
T = the absolute temperature of the gas (K)

You can also manipulate the ideal gas equation to figure out how changes in each of its variables affect the other variables. The following equation, often called the combined gas law, can only be used when the number of moles is held constant.

$$\frac{P_1V_1}{T_1} = \frac{P_2V_2}{T_2}$$

P = the pressure of the gas (atm)
V = the volume of the gas (L)
T = the absolute temperature of the gas (K)

You should be comfortable with the following simple relationships:

- If the volume is constant: As pressure increases, temperature increases; as temperature increases, pressure increases.
- If the temperature is constant: As pressure increases, volume decreases; as volume increases, pressure decreases. That's Boyle's law.
- If the pressure is constant: As temperature increases, volume increases; as volume increases, temperature increases. That's Charles's law.

DALTON'S LAW

Dalton's law states that the total pressure of a mixture of gases is just the sum of all the partial pressures of the individual gases in the mixture.

Dalton's Law

$$P_{total} = P_a + P_b + P_c + \ldots$$

You should also note that the partial pressure of a gas is directly proportional to the number of moles of that gas present in the mixture. So if 25 percent of the gas in a mixture is helium, then the partial pressure due to helium will be 25 percent of the total pressure.

Partial Pressure

$$P_a = (P_{total})(X_a)$$

$$X_a = \frac{\text{moles of gas A}}{\text{total moles of gas}}$$

DEVIATIONS FROM IDEAL BEHAVIOR

At low temperature and/or high pressure, gases behave in a less-than-ideal manner. That's because the assumptions made in kinetic molecular theory become invalid under conditions where gas molecules are packed too tightly together.

Two things happen when gas molecules are packed too tightly.

- *The volume of the gas molecules becomes significant.*
  The ideal gas equation does not take the volume of gas molecules into account, so the actual volume of a gas under nonideal conditions will be larger than the volume predicted by the ideal gas equation.
- *Gas molecules attract one another and stick together.*
  The ideal gas equation assumes that gas molecules never stick together. When a gas is packed tightly together, intermolecular forces become significant, causing some gas molecules to stick together. When gas molecules stick together, there are fewer particles bouncing around and creating pressure, so the real pressure in a nonideal situation will be smaller than the pressure predicted by the ideal gas equation.

DENSITY

You may be asked about the density of a gas. The density of a gas is measured in the same way as the density of a liquid or solid: in mass per unit of volume.

Density of a Gas

$$D = \frac{m}{V}$$

D = density
m = mass of gas, usually in grams
V = volume occupied by a gas, usually in liters

The density of any gas sample can also be determined by combining the density equation with the ideal gas law.

If $D = \frac{m}{V}$, then $V = \frac{m}{D}$.

Substituting that into the ideal gas law:

$$\frac{Pm}{D} = nRT$$

A little rearrangement yields:

$$D = \frac{Pm}{nRT}$$

The term (m/n) describes mass per mole, which is how molar mass (MM) is measured. Thus:

$$D = \frac{P(MM)}{RT}$$

If you are given the density of a gas and need to find the molar mass, this can also be rewritten as:

$$MM = \frac{DRT}{P}$$

SOLUTIONS

Molarity

Molarity (M) expresses the concentration of a solution in terms of volume. It is the most widely used unit of concentration, turning up in calculations involving equilibrium, acids and bases, and electrochemistry, among others.

When you see a chemical symbol in brackets on the test, that means they are talking about molarity. For instance, "[Na^+]" is the same as "the molar concentration (molarity) of sodium ions."

$$\text{Molarity } (M) = \frac{\text{moles of solute}}{\text{liters of solution}}$$

Mole Fraction

Mole fraction (X_s) gives the fraction of moles of a given substance (S) out of the total moles present in a sample.

$$\text{Mole Fraction } (X_s) = \frac{\text{moles of substance S}}{\text{total number of moles in solution}}$$

Solutes and Solvents

There is a basic rule for remembering which solutes will dissolve in which solvents.

Like dissolves like

That means that polar or ionic solutes (such as salt) will dissolve in polar solvents (such as water). That also means that nonpolar solutes (such as oils) are best dissolved in nonpolar solvents. When an ionic substance dissolves, it breaks up into ions. That's **dissociation.** Free ions in a solution are called electrolytes because they can conduct electricity.

The more ions that are present in an ionic compound, the greater the conductivity of that compound will be when those ions are dissociated. For instance, a solution of magnesium chloride will dissociate into three ions (one Mg^{2+} and two Cl^-). A solution of sodium chloride will dissociate into just two ions (Na^+ and Cl^-). Thus, a solution of magnesium chloride will conduct electricity better than a solution of sodium chloride if both solutions have identical concentrations.

SOLUTION SEPARATION

We can use intermolecular forces and the various Coulombic attractions that occur between ions and polar molecules in order to help separate various substances out from each other. There are several ways to do this.

Paper Chromatography

Chromatography is the separation of a mixture by passing it in solution through a medium in which the components of the solution move at different rates. There are several major types of chromatography. The first is paper chromatography, in which paper is the medium through which the solution passes.

Many chemical solutions, such as the ink found in most pens, are a mixture of a number of covalent substances. Each of these substances has its own polarity value, and thus has a different affinity depending on the solvent. One of the most common paper chromatography experiments involves the separation of pigments in black ink. Black ink is usually made up of substances of several different colors, which when combined create black.

In paper chromatography, a piece of filter paper is suspended above a solvent so that the very bottom of the paper is touching the solvent. The ink in question is dotted onto a line at the bottom of the filter paper that starts out just above the solvent level. As the solvent climbs the paper, the various substances inside the ink will be attracted to the polar water molecules. The more polar the substance is, the more it will be attracted to the water molecules, and the further it will travel. You might end up with something that looks like this:

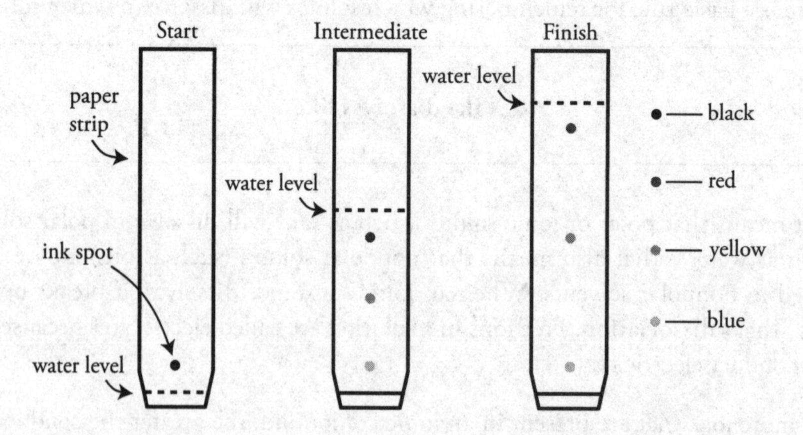

Looking at that strip, you can conclude that the ink was made of three different substances. The one that traveled the farthest with the water (the red pigment) experienced the strongest attractions and was the most polar, whereas the one that didn't travel very far from the original starting line (the blue pigment) was the least polar. Paper chromatography is the most useful with colored substances, which is why ink is used in the above example. If there were components to the ink that had no visible color, you would not be able to see them on the filter paper, and that is one major limitation of paper chromatography.

The distance the ink travels along the paper is measured via the retention (or retardation) factor, also known as the R_f value. The R_f value is calculated as such:

$$R_f = \frac{\text{Distance traveled by solute}}{\text{Distance traveled by solvent front}}$$

The stronger the attraction between the solute and the solvent front is, the larger the R_f value will be. In the diagram above, the red pigment would have the highest R_f value.

Water is not the only solvent that can be used in polar chromatography. There are many nonpolar solvents (such as cyclohexane) that can be used instead. In the case of a nonpolar solvent, the position of the various ink components in the above diagram would have been reversed—the most nonpolar substance would travel the furthest, and the most polar substance would travel the least.

Column Chromatography

Another type of chromatography is column chromatography. In this process, a column is packed with a stationary substance. Then, the solution to be separated (the analyte) is injected into the column, where it adheres to the stationary phase. After that, another solution (called the eluent) is injected into the column. As the eluent passes through the stationary phase, the analyte molecules will be attracted to it with varying degrees of strength depending on their polarity. The more attracted certain analyte molecules are to the eluent, the faster they will travel through ("elute") and leave the column.

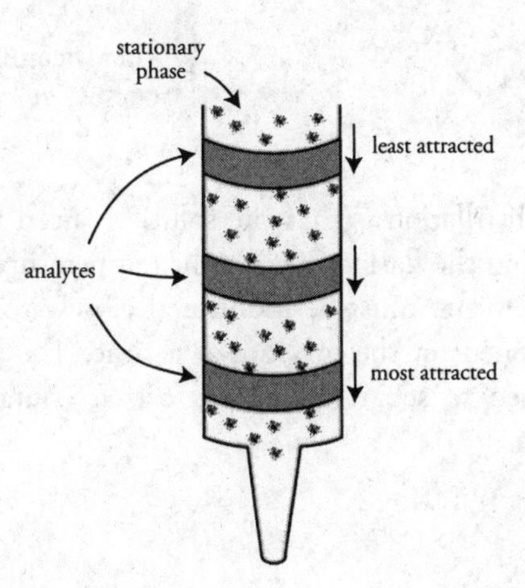

The speed at which the substances move through the column can be monitored, and if there is a sufficient polarity difference between the components they will leave the column at different times, allowing them to be separated. Generally, after collecting the eluted mixture, it can be analyzed for compositional analysis via a variety of methods.

In column chromatography, either liquids or gases can be used as the eluent, depending on the situation.

Distillation

A third method for separating solutions is distillation. Distillation takes advantage of the different boiling points of substances in order to separate them. For instance, if you have a mixture of water (BP: 100°C) and ethanol (BP: 78°C) and then heat that mixture to 85°C, the ethanol will boil but the water will not.

A condenser is a piece of glassware that consists of a smaller tube running through a larger tube. The larger tube has hose connections on it, allowing for water to be run through it. This effectively cools the inner tube and not the outer tube. At that point, the ethanol vapor, when run through the inner tube, will cool and condense back into a liquid form, which can be collected on the other side of the condenser.

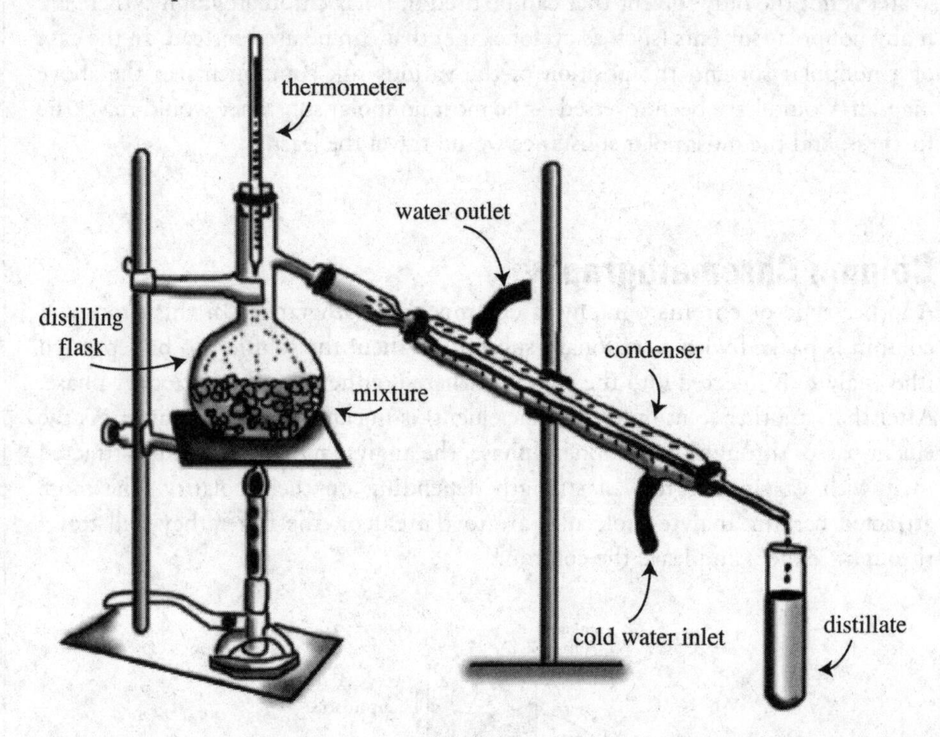

A major advantage to distillation is that the solutions need not be colored at all to separate them. Keeping the flask at a constant temperature can be a challenge, which is why the temperature must be monitored closely to ensure that you are only boiling one component of the mixture at a time. The biggest disadvantage is that it cannot be used to separate a mixture that contains substances with unknown boiling points.

CHAPTER 4 QUESTIONS

Multiple-Choice Questions

1. Why does CaF_2 have a higher melting point than NH_3?

   (A) CaF_2 is more massive and thus has stronger London disperson forces.
   (B) CaF_2 exhibits network covalent bonding, which is the strongest type of bonding.
   (C) CaF_2 is smaller and exhibits greater Coulombic attractive forces.
   (D) CaF_2 is an ionic substance and it requires a lot of energy to break up an ionic lattice.

2. Which of the following pairs of elements is most likely to create an interstitial alloy?

   (A) Titanium and copper
   (B) Aluminum and lead
   (C) Silver and tin
   (D) Magnesium and calcium

3. Why can a molecule with the structure of NBr_5 not exist?

   (A) Nitrogen only has two energy levels and is thus unable to expand its octet.
   (B) Bromine is much larger than nitrogen and cannot be a terminal atom in this molecule.
   (C) It is impossible to complete the octets for all six atoms using only valence electrons.
   (D) Nitrogen does not have a low enough electronegativity to be the central atom of this molecule.

Use the following information to answer questions 4-7.

An evacuated rigid container is filled with exactly 2.00 g of hydrogen and 10.00 g of neon. The temperature of the gases is held at 0°C and the pressure inside the container is a constant 1.0 atm.

4. What is the mole fraction of neon in the container?

   (A) 0.17
   (B) 0.33
   (C) 0.67
   (D) 0.83

5. What is the volume of the container?

   (A) 11.2 L
   (B) 22.4 L
   (C) 33.5 L
   (D) 48.8 L

6. Which gas has the higher boiling point and why?

   (A) Hydrogen, because it has a lower molar mass
   (B) Neon, because it has more electrons
   (C) Hydrogen, because it has a smaller size
   (D) Neon, because it has more protons

7. Which gas particles have a higher RMS velocity and why?

(A) Hydrogen, because it has a lower molar mass
(B) Neon, because it has a higher molar mass
(C) Hydrogen, because it has a larger atomic radius
(D) Neon, because it has a smaller atomic radius

8. Which of the following compounds would have the highest lattice energy?

(A) LiF
(B) $MgCl_2$
(C) $CaBr_2$
(D) C_2H_6

9. A liquid whose molecules are held together by only which of the following forces would be expected to have the lowest boiling point?

(A) Ionic bonds
(B) London dispersion forces
(C) Hydrogen bonds
(D) Metallic bonds

10. The six carbon atoms in a benzene molecule are shown in different resonance forms as three single bonds and three double bonds. If the length of a single carbon–carbon bond is 154 pm and the length of a double carbon–carbon bond is 133 pm, what length would be expected for the carbon–carbon bonds in benzene?

(A) 126 pm
(B) 133 pm
(C) 140 pm
(D) 154 pm

11. Which of the following lists of species is in order of increasing boiling points?

(A) H_2, N_2, NH_3
(B) N_2, NH_3, H_2
(C) NH_3, H_2, N_2
(D) NH_3, N_2, H_2

12. A mixture of gases contains 1.5 moles of oxygen, 3.0 moles of nitrogen, and 0.5 moles of water vapor. If the total pressure is 700 mmHg, what is the partial pressure of the nitrogen gas?

(A) 210 mmHg
(B) 280 mmHg
(C) 350 mmHg
(D) 420 mmHg

13. A mixture of helium and neon gases has a total pressure of 1.2 atm. If the mixture contains twice as many moles of helium as neon, what is the partial pressure due to neon?

(A) 0.2 atm
(B) 0.3 atm
(C) 0.4 atm
(D) 0.8 atm

14. Nitrogen gas was collected over water at 25°C. If the vapor pressure of water at 25°C is 23 mmHg, and the total pressure in the container is measured at 781 mmHg, what is the partial pressure of the nitrogen gas?

(A) 46 mmHg
(B) 551 mmHg
(C) 735 mmHg
(D) 758 mmHg

15. A 22.0 gram sample of an unknown gas occupies 11.2 liters at standard temperature and pressure. Which of the following could be the identity of the gas?

(A) CO_2
(B) SO_3
(C) O_2
(D) He

16. Which of the following expressions is equal to the density of helium gas at standard temperature and pressure?

(A) $\frac{1}{22.4}$ g/L

(B) $\frac{2}{22.4}$ g/L

(C) $\frac{1}{4}$ g/L

(D) $\frac{4}{22.4}$ g/L

17. In an experiment 2 moles of $H_2(g)$ and 1 mole of $O_2(g)$ were completely reacted, according to the following equation in a sealed container of constant volume and temperature:

$$2H_2(g) + O_2(g) \rightarrow 2H_2O(g)$$

If the initial pressure in the container before the reaction is denoted as P_i, which of the following expressions gives the final pressure, assuming ideal gas behavior?

(A) P_i
(B) $2P_i$
(C) $(3/2)P_i$
(D) $(2/3)P_i$

18. An ideal gas fills a balloon at a temperature of 27°C and 1 atm pressure. By what factor will the volume of the balloon change if the gas in the balloon is heated to 127°C at constant pressure?

(A) $\frac{27}{127}$

(B) $\frac{3}{4}$

(C) $\frac{4}{3}$

(D) $\frac{2}{1}$

19. A gas sample with a mass of 10 grams occupies 5.0 liters and exerts a pressure of 2.0 atm at a temperature of 26°C. Which of the following expressions is equal to the molecular mass of the gas? The gas constant, R, is 0.08 (L·atm)/(mol·K).

(A) (0.08)(299) g/mol

(B) $\frac{(299)(0.50)}{(2.0)(0.08)}$ g/mol

(C) $\frac{299}{0.08}$ g/mol

(D) (2.0)(0.08) g/mol

20. A substance is dissolved in water, forming a 0.50-molar solution. If 4.0 liters of solution contains 240 grams of the substance, what is the molecular mass of the substance?

(A) 60 g/mol
(B) 120 g/mol
(C) 240 g/mol
(D) 480 g/mol

21. How many moles of Na_2SO_4 must be added to 500 milliliters of water to create a solution that has a 2-molar concentration of the Na^+ ion? (Assume the volume of the solution does not change).

(A) 0.5 mol
(B) 1 mol
(C) 2 mol
(D) 5 mol

Use the following Lewis diagrams to answer questions 22-24.

The following three substances are kept in identical containers 25°C. All three substances are in the liquid phase:

```
    H   H                  H  :O:  H
    |   |                  |   ||  |
H—C—C—Ö—H          H—C—C—C—H
    |   |                  |       |
    H   H                  H       H
   Ethanol                  Acetone

           H   H
           |   |
     H—Ö—C—C—Ö—H
           |   |
           H   H
        Ethylene Glycol
```

22. Which substance would have the highest boiling point?

(A) Ethanol, because it is the most asymmetrical
(B) Acetone, because of the double bond
(C) Ethylene glycol, because it has the most hydrogen bonding
(D) All three substances would have very similar boiling points because their molar masses are similar.

23. Which substance would have the highest vapor pressure?

(A) Ethanol, because of the hybridization of its carbon atoms
(B) Acetone, because it exhibits the weakest intermolecular forces
(C) Ethylene glycol, because it has the most lone pairs assigned to individual atoms
(D) All three substances would have similar vapor pressure because they have a similar number of electrons.

24. Which of the substances would be soluble in water?

(A) Ethylene glycol only, because it has the longest bond lengths
(B) Acetone only, because it is the most symmetrical
(C) Ethanol and ethylene glycol only, because of their hydroxyl (–OH) groups
(D) All three substances would be soluble in water due to their permanent dipoles.

Use the following information to answer questions 25-29.

There are several different potential different Lewis diagrams for the sulfate ion, two of which are below.

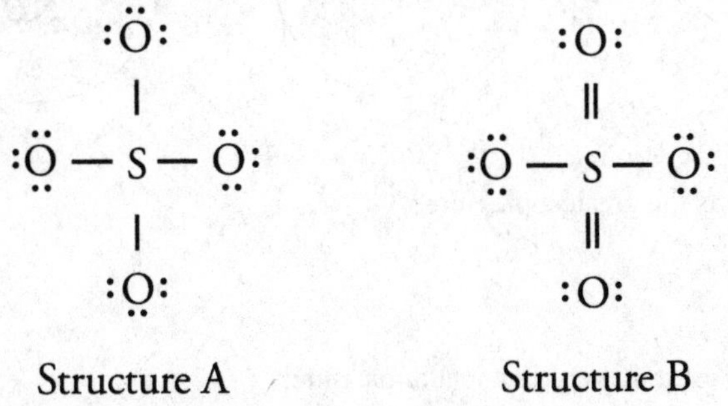

```
       :Ö:                    :O:
        |                      ||
  :Ö — S — Ö:            :Ö — S — Ö:
        |                      ||
       :O:                    :O:

   Structure A            Structure B
```

25. What is the formal charge on the sulfur atom in the structure A?

(A) –1
(B) 0
(C) +1
(D) +2

26. What is the molecular geometry in the structure A?

(A) Tetrahedral
(B) Trigonal Planar
(C) Trigonal Pyramidal
(D) Octahedral

27. What is the S–O bond order in the structure B?

(A) 1.0
(B) 1.33
(C) 1.5
(D) 1.67

28. Which of the following statements regarding the structure B is true?

(A) The double bonds must be located opposite of each other due to additional electron repulsion.
(B) It is a more polar molecule than the molecule represented by structure A.
(C) The bonds in the molecule are weaker than those in structure A.
(D) All bonds in the molecule are identical to each other.

29. Which structure is more likely to correspond with the actual Lewis diagram for the sulfate ion?

(A) Structure A; single bonds are more stable than double bonds
(B) Structure A; it has the most unshared pairs of electrons
(C) Structure B; there are more possible resonance structures
(D) Structure B; fewer atoms have formal charges

Use the following information to answer questions 30-32.

The diagram below shows three identical 1.0 L containers filled with the indicated amounts of gas. The stopcocks connecting the containers are originally closed and the gases are all at 25°C. Assume ideal behavior.

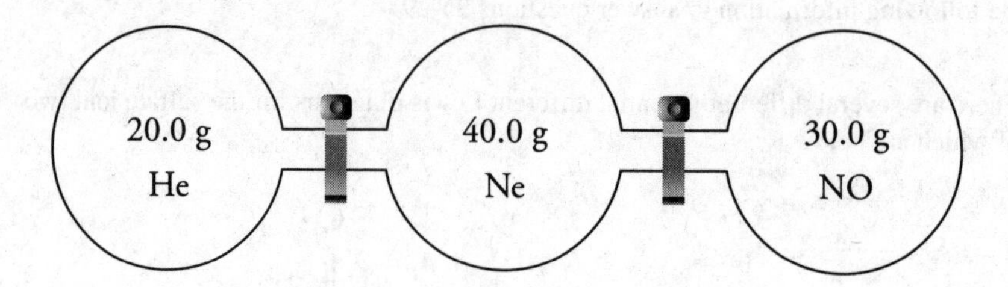

30. Which gas exerts the greatest pressure?

(A) He
(B) Ne
(C) NO
(D) All gases exert the same amount of pressure.

31. Which gas has the strongest IMFs?

(A) He
(B) Ne
(C) NO
(D) All gases have identical IMFs.

32. The stopcocks are opened. If the tubing connecting the containers has negligible volume, by what percentage will the pressure exerted by the neon gas decrease?

(A) 25%
(B) 33%
(C) 50%
(D) 67%

Free-Response Questions

1. A 250 mL Erlenmeyer flask contains a mixture of two liquids: diethyl ether and ethylamine. The flask is attached to a distillation apparatus and heated until the mixture starts to boil.

   (a) Why it important to keep the flask at a constant temperature once the mixture starts to boil?

   (b) (i) Which liquid is the primary component of the distillate? Justify your answer in terms of IMFs.

   (ii) The distillate is not a completely pure substance. Why?

   (c) After the distillate is collected, half of it is transferred into a 250 mL beaker, as shown below. If both containers are left uncovered, which liquid (if either) will evaporate first? Why?

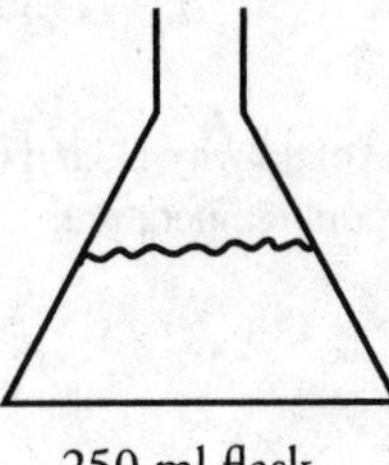

250 ml flask

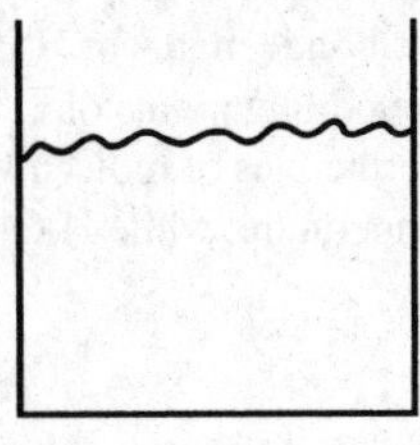

250 ml beaker

2. The carbonate ion CO_3^{2-} is formed when carbon dioxide, CO_2, reacts with slightly basic cold water.

   (a) (i) Draw the Lewis electron dot structure for the carbonate ion. Include resonance forms if they apply.

   (ii) Draw the Lewis electron dot structure for carbon dioxide.

   (b) Describe the hybridization of carbon in the carbonate ion.

   (c) (i) Describe the relative lengths of the three C–O bonds in the carbonate ion.

   (ii) Compare the average length of the C–O bonds in the carbonate ion to the average length of the C–O bonds in carbon dioxide.

3.

Substance	Boiling Point (°C)	Bond Length (Å)	Bond Strength (kcal/mol)
H_2	–253°	0.75	104.2
N_2	–196°	1.10	226.8
O_2	–182°	1.21	118.9
Cl_2	–34°	1.99	58.0

(a) Explain the differences in the properties given in the table above for each of the following pairs.

(i) The bond strengths of N_2 and O_2

(ii) The bond lengths of H_2 and Cl_2

(iii) The boiling points of O_2 and Cl_2

(b) Use the principles of molecular bonding to explain why H_2 and O_2 are gases at room temperature, while H_2O is a liquid at room temperature.

4. A student has a mixture containing three different organic substances. The Lewis diagrams of the substances are below:

```
    H  H  H  H                        H  H
    |  |  |  |                        |  |
 H—C—C—C—C—Ö—H                   H—C—C—C̈l:
    |  |  |  |   ··                   |  |  ··
    H  H  H  H                        H  H
     n-butanol                      ethyl chloride

              H  H  H   H
              |  |  |  /
           H—C—C—C—N:
              |  |  |  \
              H  H  H   H
              n-propylamine
```

(a) If the mixture was dabbed onto chromatography paper that was then placed into a nonpolar solvent, rank the R_f values for each component of the mixture from high to low after the solvent has saturated the paper. Justify your answer.

(b) If the mixture is poured into a chromatography column and then eluted with a very polar substance, which component of the mixture would leave the column first, and why?

(c) (i) The mixture is heated until it begins to boil. Which substance would be the easiest to separate via distillation, and why?

(ii) After the substance begins boiling, it continues to be heated at the same rate. Compared to the rate at which it was changing prior to boiling, will the temperature increase faster, slower, or at the same rate? Explain.

(d) (i) After the components of the mixture have been separated, they are returned to room temperature. Of the three substances, which would have the highest vapor pressure at room temperature? Justify your answer.

(ii) If the substances were heated (but not boiled), explain what would happen to their vapor pressures.

5.

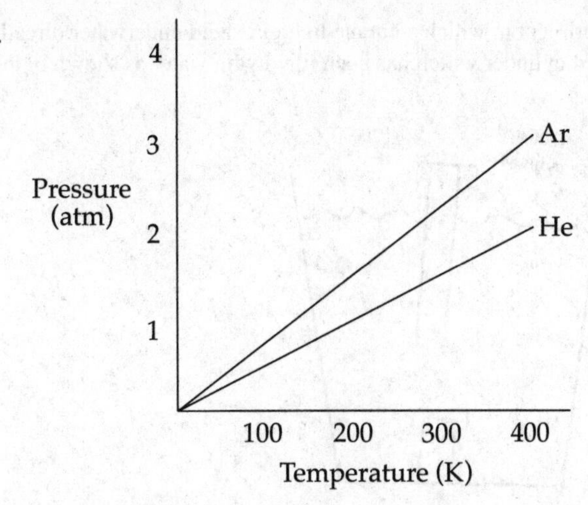

The graph above shows the changes in pressure with changing temperature of gas samples of helium and argon confined in a closed 2-liter vessel.

(a) What is the total pressure of the two gases in the container at a temperature of 200 K?
(b) How many moles of helium are contained in the vessel?
(c) How many molecules of helium are contained in the vessel?
(d) Molecules of which gas will have a greater distribution of velocities at 200 K? Justify your answer.
(e) If the volume of the container were reduced to 1 liter at a constant temperature of 300 K, what would be the new pressure of the helium gas?

6.

$$2\ KClO_3(s) \rightarrow 2\ KCl(s) + 3\ O_2(g)$$

The reaction above took place, and 1.45 liters of oxygen gas were collected over water at a temperature of 29°C and a pressure of 755 millimeters of mercury. The vapor pressure of water at 29°C is 30.0 millimeters of mercury.

(a) What is the partial pressure of the oxygen gas collected?
(b) How many moles of oxygen gas were collected?
(c) What would be the dry volume of the oxygen gas at a pressure of 760 millimeters of mercury and a temperature of 273 K?
(d) What was the mass of the $KClO_3$ consumed in the reaction?

7. Equal molar quantities of two gases, O_2 and H_2O, are confined in a closed vessel at constant temperature.

(a) Which gas, if either, has the greater partial pressure?
(b) Which gas, if either, has the greater density?
(c) Which gas, if either, has the greater concentration?
(d) Which gas, if either, has the greater average kinetic energy?
(e) Which gas, if either, will show the greater deviation from ideal behavior?

8. A student performs an experiment in which a butane lighter is held underwater directly beneath a 100-mL graduated cylinder which has been filled with water as shown in the diagram below.

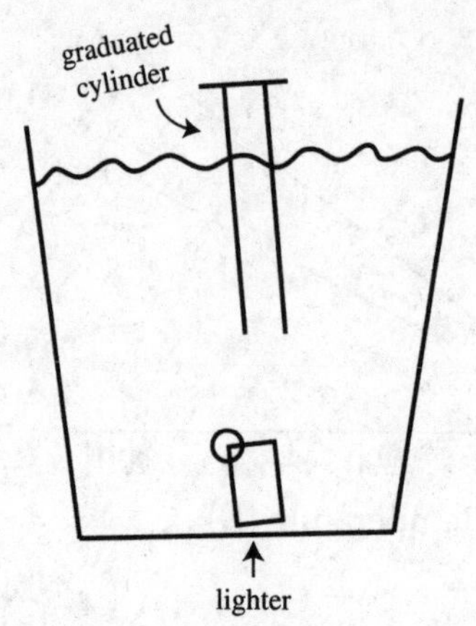

The switch on the lighter is pressed, and butane gas is released into the graduated cylinder. The student's data table for this lab is as follows:

Mass of lighter before gas release	20.432 g
Mass of lighter after gas release	20.296 g
Volume of gas collected	68.40 mL
Water Temperature	19.0°C
Atmospheric Pressure	745 mmHg

(a) Given that the vapor pressure of water at 19.0°C is 16.5 mmHg, determine the partial pressure of the butane gas collected in atmospheres.

(b) Calculate the molar mass of butane gas from the experimental data given.

(c) If the formula of butane is C_4H_{10}, determine the percent error for the student's results.

(d) The following are common potential error sources that occur during this lab. Explain whether or not each error could have been responsible for the error in the student's results.

(i) The lighter was not sufficiently dried before massing it after the gas was released.

(ii) The gas in the lighter was not held underwater long enough to sufficiently cool it to the same temperature of the water and was actually at a higher temperature than the water.

(iii) Not all of the butane gas released was collected in the graduated cylinder.

CHAPTER 4 ANSWERS AND EXPLANATIONS

Multiple-Choice

1. **D** Calcium is a metal, and fluorine is a nonmetal. Their electronegativities differ sufficiently for them to create an ionic bond, which is stronger than all other bond types except for network covalent bonding (which NH_3 does not exhibit).

2. **B** Interstitial alloys form when atoms of greatly different sizes combine. The aluminum atoms would have a chance to fit between the comparatively larger lead atoms.

3. **A** Only atoms with at least three energy levels (n = 3 and above) have empty *d*-orbitals that additional electrons can fit into, thus expanding their octet.

4. **B** Moles of $H_2 = \dfrac{2.00\text{ g}}{2.00\text{ g/mol}} = 1.00$ mol (remember, hydrogen is a diatomic)

   Moles of $\text{Ne} = \dfrac{10.00\text{ g}}{20.0\text{ g/mol}} = 0.500$ mol

   Total moles = 1.50 moles. $X_{Ne} = \dfrac{\text{moles Ne}}{\text{total moles}} = \dfrac{0.500}{1.500} = 0.33$

5. **C** At STP, 1 mol of gas takes up 22.4 L of space: (1.5)(22.4) = 33.5 L.

6. **B** Both gases only have London dispersion forces. The more electrons a gas has, the more polarizable it is and the stronger the intermolecular forces are.

7. **A** The gas molecules have the same amount of kinetic energy due to their temperature being the same. Via $KE = \frac{1}{2}mv^2$, if KE is the same, then the molecule with less mass must correspondingly have a higher velocity.

8. **B** First, C_2H_6 is not an ionic substance and thus has no lattice energy. Next, LiF is composed of ions with charges +1 and –1, and will not be as strong as the two compounds which have ions with charges of +2 and –1. Finally, $MgCl_2$ is smaller than $CaBr_2$, meaning it will have a higher lattice energy, as (according to Coulomb's law) atomic radius is inversely proportional with bond energy.

9. B A liquid with a low boiling point must be held together by weak intermolecular forces, of which London dispersion forces are the weakest kind.

10. C Resonance is used to describe a situation that lies between single and double bonds, so the bond length would also be expected to be in between that of single and double bonds. The best answer here is 140 pm, (C).

11. A H_2 experiences only London dispersion forces and has the lowest boiling point.
N_2 also experiences only London dispersion forces, but it is larger than H_2 and has more electrons, so it has stronger interactions with other molecules.

NH_3 is polar and undergoes hydrogen bonding, so it has the strongest intermolecular interactions and the highest boiling point.

12. D From Dalton's law, the partial pressure of a gas depends on the number of moles of the gas that are present.

The total number of moles of gas present is

$$1.5 + 3.0 + 0.5 = 5.0 \text{ total moles}$$

If there are 3 moles of nitrogen, then $\frac{3}{5}$ of the pressure must be due to nitrogen.

$$\left(\frac{3}{5}\right)(700 \text{ mmHg}) = 420 \text{ mmHg}$$

13. C From Dalton's law, the partial pressure of a gas depends on the number of moles of the gas that are present. If the mixture has twice as many moles of helium as neon, then the mixture must be $\frac{1}{3}$ neon. So $\frac{1}{3}$ of the pressure must be due to neon.

$$\left(\frac{1}{3}\right)(1.2 \text{ atm}) = 0.4 \text{ atm.}$$

14. D From Dalton's law, the partial pressures of nitrogen and water vapor must add up to the total pressure in the container. The partial pressure of water vapor in a closed container will be equal to the vapor pressure of water, so the partial pressure of nitrogen is:

$$781 \text{ mmHg} - 23 \text{ mmHg} = 758 \text{ mmHg}$$

15. A Use the following relationship:

$$\text{Moles} = \frac{\text{liters}}{22.4\ \text{L/mol}}$$

$$\text{Moles of unknown gas} = \frac{11.2\ \text{L}}{22.4\ \text{L/mol}} = 0.500\ \text{moles}$$

$$\text{MW} = \frac{\text{grams}}{\text{mole}}$$

$$\text{MW of unknown gas} = \frac{22.0\ \text{g}}{0.500\ \text{mole}} = 44.0\ \text{grams/mole}$$

That's the molecular weight of CO_2.

16. D Density is measured in grams per liter. One mole of helium gas has a mass of 4 grams and occupies a volume of 22.4 liters at STP, so the density of helium gas at STP is $\frac{4}{22.4}$ g/L.

17. D There are initially 3.0 moles of gas in the container. If they react completely, 2.0 moles of gas are produced. 2.0 moles of gas will exert exactly $\frac{2}{3}$ as much pressure as three moles of gas.

18. C From the ideal gas laws, for a gas sample at constant pressure:

$$\frac{V_1}{T_1} = \frac{V_2}{T_2}$$

Solving for V_2 we get $V_2 = V_1\frac{T_2}{T_1}$.

So V_1 is multiplied by a factor of $\frac{T_2}{T_1}$.

Remember to convert Celsius to Kelvin, $\frac{127°\text{C} + 273}{27°\text{C} + 273} = \frac{400\ \text{K}}{300\ \text{K}} = \frac{4}{3}$

19. A $\text{MM} = \frac{DRT}{\text{P}}$. The density of this gas is $\frac{10.0\ \text{g}}{5.0\ \text{L}}$, or $\frac{2.0\ \text{g}}{\text{L}}$.

Plugging that in yields $\text{MM} = \frac{(2.0)(0.08)(299)}{(2.0)}$. The value of 2.0 cancels out, leaving the answer (0.08)(299).

20. **B** First find the number of moles.

Moles = (molarity)(volume)
Moles of substance = (0.50 M)(4.0 L) = 2 moles

$$\text{Moles} = \frac{\text{grams}}{\text{MW}}$$

$$\text{So MW} = \frac{240 \text{ g}}{2 \text{ mol}} = 120 \text{ g/mol.}$$

21. **A** Let's find out how many moles of Na^+ we have to add.

Moles = (molarity)(volume)
Moles of Na^+ = (2 *M*)(0.5 L) = 1 mole

Because we get 2 moles of Na^+ ions for every mole of Na_2SO_4 we add, we need to add only 0.5 moles of Na_2SO_4.

22. **C** Hydrogen bonds are the strongest types of intermolecular forces when dealing with molecules of a similar size. Ethylene glycol has twice as many hydrogen bonds as ethanol (acetone has none), and so it would have the highest boiling point.

23. **B** Vapor pressure arises from molecules breaking free from the intermolecular forces holding them together. Acetone, which has no hydrogen bonding, thus has the weakest intermolecular forces of the three and thus would have the highest vapor pressure.

24. **D** Water is very polar, and using the concept of "like dissolves like," any substance with polar molecules would be soluble in water. As all three molecules are polar, all three liquids would be soluble in water.

25. **D** Valence electrons – assigned electrons = formal charge: 6 – 4 = +2.

26. **A** Four charge clouds and no lone pairs means tetrahedral geometry.

27. **C** Six total bonds divided by four locations gives a bond order of 1.5.

28. **D** In any molecule displaying resonance, all bonds are identical.

29. **D** The formal charge tables for each diagram are below (note: for structure B, the double-bonded oxygens are the first two, and the single bonded are the last two)

Structure A		Structure B
S O O O O		S O O O O
6 6 6 6 6	**Valence e^-**	6 6 6 6 6
4 7 7 7 7	**Assigned e^-**	6 6 6 7 7
+2 -1 -1 -1 -1	**Formal charge**	**0 0 0 -1 -1**

The total formal charge on each potential structure is –2, which is correct as that is the charge on a sulfate ion. However, the right-hand structure has fewer atoms with formal charges, making it the more likely structure.

30. **A** Pressure is directly dependent on the number of moles. In their respective containers, there are 5 moles of He, 2 moles of Ne, and 1 mole of NO. As there are the most moles of He, the He must exert the greatest pressure.

31. **D** One of the precepts of kinetic molecular theory is that gas molecules exert no forces on each other, thus, in all containers there are no IMFs present.

32. **D** $P_1V_1 = P_2V_2$

$P_1(1.0 \text{ L}) = P_2(3.0 \text{ L})$

$P_2/P_1 = 1.0/3.0$

Thus, the pressure of the neon gas is 33% of what it was originally, meaning a decrease of 67%. Note that the same calculation could be used for any of the gases; each gas is expanding to take up three times as much space as it has originally, and thus exerts one-third as much pressure.

Free-Response

1. (a) If the temperature were to be increased, it may go higher than the boiling point of both substances, not just the one with the lower boiling point. This would create a very impure distillate.

   (b) (i) The diethyl ether would be in the distillate. Ethylamine contains hydrogen bonding, which is the strongest type of IMF, while diethyl ether does not.

   (ii) Even below the boiling point of ethylamine, some of the ethylamine will spontaneously convert into a gas. The molecules within both liquids are constantly moving, and even if the ethylamine does not boil, some of its molecules will evaporate, convert into a gas, and then be condensed, becoming part of the distillate.

   (c) The liquid in the beaker will evaporate first. Evaporation occurs when molecules in the liquid phase have sufficient kinetic energy to break free of the IMFs within the liquid. In order to break free, the molecules must first hit the surface of the liquid. The surface area of the liquid in the beaker is greater than the surface area of the liquid in the test tube, thus, the molecules in the beaker are more likely to hit the surface and evaporate.

2. (a) (i)

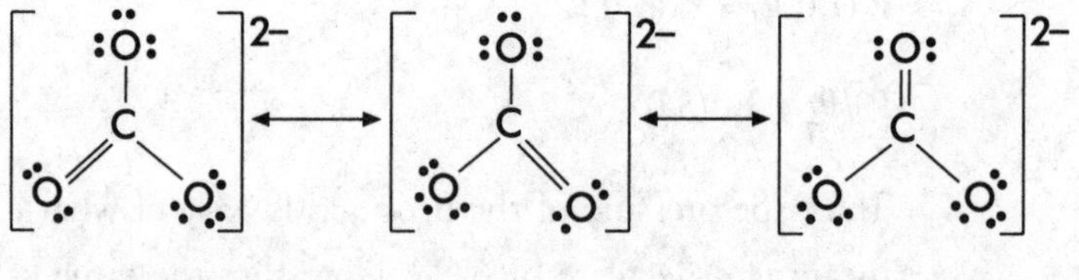

   (ii) O=C=O

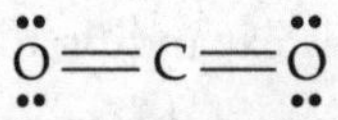

   (b) The central carbon atom forms three sigma bonds with oxygen atoms and has no free electron pairs, so its hybridization must be sp^2.

   (c) (i) All three bonds will be the same length because when a molecule exhibits resonance, all the bonds are identical to each other, being somewhere in character between single bonds and double bonds.

(ii) The C–O bonds in the carbonate ion have resonance forms between single and double bonds, while the C–O bonds in carbon dioxide are both double bonds.

The bonds in the carbonate ion will be shorter than single bonds and longer than double bonds, so the carbonate bonds will be longer than the carbon dioxide bonds.

3. (a) (i) The bond strength of N_2 is larger than the bond strength of O_2 because N_2 molecules have triple bonds and O_2 molecules have double bonds. Triple bonds are stronger and shorter than double bonds.

(ii) The bond length of H_2 is smaller than the bond length of Cl_2 because hydrogen is a smaller atom than chlorine, allowing the hydrogen nuclei to be closer together.

(iii) Liquid oxygen and liquid chlorine are both nonpolar substances that experience only London dispersion forces of attraction. These forces are greater for Cl_2 because it has more electrons (which makes it more polarizable), so Cl_2 has a higher boiling point than O_2.

(b) H_2 and O_2 are both nonpolar molecules that experience only London dispersion forces, which are too weak to form the bonds required for a substance to be liquid at room temperature.

H_2O is a polar substance whose molecules form hydrogen bonds with each other. Hydrogen bonds are strong enough to form the bonds required in a liquid at room temperature.

4. (a) All three components are polar, but ethyl chloride has no hydrogen bonding and thus has the weakest dipoles, meaning it would travel the furthest and have the highest R_f value.

Between n-butanol and n-propylamine, both have H-bonds but the butanol has more electrons, meaning its London dispersion forces are stronger. It thus has the highest polarity and would have the smallest R_f value. So: ethyl chloride > n-propylamine > n-butanol.

(b) n-butanol is the most polar and would be most attracted to a polar eluent, and thus would leave the column first.

(c) (i) Ethyl chloride has the weakest overall IMFs, and thus would have the lowest boiling point and be the easiest to separate out.

(ii) Prior to boiling, molecules in all three substances were speeding up as heat was added. The increased velocity caused the temperature increase. Once the ethyl chloride starts boiling, though, the heat that would ordinarily be causing the molecules to speed up is instead breaking the IMFs. While the molecules of both the n-butanol and n-propylamine would still be experiencing a velocity increase, those of the ethyl chloride would not. Thus, the overall rate of the temperature change would be less than it was prior to the ethyl chloride starting to boil.

(d) (i) The substance with the weakest IMFs would allow the largest number of molecules to escape the liquid phase and turn into a gas, which is the cause of vapor pressure. As such, the ethyl chloride should have the highest vapor pressure of the three substances.

(ii) As temperature increases, so does the molecular velocity. The faster the molecules are going, the more energy they have, and the more likely they are to be able to overcome the IMFs and escape the liquid phase. For all substances, as temperature increases, so does vapor pressure.

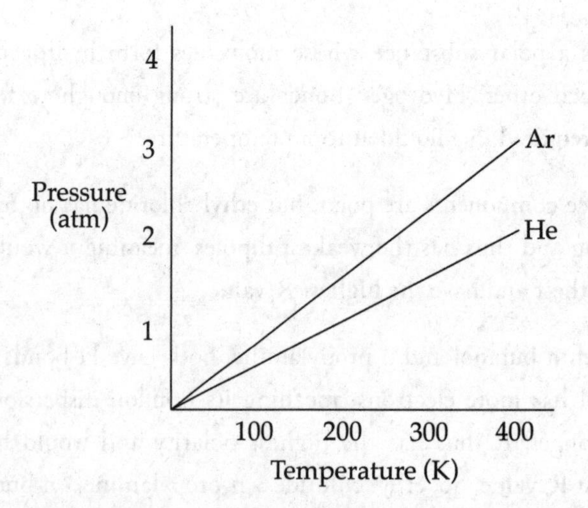

5. (a) Read the graph, and add the two pressures.

$P_{\text{Total}} = P_{\text{He}} + P_{\text{Ar}}$

$P_{\text{Total}} = (1 \text{ atm}) + (1.5 \text{ atm}) = 2.5 \text{ atm}$

(b) Read the pressure (1 atm) at 200 K, and use the ideal gas equation.

$$n = \frac{PV}{RT} = \frac{(1.0\text{ atm})(2.0\text{ L})}{(0.082\text{ L}\cdot\text{atm/mol}\cdot\text{K})(200\text{ K})} = 0.12\text{ moles}$$

(c) Use the definition of a mole.

Molecules = (moles)(6.02×10^{23})

Molecules (atoms) of helium = (0.12)(6.02×10^{23}) = 7.2×10^{22}

(d) As both gases are at the same temperature and thus have the same kinetic energy, the molecules of helium will have a higher average velocity because they have less mass. The higher the average velocity of a gas is, the wider the distribution curve is for the velocities of the individual gas molecules.

(e) Use the following relationship:

$$\frac{P_1V_1}{T_1} = \frac{P_2V_2}{T_2}$$

Since T is a constant, the equation becomes:

$$P_1V_1 = P_2V_2$$

$$(1.5\text{ atm})(2.0\text{ L}) = P_2(1.0\text{ L})$$

$$P_2 = 3.0\text{ atm}$$

6. (a) Use Dalton's law.

$$P_{Total} = P_{Oxygen} + P_{Water}$$

$$(755\text{ mmHg}) = (P_{Oxygen}) + (30.0\text{ mmHg})$$

$$P_{Oxygen} = 725\text{ mmHg}$$

(b) Use the ideal gas law. Don't forget to convert to the proper units.

$$n = \frac{PV}{RT} = \frac{\left(\frac{725}{760}\text{ atm}\right)(1.45\text{ L})}{(0.082\text{ L}\cdot\text{atm/mol}\cdot\text{K})(302\text{ K})} = 0.056\text{ moles}$$

(c) At STP, moles of gas and volume are directly related.

Volume = (moles)(22.4 L/mol)

Volume of O_2 = (0.056 mol)(22.4 L/mol) = 1.25 L

(d) We know that 0.056 moles of O_2 were produced in the reaction.

From the balanced equation, we know that for every 3 moles of O_2 produced, 2 moles of $KClO_3$ are consumed. So there are $\frac{2}{3}$ as many moles of $KClO_3$ as O_2.

$$\text{Moles of } KClO_3 = \left(\frac{2}{3}\right)(\text{moles of } O_2)$$

$$\text{Moles of } KClO_3 = \left(\frac{2}{3}\right)(0.056 \text{ mol}) = 0.037 \text{ moles}$$

$$\text{Grams} = (\text{moles})(\text{MW})$$

$$\text{Grams of } KClO_3 = (0.037 \text{ mol})(122 \text{ g/mol}) = 4.51 \text{ g}$$

7. (a) The partial pressures depend on the number of moles of gas present. Because the number of moles of the two gases are the same, the partial pressures are the same.

   (b) O_2 has the greater density. Density is mass per unit volume. Both gases have the same number of moles in the same volume, but oxygen has heavier molecules, so it has greater density.

   (c) Concentration is moles per volume. Both gases have the same number of moles in the same volume, so their concentrations are the same.

   (d) According to kinetic-molecular theory, the average kinetic energy of a gas depends only on the temperature. Both gases are at the same temperature, so they have the same average kinetic energy.

   (e) H_2O will deviate most from ideal behavior. Ideal behavior for gas molecules assumes that there will be no intermolecular interactions.

   H_2O is polar, and O_2 is not. H_2O undergoes hydrogen bonding, while O_2 does not. So H_2O has stronger intermolecular interactions, which will cause it to deviate more from ideal behavior.

8. (a) 745 mmHg – 16.5 mmHg = 729 mmHg

   $$\frac{729 \text{ mmHg}}{760 \text{ mmHg}} = 0.959 \text{ atm}$$

   (b) To determine the mass of the butane, subtract the mass of the lighter after the butane was released from the mass of the lighter before the butane was released.

   $$20.432 \text{ g} - 20.296 \text{ g} = 0.136 \text{ g}$$

To determine the moles of butane, use the ideal gas law, making any necessary conversions first.

$$PV = nRT$$
$$(0.959 \text{ atm})(0.06840 \text{ L}) = n(0.0821 \text{ atm·L/mol·K})(292 \text{ K})$$
$$n = 2.74 \times 10^{-3} \text{ mol}$$

Molar mass is defined as grams per mole, so
$0.136 \text{ g}/2.74 \times 10^{-3} \text{ mol} = 49.6 \text{ g/mol}$

(c) Actual molar mass of butane:

$(12.00 \text{ g/mol} \times 4) + (1.01 \text{ g/mol} \times 10) = 58.08 \text{ g/mol}$

Percent error is:

$$\frac{|\text{Actual value} - \text{experimental value}|}{\text{Actual value}} \times 100\%$$

So:

$$\frac{|58.08 - 49.6|}{58.08} \times 100\% = 14.5\% \text{ error}$$

(d) (i) If the lighter is not sufficiently dried, then the mass of the butane calculated will be artificially low. That means the numerator in the molar mass calculation will be too low, which would lead to an experimental molar mass that is too low. This is consistent with the student's error.

(ii) If the temperature of the butane is higher than the water temperature, the calculated moles of butane will be artificially high. This means the denominator in the molar mass calculation will be too high, which would lead to an experimental molar mass that is too low. This is consistent with the student's error.

(iii) If some butane gas escaped without going into the graduated cylinder, the volume of butane gas collected will be artificially low. That will make the calculation for moles of butane too low, which in turns means the denominator of the molar mass calculation will be too low. This would lead to an experimental molar mass that is too high. This is NOT consistent with the student's error.

Chapter 5
Big Idea #3: Chemical Reactions, Energy Changes, and Redox Reactions

Changes in matter involve the rearrangement and/or reorganization of atoms and/or the transfer of electrons.

TYPES OF REACTIONS

Reactions may be classified into several categories.

1. Synthesis Reactions: When elements or simple compounds are combined to form a single, more complex compound.

   $$2Mg(s) + O_2(g) \rightarrow 2MgO(s)$$

2. Decomposition: The opposite of a synthesis. A reaction where a single compound is split into two or more elements or simple compounds, usually in the presence of heat.

   $$HgO(s) + \text{Heat} \rightarrow Hg(s) + \tfrac{1}{2}O_2(g)$$

3. Acid-Base Reaction: A reaction when an acid (i.e. H^+) reacts with a base (i.e. OH^-) to form water and a salt.

   $$HCl(aq) + NaOH(aq) \rightarrow NaCl(aq) + H_2O(l)$$

4. Oxidation-Reduction (Redox) Reaction: A reaction that results in the change of the oxidation states of some participating species.

   $$Cu^{2+}(aq) + 2e^- \rightarrow Cu(s)$$

5. Hydrocarbon Combustion: When a covalent substance containing carbon and hydrogen (and sometimes oxygen) is ignited, it reacts with the oxygen in the atmosphere. The products of a hydrocarbon combustion are always CO_2 and H_2O. The combustion of butane, a fuel found in many lighters, is an example of a combustion reaction:

   $$C_4H_{10} + \frac{13}{2}O_2 \rightarrow 4CO_2 + 5H_2O$$

   If elements other than carbon and hydrogen are present in the substance being combusted, they, too, often combine with oxygen to form various gases.

   $$2C_2H_5S + \frac{17}{2}O_2 \rightarrow 4CO_2 + 5H_2O + 2SO_2$$

   It is not always possible to predict the exact formula of the non-carbon or hydrogen containing compounds that are created, but carbon will always lead to CO_2 and hydrogen, to H_2O.

6. Precipitation: When two aqueous solutions mix, sometimes a new cation/anion pairing can create an insoluble salt. This type of reaction is called a precipitation reaction. When potassium carbonate and magnesium nitrate mix, a precipitate of magnesium carbonate will form as follows:

   $$K_2CO_3(aq) + Mg(NO_3)_2(aq) \rightarrow 2KNO_3(aq) + MgCO_3(s)$$

   That can also be written as a net ionic equation. In solution, the potassium and nitrate ions do not actually take part in the reaction. They start out as free ions and end up as free ions. We call those ions **spectator ions**. The only thing that is changing is that carbonate and magnesium ions are bonding to form magnesium carbonate. The net ionic equation is:

$$CO_3^{2-} + Mg^{2+} \rightarrow MgCO_3(s)$$

Finally, this can be represented using particulate diagrams, as shown below:

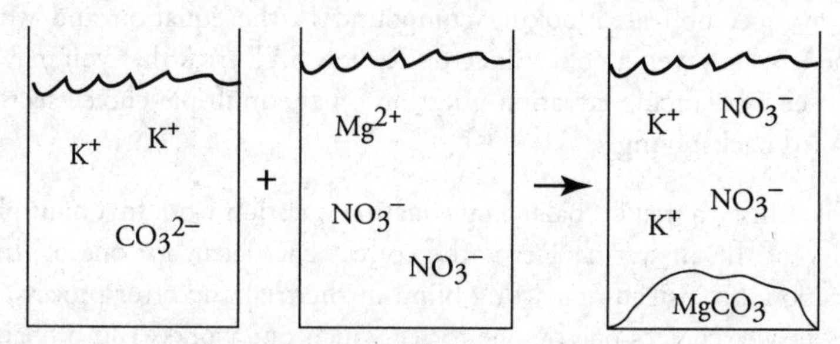

You may have heard that you need to memorize solubility rules—that is, what ions will form insoluble precipitates when you combine them. You DO NOT need to memorize most of these; for the most part, the AP Exam will provide you with them as needed. The only ones you need to know are:

1. Compounds with an alkali metal cation (Na^+, Li^+, K^+, etc) or an ammonium cation (NH_4^+) are always soluble.
2. Compounds with a nitrate (NO_3^-) anion are always soluble.

That's it! You should know how to interpret any solubility rules that you are given on the test, but you need not memorize anything beyond those two.

Ionic substances that dissolve in water do so because of the attraction of the ions to the dipoles of the water molecules exceeds the attraction of the ions to each other. Let's take NaBr for example:

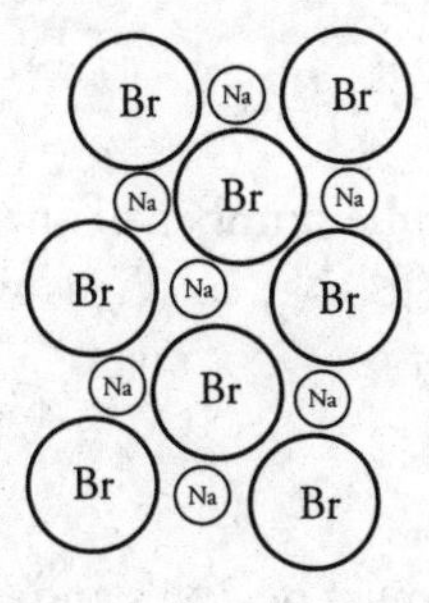

When it dissolves in water, the positive Na^+ cations are attracted to the negative (oxygen) ends of the water molecules. The negative Br^- anions are attracted to the positive (hydrogen) ends of the water molecules. Thus, when the substance dissolves it looks like this on the particulate level:

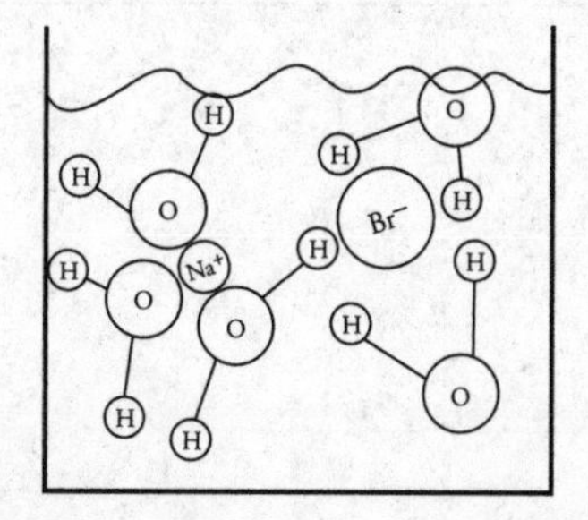

CHEMICAL EQUATIONS

Balancing Chemical Equations

Normally, balancing a chemical equation is a trial-and-error process. You start with the most complicated-looking compound in the equation and work from there. There is, however, an old Princeton Review SAT trick that you may want to try if you see a balancing equation question on the multiple-choice section. The trick is called **backsolving.**

It works like this: To make a balancing equation question work in a multiple-choice format, one of the answer choices is the correct coefficient for one of the species in the reaction. So instead of starting blind in the trial-and-error process, you can insert the answer choices one by one to see which one works. You probably won't have to try all four, and if you start in the middle and the number doesn't work, it might be obviously too small or large, eliminating other choices before you have to try them. Let's try it.

A Note About Backsolving

Backsolving is more efficient than the methods that you're accustomed to. If you use the answer choices that you're given, you streamline the trial-and-error process and allow yourself to use POE as you work on the problem.

$$...NH_3 + ...O_2 \rightarrow ...N_2 + ...H_2O$$

1. If the equation above were balanced with lowest whole number coefficients, the coefficient for NH_3 would be

(A) 1
(B) 2
(C) 3
(D) 4

Here's How to Crack It!

Start at (C) because it's the middle number. If there are $3NH_3$'s, then there can't be a whole number coefficient for N_2, so (C) is wrong, and so is the other odd number answer, (A).

Try (D).

If there are $4NH_3$'s, then there must be $2N_2$'s and $6H_2O$'s.

If there are $6H_2O$'s, then there must be $3O_2$'s, and the equation is balanced with lowest whole number coefficients.

Chemical Equations and Calculations

Because of the math involved, stoichiometry like this will most likely appear only as a free-response question. In those instances, expect the following: You will be given a balanced chemical equation and told that you have some number of grams (or liters of gas, or molar concentration, and so on) of reactant. Then you will be asked what number of grams (or liters of gas, or molar concentration, and so on) of products are generated.

In these cases, follow this simple series of steps.

1. Convert whatever quantity you are given into moles.
2. If you are given information about two reactants, you may have to use the equation coefficients to determine which one is the limiting reagent. Remember, the limiting reagent is not necessarily the reactant that you have the least of; it is the reactant that runs out first.
3. Use the balanced equation to determine how many moles of the desired product are generated.
4. Convert moles of product to the desired unit.

Let's try one.

$$2HBr(aq) + Zn(s) \rightarrow ZnBr_2(aq) + H_2(g)$$

2. A piece of solid zinc weighing 98 grams was added to a solution containing 324 grams of HBr. What is the volume of H_2 produced at standard temperature and pressure if the reaction above runs to completion?

Here's How to Crack It!

1. Convert whatever quantity you are given into moles.

$$\text{Moles of Zn} = \frac{\text{grams}}{\text{molar mass}} = \frac{(98\text{ g})}{(65.4\text{ g/mol})} = 1.5\text{ mol}$$

$$\text{Moles of HBr} = \frac{\text{grams}}{\text{molar mass}} = \frac{(324\text{ g})}{(80.9\text{ g/mol})} = 4.0\text{ mol}$$

2. Use the balanced equation to find the limiting reagent.
From the balanced equation, 2 moles of HBr are used for every mole of Zn that reacts, so when 1.5 moles of Zn react, 3 moles of HBr are consumed, and there will be HBr left over when all of the Zn is gone. That makes Zn the limiting reagent.
3. Use the balanced equation to determine how many moles of the desired product are generated.
1 mole of H_2 is produced for every mole of Zn consumed, so if 1.5 moles of Zn are consumed, then 1.5 moles of H_2 are produced.
4. Convert moles of product to the desired unit.
The H_2 gas is at STP, so we can convert directly from moles to volume.
Volume of H_2 = (moles)(22.4 L/mol) = (1.5 mol)(22.4 L/mol) = 33.6 L $\approx$ 34 L

Let's try another one using the same reaction.

$$2HBr(aq) + Zn(s) \rightarrow ZnBr_2(aq) + H_2(g)$$

3. A piece of solid zinc weighing 13.1 grams was placed in a container. A 0.10-molar solution of HBr was slowly added to the container until the zinc was completely dissolved. What was the volume of HBr solution required to completely dissolve the solid zinc?

Here's How to Crack It!

1. Convert whatever quantity you are given into moles.

$$\text{Moles of Zn} = \frac{\text{grams}}{\text{MW}} = \frac{(13.1 \text{ g})}{(65.4 \text{ g/mol})} = 0.200 \text{ mol}$$

2. Use the balanced equation to find the limiting reagent.
3. Use the balanced equation to determine how many moles of the desired product are generated.
In this case, we're using the balanced reaction to find out how much of one reactant is required to consume the other reactant.
We can see from the balanced equation that it takes 2 moles of HBr to react completely with 1 mole of Zn, so it will take 0.400 moles of HBr to react completely with 0.200 moles of Zn.
4. Convert moles of product to the desired unit.
Moles of HBr = (molarity)(volume)

$$\text{Volume of HBr} = \frac{\text{moles}}{\text{molarity}} = \frac{(0.400 \text{ mol})}{(0.10 \text{ mol/L})} = 4.0 \text{ L}$$

When you perform calculations, always include units. Including units in your calculations will help you (and the person scoring your test) keep track of what you are doing. While using the proper units in an answer is not typically worth any points in and of itself (with a few exceptions), AP graders may take off points if your units are missing or incorrect.

Combustion Analysis

When a hydrocarbon is combusted, producing carbon dioxide and water, the formula of that hydrocarbon can often be determined using a little math. One of the most important concepts in chemistry is the **law of conservation of mass.** This states that matter can be neither created nor destroyed during a chemical reaction. Thus, when a hydrocarbon is combusted, all of the carbon in that hydrocarbon will end up in CO_2, and all of the hydrogen will end up in H_2O. With that in mind, we can often determine the empirical formula of a hydrocarbon by determining the mass of the products it forms.

4. A compound containing hydrogen, carbon, and oxygen is combusted. 44.0 g of CO_2 and 27.0 g of H_2O are produced. Which of the following is a possible empirical formula for that compound?

(A) CH_4O
(B) C_2H_6O
(C) $C_2H_3O_2$
(D) $C_3H_5O_2$

Here's How to Crack It!

If 44.0 grams of CO_2 are created, that is 1.00 mole of CO_2, and thus, 1.00 mole of C as well (as there is one mole of carbon in each mole of CO_2). Through conservation of mass, this means there was 1.00 mole of C in the original compound.

27.0 g of H_2O means 1.5 moles of H_2O were created. Because there are two hydrogen atoms in every H_2O, that means there were 3.0 moles of hydrogen in the water which was created, and thus, 3.00 moles of H in the original compound. The C:H ratio in the original compound must be 1:3, which is consistent with (B).

GRAVIMETRIC ANALYSIS

A common way to use precipitation reactions in order to make quantitative determinations about the identity of an unknown sample is through gravimetric analysis. In this process, a soluble sample of an unknown compound is mixed with another solution that will cause one of the ions from the unknown sample to fully precipitate out. If the identity of the precipitate is known, stoichiometric conversions will reveal the mass of the ion in the unknown, which allows for a mass percent calculation.

Due to the large amount of math involved, it's easiest to demonstrate.

Example:

A 4.33 g sample of an unknown alkali hydroxide compound is dissolved completely in water. A sufficient solution of copper (II) nitrate is added to the hydroxide solution such that it will fully precipitate copper (II) hydroxide via the following reaction:

$Cu^{2+} + 2\ OH^- \rightarrow Cu(OH)_2\ (s)$

After the precipitate is filtered and dried, its mass is found to be 3.81 g. Is the original alkali hydroxide sample most likely LiOH, NaOH, or KOH?

First, you convert the mass of the precipitate to moles:

$$3.81\text{g } Cu(OH)_2 \times \frac{1 \text{ mol } Cu(OH)_2}{97.55\text{g } Cu(OH)_2} = 0.0391 \text{ mol } Cu(OH)_2$$

Next, you need to determine the moles of hydroxide in the precipitate and convert that to grams.

$$0.0391 \text{ mol } Cu(OH)_2 \times \frac{2 \text{ mol } OH^-}{1 \text{ mol } Cu(OH)_2} \times \frac{17.02 \text{ g } OH^-}{1 \text{ mol } OH^-} = 1.33 \text{ g } OH^-$$

If there are 1.33 g of OH^- present in the precipitate, then there were also 1.33 g of OH^- in the original sample. The final calculation is to determine the mass percent of the OH^- in the original sample:

$$\frac{1.33 \text{ g } OH^-}{4.33 \text{ g sample}} \times 100\% = 30.7\%$$

To determine the identity of the sample, you have to know the mass percent of hydroxide in all of the possible compounds. To do that, you divide the mass of one mole of hydroxide by the molar mass of the compound.

LiOH: $\frac{17.02\ g}{23.96\ g} \times 100\% = 71.0\%$

NaOH: $\frac{17.02\ g}{40.02\ g} \times 100\% = 42.5\%$

KOH: $\frac{17.02\ g}{56.12\ g} \times 100\% = 30.3\%$

The hydroxide mass percent in KOH is closest to the mass percent determined via gravimetric analysis, so that is the most likely identity of the original unknown.

ENTHALPY

Enthalpy Change, ΔH

The enthalpy of a substance is a measure of the energy that is released or absorbed by the substance when bonds are broken and formed during a reaction.

The Basic Rules of Enthalpy

When bonds are ***formed***, energy is ***released***.

When bonds are ***broken***, energy is ***absorbed***.

The change in enthalpy, ΔH, that takes place over the course of a reaction can be calculated by subtracting the enthalpy of the reactants from the enthalpy of the products.

Enthalpy Change

$$\Delta H = H_{products} - H_{reactants}$$

If the products have stronger bonds than the reactants, then the products have lower enthalpy than the reactants and are more stable; in this case, energy is released by the reaction, or the reaction is **exothermic.**

If the products have weaker bonds than the reactants, then the products have higher enthalpy than the reactants and are less stable; in this case, energy is absorbed by the reaction, or the reaction is **endothermic.**

All substances are the most stable in their lowest energy state. This means that, in general, exothermic processes are more likely to be favorable than endothermic processes.

ENERGY DIAGRAMS

Exothermic and Endothermic Reactions

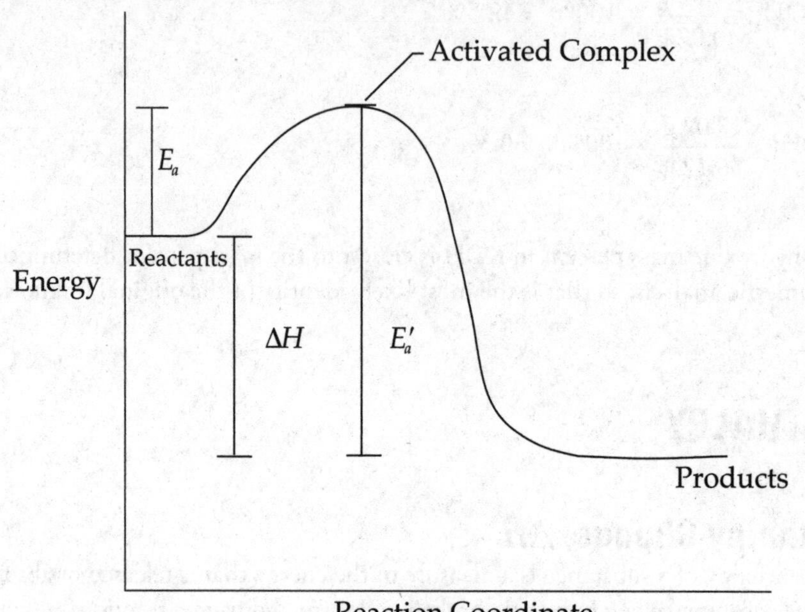

EXOTHERMIC REACTION

The diagram above shows the energy change that takes place during an exothermic reaction. The reactants start with a certain amount of energy (read the graph from left to right). For the reaction to proceed, the reactants must have enough energy to reach the transition state, where they are part of an activated complex. This is the highest point on the graph above. The amount of energy needed to reach this point is called the **activation energy, E_a**. At this point, all reactant bonds have been broken, but no product bonds have been formed, so this is the point in the reaction with the highest energy and lowest stability. The energy needed for the reverse reaction is shown as line E_a'.

Moving to the right past the activated complex, product bonds start to form, and we eventually reach the energy level of the products.

This diagram represents an exothermic reaction, so the products are at a lower energy level than the reactants and ΔH is negative.

The diagram below shows an endothermic reaction.

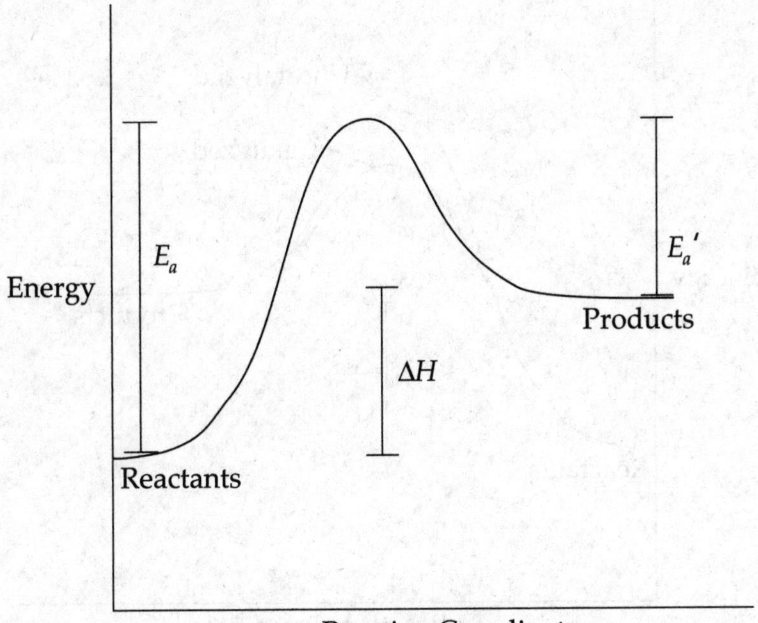

ENDOTHERMIC REACTION

In this diagram, the energy of the products is greater than the energy of the reactants, so ΔH is positive.

Reaction diagrams can be read in both directions, so the reverse reaction for an exothermic reaction is endothermic and vice versa.

CATALYSTS AND ENERGY DIAGRAMS

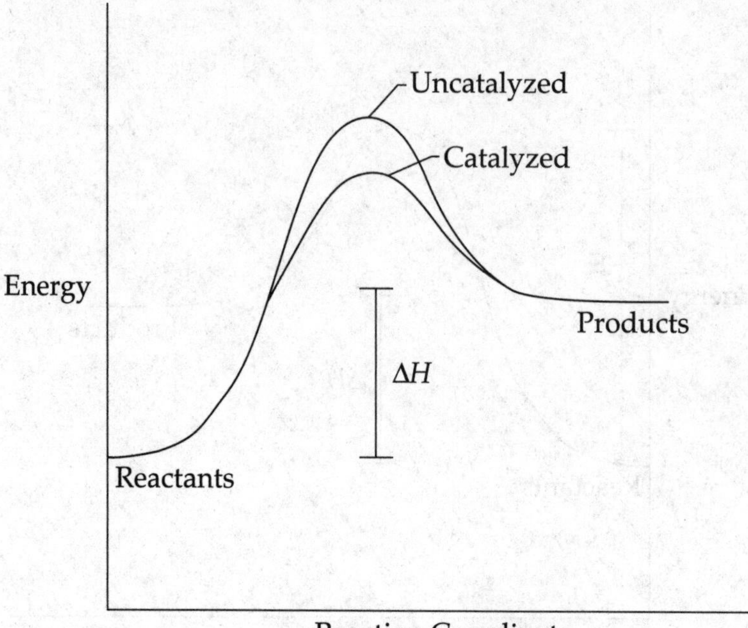

A catalyst speeds up a reaction by providing the reactants with an alternate pathway that has a lower activation energy, as shown in the diagram above.

Another helpful mnemonic for oxidation-reduction reactions is OIL RIG. Oxidation Is Loss Reduction Is Gain

Notice that the only difference between the catalyzed reaction and the uncatalyzed reaction is that the energy of the activated complex is lower for the catalyzed reaction. A catalyst lowers the activation energy, but it has no effect on the energy of the reactants, the energy of the products, or ΔH for the reaction.

Also note that a catalyst lowers the activation energy for both the forward and the reverse reaction, so it has no effect on the equilibrium conditions.

OXIDATION STATES

The **oxidation state** (or oxidation number) of an atom indicates the number of electrons that it gains or loses when it forms a bond. For instance, upon forming a bond with another atom, oxygen generally gains two electrons, which are negatively charged, so the oxidation state of oxygen in a bond is –2. Similarly, sodium generally loses one electron when it bonds to another atom, so its oxidation state in a bond is +1. Here are three important things you have to keep in mind when dealing with oxidation numbers.

- The oxidation state of an atom that is not bonded to an atom of another element is zero. That means either an atom that is not bonded to any other atom or an atom that is bonded to another atom of the same element (like the oxygen atoms in O_2).

- The oxidation numbers for all the atoms in a molecule must add up to zero.
- The oxidation numbers for all the atoms in a polyatomic ion must add up to the charge on the ion.

Most elements have different oxidation numbers that can vary depending on the molecule that they are a part of. The following chart shows some elements that consistently take on the same oxidation numbers.

Element	Oxidation Number
Hydrogen	+1
Alkali metals (Li, Na, ...)	+1
Alkaline earths (Be, Mg, ...)	+2
Oxygen	–2
Halogens (F, Cl, ...)	–1

Transition metals can have several oxidation states, which are differentiated from one another by a Roman numeral in the name of the compound. For example, in copper (II) sulfate ($CuSO_4$), the oxidation state for copper is +2, and in lead (II) oxide (PbO), the oxidation state for lead is +2. However, copper can also have an oxidation number of +1, and Pb can range from –4 to +4. In general, transition metals are characterized by variable oxidation states, as are the group 14 metals, tin and lead. Two exceptions to this rule are silver, which always takes on an oxidation state of +1, and zinc, which always takes on an oxidation state of +2.

You should be familiar with the following polyatomic ions and their charges.

Hydroxide	OH^-
Nitrate	NO_3^-
Acetate	$C_2H_3O_2^-$
Cyanide	CN^-
Permanganate	MnO_4^-
Carbonate	CO_3^{2-}
Sulfate	SO_4^{2-}
Dichromate	$Cr_2O_7^{2-}$
Phosphate	PO_4^{3-}
Ammonium	NH_4^+

OXIDATION-REDUCTION REACTIONS

In an oxidation-reduction (or redox, for short) reaction, electrons are exchanged by the reactants, and the oxidation states of some of the reactants are changed over the course of the reaction. Look at the following reaction:

$$Fe + 2HCl \rightarrow FeCl_2 + H_2$$

The oxidation state of Fe changes from 0 to +2.

The oxidation state of H changes from +1 to 0.

Here's a mnemonic device that might be useful.

LEO the lion says GER
LEO: you Lose Electrons in Oxidation
GER: you Gain Electrons in Reduction

- *When an atom gains electrons, its oxidation number decreases, and it is said to have been reduced.*

In the reaction above, H was reduced.

- *When an atom loses electrons, its oxidation number increases, and it is said to have been oxidized.*

In the reaction above, Fe was oxidized.

Oxidation and reduction go hand in hand. If one atom is losing electrons, another atom must be gaining them.

An oxidation-reduction reaction can be written as two **half-reactions**: one for the reduction and one for the oxidation. For example, the reaction

$$Fe + 2HCl \rightarrow FeCl_2 + H_2$$

can be written as

$Fe \rightarrow Fe^{2+} + 2e^-$	Oxidation
$2H^+ + 2e^- \rightarrow H_2$	Reduction

Redox Titrations

A titration involves the slow addition of a solution at a known concentration to another solution of unknown concentration in order to determine the concentration of the unknown solution. To determine the endpoint of a titration, a color change is often used. Titrations are frequently used in acid-base reactions (more on that in Chapter 8), but redox reactions can also be used in titration situations.

Potassium permanganate, $KMnO_4$, is frequently used in redox titrations. The manganese ion has an oxidation state of +7 in the permanganate ion (MnO_4^-), and a solution containing permanganate ions is a deep purple color. The manganese in MnO_4^- reduces to Mn^{2+} (thus changing its oxidation state to +2) when mixed with compounds that can be oxidized. Mn^{2+} is a colorless ion.

Frequently, when potassium permanganate is titrated into a colorless solution that contains a compound that can be oxidized, the end of the titration is marked by the solution turning pink. Initially, the permanganate ions take electrons from the oxidized compound and reduce to Mn^{2+}. However, once the compound that is being oxidized runs out, there are no electrons left for the MnO_4^- to take, and thus any extra permanganate ion that is added remains unreduced and retains its purple color, which when diluted sufficiently appears pink.

Example:

A 50.0 mL solution of sodium oxalate, $Na_2C_2O_4$ is poured into an Erlenmeyer flask. An acidified solution of 0.135 *M* potassium permanganate is titrated into the flask while the solution is swirled on a stir station and heated. The solution in the flask is originally colorless, but turns pink after 14.56 mL of potassium permanganate is added. Given the following half-reactions, what is the concentration of the oxalate solution?

Reduction: $8H^+ + MnO_4^- + 5e^- \rightarrow Mn^{2+} + 4H_2O\ (l)$

Oxidation: $C_2O_4^{2-} \rightarrow 2\ CO_2(g) + 2e^-$

The first step to solving this problem is to balance the half-reactions. In this case, the oxidation half-reaction must be multiplied by 5 in order to make the electrons balance with those in the reduction half-reaction, and the reduction half-reaction must be multiplied by 2. Thus, the full reaction (which omits the sodium and potassium spectator atoms) is:

$$16H^+ + 5C_2O_4^{2} + 2MnO_4^- \rightarrow 2Mn^{2+} + 8H_2O\ (l) + 5CO_2(g)$$

So, for every mole of permanganate that reacts, five moles of oxalate are required. The number of moles of permanganate can be easily determined:

$$M = \frac{n}{V} \qquad 0.135\ \text{M}\ MnO_4^- = \frac{n}{0.01456\ \text{L}\ MnO_4^-} \qquad n = 0.00197\ MnO_4^-$$

That can then be converted to moles of oxalate:

$$0.00197\ MnO_4^- \times \frac{5\ \text{mol}\ C_2O_4^{2-}}{2\ \text{mol}\ MnO_4^-} = 0.00493\ \text{mol}\ C_2O_4^{2-}$$

Finally, the concentration of the oxalate solution can be determined:

$$\frac{0.00493\ \text{mol}\ C_2O_4^{2-}}{0.0500\ \text{L}\ C_2O_4^{2-}} = 0.0986\ M = [C_2O_4^{2-}] = [Na_2C_2O_4^{2-}]$$

You will not be expected to memorize any of the various color changes that can occur throughout a redox titration. However, you should understand the concept of using a color change to determine an endpoint if you are provided with the colors of the various compounds that are present during a titration.

Reduction Potentials

Every half-reaction has an electric potential, or voltage, associated with it. You will be given the necessary values for the standard reduction potential of half-reactions for any question in which they are required. Potentials are always given as reduction half-reactions, but you can read them in reverse and flip the sign on the voltage to get oxidation potentials.

Look at the reduction potential for Zn^{2+}.

$$Zn^{2+} + 2e^- \rightarrow Zn(s) \qquad E^\circ = -0.76\text{ V}$$

Read the reduction half-reaction in reverse and change the sign on the voltage to get the oxidation potential for Zn.

$$Zn(s) \rightarrow Zn^{2+} + 2e^- \qquad E^\circ = +0.76\text{ V}$$

The larger the potential for a half-reaction, the more likely it is to occur.

$$F_2(g) + 2e^- \rightarrow 2F^- \qquad E^\circ = +2.87\text{ V}$$

$F_2(g)$ has a very large reduction potential, so it is likely to gain electrons and be reduced.

$$Li(s) \rightarrow Li^+ + e^- \qquad E^\circ = +3.05\text{ V}$$

$Li(s)$ has a very large oxidation potential, making it very likely to lose electrons and be oxidized.

You can calculate the potential of a redox reaction if you know the potentials for the two half-reactions that constitute it. There are two important things to remember when calculating the potential of a redox reaction.

- Add the potential for the oxidation half-reaction to the potential for the reduction half-reaction.
- Never multiply the potential for a half-reaction by a coefficient.

Let's look at the following reaction:

$$Zn + 2Ag^+ \rightarrow Zn^{2+} + 2Ag(s)$$

The two half-reactions are:

Oxidation:	$Zn \rightarrow Zn^{2+} + 2e^-$	$E^\circ = +0.76\text{ V}$
Reduction:	$Ag^+ + e^- \rightarrow Ag(s)$	$E^\circ = +0.80\text{ V}$

$$E = E_{\text{oxidation}} + E_{\text{reduction}}$$

$$E = 0.76\text{ V} + 0.80\text{ V} = 1.56\text{ V}$$

Notice that we ignored that silver has a coefficient of 2 in the balanced equation.

The relative reduction strengths of two different metals can also be determined qualitatively. In the above reaction, if $Zn(s)$ were placed in a solution containing

Ag^+ ions, the silver ions have a high enough reduction potential that they would take electrons from the zinc and start forming solid silver, which would precipitate out on the surface of the zinc.

However, if Ag(*s*) were placed in a solution containing Zn^{2+} ions, zinc does not have a high enough reduction potential to take electrons from silver and so no reaction would occur. So, when a solid metal is placed into a metallic solution and a new solid starts to form, the reduction potential of the metal in solution is greater than that of the solid metal. If no solid forms, the reduction potential of the solid metal is higher.

GALVANIC CELLS

In a **galvanic cell** (also called a voltaic cell), a favored redox reaction is used to generate a flow of current.

Look at the following favored redox reaction:

	$Zn(s) + Cu^{2+}(aq) \rightarrow Zn^{2+}(aq) + Cu(s)$	$E° = 1.10$ V
Oxidation:	$Zn(s) \rightarrow Zn^{2+}(aq) + 2e^-$	$E° = 0.76$ V
Reduction:	$Cu^{2+}(aq) + 2e^- \rightarrow Cu(s)$	$E° = 0.34$ V

Going Platinum
Because it is unreactive with most solutions, platinum is often used as a conductor in an electrolytic cell.

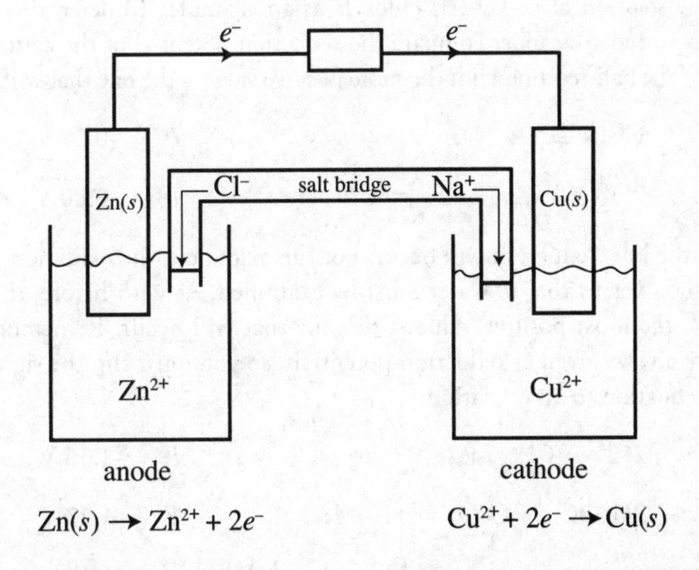

In a galvanic cell, the two half-reactions take place in separate chambers, and the electrons that are released by the oxidation reaction pass through a wire to the chamber where they are consumed in the reduction reaction. That's how the current is created. **Current** is defined as the flow of electrons from one place to another.

There's a mnemonic device to remember that.

AN OX
RED CAT

In any electric cell (either a galvanic cell or an electrolytic cell, which we'll discuss a little later) oxidation takes place at the electrode called the **anode.** Reduction takes place at the electrode called the **cathode.**

The salt bridge maintains the electrical neutrality in the cell. At the cathode, where reduction occurs and the solution is becoming less positively charged, the positive cations from the salt bridge solution flow into the half-cell. At the anode, where oxidation occurs and the solution is becoming more positively charged, the negative anions from the salt bridge solution flow into the half-cell. If the salt bridge were to be removed, the solutions in the half-cells would become electrically imbalanced, and the voltage of the cell would drop to 0. In the previous diagram, the salt bridge is filled with aqueous NaCl. As the cell functions, the Na^+ cations flow to the cathode and the Cl^- anions flow to the anode.

Under standard conditions (when all concentrations are 1 *M*), the voltage of the cell is the same as the total voltage of the redox reaction. To determine what would happen to the cell potential under nonstandard conditions, we must first learn more about equilibrium, which will be covered in Chapter 8.

ELECTROLYTIC CELLS

In an electrolytic cell, an outside source of voltage is used to force an unfavored redox reaction to take place. Most electrolytic cells occur in aqueous solutions which are created when a chemical dissolves in water; either the ions or the water molecules can be reduced or oxidized.

Let's look at a solution of nickel (II) chloride as an example. To determine which substance is reduced, you must compare the reduction potential of the cation with that of water. The half-reaction with the more positive value is the one that will occur.

$$Ni^{2+} + 2e^- \rightarrow Ni(s) \qquad E° = -0.25\text{ V}$$

$$2H_2O(l) + 2e^- \rightarrow H_2(g) + 2OH^- \qquad E° = -0.80\text{ V}$$

In this case, the Ni^{2+} reduction will occur. For the oxidation, the oxidation potential of the anion versus that of water must be examined. As with before, the half-reaction with the most positive value is the one that will occur. Remember that potentials are always given as reduction potentials, so you must flip the sign when you flip the equation to an oxidation.

$$2Cl^- \rightarrow Cl_2(g) + 2e^- \qquad E° = -1.36\text{ V}$$

$$2H_2O(l) \rightarrow O_2(g) + 4H^+ + 4e^- \qquad E° = -1.23\text{ V}$$

So, in this case, the water itself would be reduced. When the equations are balanced for electron transfer, the net ionic equation looks like this:

$$2Ni^{2+} + 2H_2O(l) \rightarrow 2Ni(s) + O_2(g) + 4H^+$$

$$E = -0.25\text{ V} + -1.24\text{V} = -1.49\text{ V}$$

The anode and cathode in an electrolytic reaction are usually just metal bars that conduct current and do not take part in the reaction. In the above reaction, solid nickel is being created at the cathode, and oxygen gas is being evolved at the anode. The sign of your total cell potential (*E*) for an electrolytic reaction is always negative.

Occasionally, a current will either be run through a molten compound or pure water. In this case, you do not need to determine which redox reactions are taking place as you will only have one choice for each.

The AN OX/RED CAT rule applies to the electrolytic cell in the same way that it applies to the galvanic cell.

Electroplating

Electrolytic cells are used for electroplating. You may see a question on the test that gives you an electrical current and asks you how much metal "plates out."

There are roughly four steps for figuring out electrolysis problems.

1. If you know the current and the time, you can calculate the charge in coulombs.

Current

$$I = \frac{q}{t}$$

I = current (amperes, A)
q = charge (coulombs, C)
t = time (seconds, s)

2. Once you know the charge in coulombs, you know how many electrons were involved in the reaction.

$$\text{Moles of electrons} = \frac{\text{coulombs}}{96{,}500 \text{ coulombs/mol}}$$

3. When you know the number of moles of electrons and you know the half-reaction for the metal, you can find out how many moles of metal plated out. For example, from this half-reaction for gold,

$$Au^{3+} + 3e^{-} \rightarrow Au(s),$$

you know that for every 3 moles of electrons consumed, you get 1 mole of gold.

4. Once you know the number of moles of metal, you can use what you know from stoichiometry to calculate the number of grams of metal.

For instance, if a current of 2.50 A is run through a solution of iron (III) chloride for 15 minutes, it would cause the following mass of iron to plate out:

$$15 \text{ minutes} \times \frac{60 \text{ seconds}}{1 \text{ minute}} \times \frac{2.50 \text{ C}}{1.0 \text{ second}} \times \frac{1 \text{ mol } e^-}{96{,}500 \text{ C}} \times \frac{1 \text{ mol Fe}}{3 \text{ mol } e^-} \times \frac{55.85 \text{ g}}{1 \text{ mol Fe}} = 7.23 \times 10^{-3} \text{ g Fe}$$

It is particularly important to keep track of your units in an electroplating problem; there are many different conversions before you come up with your final answer. In general, as long as all of your conversions are set up correctly your final answer will have the correct units.

CHAPTER 5 QUESTIONS

Multiple-Choice Questions

Use the following solubility rules to answer questions 1-4.

Salts containing halide anions are soluble except for those containing Ag^+, Pb^{2+}, and Hg_2^{2+}.

Salts containing carbonate anions are insoluble except for those containing alkali metals or ammonium.

1. If solutions of iron (III) nitrate and sodium carbonate are mixed, what would be the formula of the precipitate?

(A) Fe_3CO_3
(B) $Fe_2(CO_3)_3$
(C) $NaNO_3$
(D) No precipitate would form.

2. If solutions containing equal amounts of $AgNO_3$ and KCl are mixed, what is the identity of the spectator ions?

(A) Ag^+, NO_3^-, K^+, and Cl^-
(B) Ag^+ and Cl^-
(C) K^+ and Ag^+
(D) K^+ and NO_3^-

3. If equimolar solutions of $Pb(NO_3)_2$ and NaCl are mixed, which ion will not be present in significant amounts in the resulting solution after equilibrium is established?

(A) Pb^{2+}
(B) NO_3^-
(C) Na^+
(D) Cl^-

4. Choose the correct net ionic equation representing the reaction that occurs when solutions of potassium carbonate and copper (I) chloride are mixed.

(A) $K_2CO_3(aq) + 2CuCl(aq) \rightarrow 2KCl(aq) + Cu_2CO_3(s)$
(B) $K_2CO_3(aq) + 2CuCl(aq) \rightarrow 2KCl(s) + Cu_2CO_3(aq)$
(C) $CO_3^{2-} + 2Cu^+ \rightarrow Cu_2CO_3(s)$
(D) $CO_3^{2-} + Cu^{2+} \rightarrow CuCO_3(s)$

5. A strip of metal X is placed into a solution containing Y^{2+} ions and no reaction occurs. When metal X is placed in a separate solution containing Z^{2+} ions, metal Z starts to form on the strip. Which of the following choices organizes the reduction potentials for metals X, Y, and Z from greatest to least?

(A) $X > Y > Z$
(B) $Y > Z > X$
(C) $Z > X > Y$
(D) $Y > X > Z$

6. Which of the following is true for an endothermic reaction?

(A) The strength of the bonds in the products exceeds the strength of the bonds in the reactants.
(B) The activation energy is always greater than the activation energy for an exothermic reaction.
(C) Energy is released over the course of the reaction.
(D) A catalyst will increase the rate of the reaction by increasing the activation energy.

7. In which of the following compounds is the oxidation number of chromium the greatest?

(A) CrO_4^{2-}
(B) CrO
(C) Cr^{3+}
(D) Cr(*s*)

8. For an endothermic reaction, which of the following is true regarding the energy level of the activated complex?

(A) It is above the energy level of the reactants, but below the energy level of the products.
(B) It is below the energy level of the reactants, but above energy level of the products.
(C) It is above the energy level of both the products and reactants.
(D) It is below the energy level of both the products and reactants.

9. Based on the particulate drawing of the products for the reaction below, which reactant is limiting for the following reaction and why?

$$2K_3PO_4(aq) + 3MgI_2(aq) \rightarrow Mg_3(PO_4)_2(s) + 6KI(aq)$$

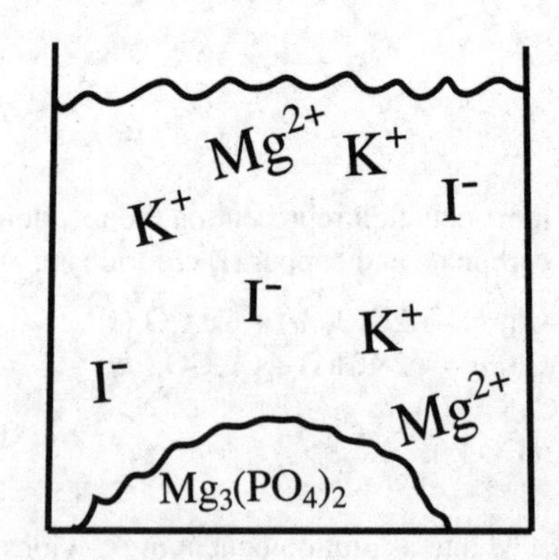

(A) The K_3PO_4, because there are no excess PO_4^{3-} ions after the reaction
(B) The MgI_2, because there are excess Mg^{2+} cations remaining after the reaction
(C) The K_3PO_4, because it contains a cation that cannot form a precipitate
(D) The MgI_2, because it requires more of itself to create the products

10. What is the mass of oxygen in 148 grams of calcium hydroxide ($Ca(OH)_2$)?

(A) 24 grams
(B) 32 grams
(C) 48 grams
(D) 64 grams

11. An ion containing only oxygen and chlorine is 31% oxygen by mass. What is its empirical formula?

(A) ClO^-
(B) ClO_2^-
(C) ClO_3^-
(D) ClO_4^-

Use the following information to answer questions 12-15.

When heated in a closed container in the presence of a catalyst, potassium chlorate decomposes into potassium chloride and oxygen gas via the following reaction:

$$2KClO_3(s) \rightarrow 2KCl(s) + 3O_2(g)$$

12. If 12.25 g of potassium chlorate decomposes, how many grams of oxygen gas will be generated?

(A) 1.60 g
(B) 3.20 g
(C) 4.80 g
(D) 18.37 g

13. Approximately how many liters of oxygen gas will be evolved at STP?

(A) 2.24 L
(B) 3.36 L
(C) 4.48 L
(D) 22.4 L

14. If the temperature of the gas is doubled while the volume is held constant, what will happen to the pressure exerted by the gas and why?

(A) It will also double because the gas molecules will be moving faster.
(B) It will also double because the gas molecules are exerting a greater force on each other.
(C) It will be cut in half because the molecules will lose more energy when colliding.
(D) It will increase by a factor of 4 because the kinetic energy will be four times greater.

15. Why is a catalyst present during the reaction?

(A) A catalyst is necessary for all decomposition reactions to occur.
(B) A catalyst reduces the bond energy in the reactants, making them easier to activate.
(C) A catalyst reduces the energy differential between the reactants and the products.
(D) A catalyst lowers the activation energy of the overall reaction and speeds it up.

16. A sample of a hydrate of $CuSO_4$ with a mass of 250 grams was heated until all the water was removed. The sample was then weighed and found to have a mass of 160 grams. What is the formula for the hydrate?

(A) $CuSO_4 \cdot 10H_2O$
(B) $CuSO_4 \cdot 7H_2O$
(C) $CuSO_4 \cdot 5H_2O$
(D) $CuSO_4 \cdot 2H_2O$

17. $$CaCO_3(s) + 2H^+(aq) \rightarrow Ca^{2+}(aq) + H_2O(l) + CO_2(g)$$

If the reaction above took place at standard temperature and pressure and 150 grams of $CaCO_3(s)$ were consumed, what was the volume of $CO_2(g)$ produced at STP?

(A) 11 L
(B) 22 L
(C) 34 L
(D) 45 L

18. A gaseous mixture at 25°C contained 1 mole of CH_4 and 2 moles of O_2 and the pressure was measured at 2 atm. The gases then underwent the reaction shown below.

$$CH_4(g) + 2O_2(g) \rightarrow CO_2(g) + 2H_2O(g)$$

What was the pressure in the container after the reaction had gone to completion and the temperature was allowed to return to 25°C?

(A) 1 atm
(B) 2 atm
(C) 3 atm
(D) 4 atm

19. A hydrocarbon was found to be 20% hydrogen by weight. If 1 mole of the hydrocarbon has a mass of 30 grams, what is its molecular formula?

(A) CH_2
(B) CH_3
(C) C_2H_4
(D) C_2H_6

20. $$Cr_2O_7^{2-} + 6I^- + 14H^+ \rightarrow 2Cr^{3+} + 3I_2 + 7H_2O$$

Which of the following statements about the reaction given above is NOT true?

(A) The oxidation number of chromium changes from +6 to +3.
(B) The oxidation number of iodine changes from –1 to 0.
(C) The oxidation number of hydrogen changes from +1 to 0.
(D) The oxidation number of oxygen remains the same.

21. Oxygen takes the oxidation state –1 in hydrogen peroxide, H_2O_2. The equation for the decomposition of H_2O_2 is shown below.

$$2H_2O_2 \rightarrow 2H_2O + O_2$$

Which of the following statements about the reaction shown above is true?

(A) Oxygen is reduced, and hydrogen is oxidized.
(B) Oxygen is oxidized, and hydrogen is reduced.
(C) Oxygen is both oxidized and reduced.
(D) Hydrogen is both oxidized and reduced.

22. $Cu^{2+} + 2e^- \rightarrow Cu$ $E° = +0.3$ V

$Fe^{2+} + 2e^- \rightarrow Fe$ $E° = -0.4$ V

Based on the reduction potentials given above, what is the reaction potential for the following reaction?

$$Fe^{2+} + Cu \rightarrow Fe + Cu^{2+}$$

(A) –0.7 V
(B) –0.1 V
(C) +0.1 V
(D) +0.7 V

23. $Cu^{2+} + 2e^- \rightarrow Cu$ $E° = +0.3$ V
$Zn^{2+} + 2e^- \rightarrow Zn$ $E° = -0.8$ V
$Mn^{2+} + 2e^- \rightarrow Mn$ $E° = -1.2$ V

Based on the reduction potentials given above, which of the following reactions will be favored?

(A) $Mn^{2+} + Cu \rightarrow Mn + Cu^{2+}$
(B) $Mn^{2+} + Zn \rightarrow Mn + Zn^{2+}$
(C) $Zn^{2+} + Cu \rightarrow Zn + Cu^{2+}$
(D) $Zn^{2+} + Mn \rightarrow Zn + Mn^{2+}$

24. Molten $AlCl_3$ is electrolyzed with a constant current of 5.00 amperes over a period of 600.0 seconds. Which of the following expressions is equal to the maximum mass of Al(*s*) that plates out? (1 faraday = 96,500 coulombs)

(A) $\dfrac{(600)(5.00)}{(96{,}500)(3)(27.0)}$ grams

(B) $\dfrac{(600)(5.00)(3)(27.0)}{(96{,}500)}$ grams

(C) $\dfrac{(600)(5.00)(27.0)}{(96{,}500)(3)}$ grams

(D) $\dfrac{(96{,}500)(3)(27.0)}{(600)(5.00)}$ grams

Use the following information to answer questions 25-28.

A voltaic cell is created using the following half-cells:

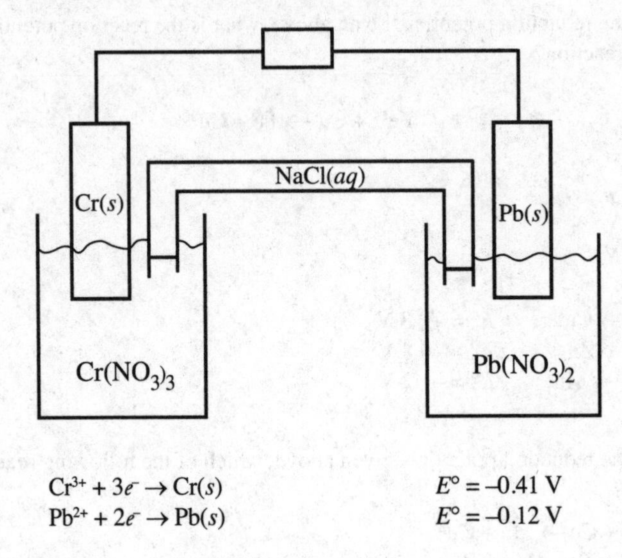

$Cr^{3+} + 3e^- \rightarrow Cr(s)$ $E° = -0.41$ V

$Pb^{2+} + 2e^- \rightarrow Pb(s)$ $E° = -0.12$ V

The concentrations of the solutions in each half-cell are 1.0 *M*.

25. Which of the following occurs at the cathode?

(A) Cr^{3+} is reduced to Cr(*s*).
(B) Pb^{2+} is reduced to Pb(*s*).
(C) Cr(*s*) is oxidized to Cr^{3+}.
(D) Pb(*s*) is oxidized to Pb^{2+}.

26. If the Cr^{3+} solution was replaced with a solution at a higher concentration, what would happen to the *E*° value for the cell?

(A) It would increase because of the presence of extra Cr^{3+} ions, which are able to donate more electrons.
(B) It would increase because the reaction would become more favored.
(C) It would decrease because the additional Cr^{3+} ions would drive the equation to the left.
(D) It would decrease because the stoichiometric ratio of Cr^{3+} ions to Pb^{2+} ions is less than 1 in the balanced reaction.

27. Which of the following best describes the activity in the salt bridge as the reaction progresses?

(A) Electrons flow through the salt bridge from the Pb/Pb^{2+} half-cell to the Cr/Cr^{3+} half-cell.
(B) Pb^{2+} flows to the Cr/Cr^{3+} half-cell, and Cr^{3+} flows to the Pb/Pb^{2+} half-cell.
(C) Na^+ flows to the Cr/Cr^{3+} half-cell, and Cl^- flows to the Pb/Pb^{2+} half-cell.
(D) Na^+ flows to the Pb/Pb^{2+} half-cell, and Cl^- flows to the Cr/Cr^{3+} half-cell.

28. Based on the given reduction potentials, which of the following would lead to a reaction?

(A) Placing some Cr(*s*) in a solution containing Pb^{2+} ions
(B) Placing some Pb(*s*) in a solution containing Cr^{3+} ions
(C) Placing some Cr(*s*) in a solution containing Cr^{3+} ions
(D) Placing some Pb(*s*) in a solution containing Pb^{2+} ions

Use the following information to answer questions 29-31.

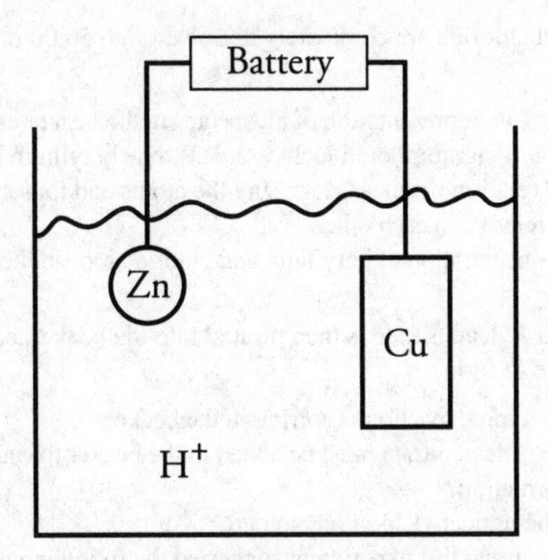

Pennies are made primarily of zinc, which is coated with a thin layer of copper through electroplating, using a setup like the one above. The solution in the beaker is a strong acid (which produces H^+ ions), and the cell is wired so that the copper electrode is the anode and zinc penny is the cathode. Use the following reduction potentials to answer questions 29-31.

Half-Reaction	Standard Reduction Potential
$Cu^{2+} + 2e^- \rightarrow Cu\ (s)$	+0.34 V
$2\ H^+ + 2e^- \rightarrow H_2\ (g)$	0.00 V
$Ni^{2+} + 2e^- \rightarrow Ni\ (s)$	-0.25 V
$Zn^{2+} + 2e^- \rightarrow Zn\ (s)$	-0.76 V

29. When the cell is connected, which of the following reactions takes place at the anode?

(A) $Cu^{2+} + 2e^- \rightarrow Cu\ (s)$
(B) $Cu\ (s) \rightarrow Cu^{2+} + 2e^-$
(C) $2H^+ + 2e^- \rightarrow H_2\ (g)$
(D) $H_2\ (g) \rightarrow 2H^+ + 2e^-$

30. What is the required voltage to make this cell function?

(A) 0.34 V
(B) 0.42 V
(C) 0.76 V
(D) 1.10 V

31. If, instead of copper, a nickel bar were to be used, could nickel be plated onto the zinc penny effectively? Why or why not?

(A) Yes, nickel's SRP is greater than that of zinc, which is all that is required for nickel to be reduced at the cathode
(B) Yes, nickel is able to take electrons from the H^+ ions in solution, allowing it to be reduced
(C) No, nickel's SRP is lower than that of H^+ ions, which means the only product being produced at the cathode would be hydrogen gas
(D) No, nickel's SRP is negative, meaning it cannot be reduced in an electrolytic cell

Free-Response Questions

1. 2.54 g of beryllium chloride are completely dissolved into 50.00 mL of water inside a beaker.

    (a) Draw a particulate representation of all species in the beaker after the solute has dissolved. Your diagram should include at least one beryllium ion, one chloride ion, and four water molecules. Make sure the atoms and ions are correctly sized and oriented relative to each other.
    (b) What is the concentration of beryllium and chloride ions in the beaker?

    A solution of 0.850 *M* lead nitrate is then titrated into the beaker, causing a precipitate of lead (II) chloride to form.

    (c) Identify the net ionic reaction occurring in the beaker.
    (d) What volume of lead nitrate must be added to the beaker to cause the maximum precipitate formation?
    (e) What is the theoretical yield of precipitate?
    (f) Students performing this experiment suggested the following techniques to separate the precipitate from the water. Their teacher rejected each idea. Explain why the teacher may have done so, and name the inherent errors of
        (i) boiling off the water.
        (ii) decanting (pouring off) the water.

2. Hydrogen peroxide, H_2O_2, is a common disinfectant. Pure hydrogen peroxide is a very strong oxidizer, and as such, it is diluted with water to low percentages before being bottled and sold. One method to determine the exact concentration of H_2O_2 in a bottle of hydrogen peroxide is to titrate a sample with a solution of acidified potassium permanganate. This causes the following redox reactions to occur:

    Reduction: $8H^+ + MnO_4^- + 5e^- \rightarrow Mn^{2+} + 4H_2O(l)$

    Oxidation: $H_2O_2(aq) \rightarrow 2H^+ + O_2(g) + 2e^-$

    During a titration, a student measures out 5.0 mL of hydrogen peroxide solution into a graduated cylinder, and the pours it into a flask, diluting it to 50.0 mL with water. The student then titrates 0.150 *M* potassium permanganate solution into the flask with constant stirring.

    (a) Write out the full, balanced redox reaction that is taking place during the titration.
    (b) List two observations that the student will see as the titration progresses that are indicative of chemical reactions.

Diagrams of the permanganate in the buret at the start and end of the titration are as follows:

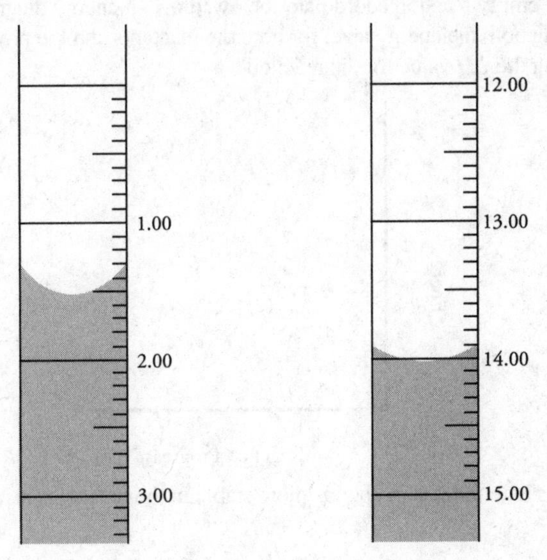

(c) (i) What volume of $KMnO_4$ was titrated?
(ii) What is the concentration of hydrogen peroxide in the original sample?

(d) How would the precision of the student's results have changed if the hydrogen peroxide sample were measured out in a 50 mL beaker instead of a graduated cylinder?

(e) How would each of the following errors affect the student's final calculated hydrogen peroxide concentration?
(i) Not filling the buret tip with solution prior to the titration
(ii) Not rinsing down the sides of the flask during titration

3. Acetylene (C_2H_2) is a fuel that is commonly used in metallurgical applications, particularly welding.

(a) Draw the Lewis diagram for acetylene.

(b) Write out and balance the reaction that occurs when acetylene is combusted in the atmosphere.

The combustion of acetylene is a very exothermic process.

(c) On the empty reaction coordinates below, draw an energy diagram for ethylene, labeling both the energy level for both the reactants and the products, as well as labeling the ΔH value for the reaction.

(d) For this reaction, are the bonds more stable in the reactants or in the products? Why?

4. $$2Mg(s) + 2CuSO_4(aq) + H_2O(l) \rightarrow 2MgSO_4(aq) + Cu_2O(s) + H_2(g)$$

(a) If 1.46 grams of Mg(*s*) are added to 500 milliliters of a 0.200-molar solution of $CuSO_4$, what is the maximum molar yield of $H_2(g)$?

(b) When all of the limiting reagent has been consumed in (a), how many moles of the other reactant (not water) remain?

(c) What is the mass of the Cu_2O produced in (a)?

(d) What is the value of $[Mg^{2+}]$ in the solution at the end of the experiment? (Assume that the volume of the solution remains unchanged.)

5. A student performs an experiment in which a bar of unknown metal M is placed in a solution with the formula MNO_3. The metal is then hooked up to a copper bar in a solution of $CuSO_4$ as shown below. A salt bridge that contains aqueous KCl links the cell together.

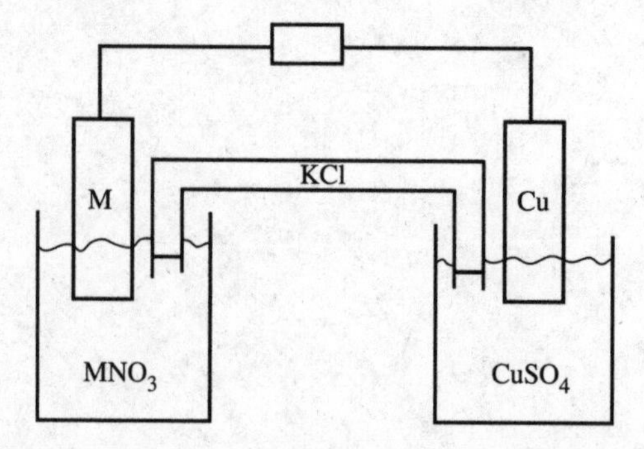

The cell potential is found to be +0.74 V. Separately, when a bar of metal M is placed in the copper sulfate solution, solid copper starts to form on the bar. When a bar of copper is placed in the MNO_3 solution, no visible reaction occurs.

The following gives some reduction potentials for copper:

Half-reaction	$E°$
$Cu^{2+} + 2e^- \rightarrow Cu(s)$	0.34 V
$Cu^{2+} + e^- \rightarrow Cu^+$	0.15 V
$Cu^+ + e^- \rightarrow Cu(s)$	0.52 V

(a) Write the net ionic equation that takes place in the Cu/M cell.
(b) What is the standard reduction potential for metal M?
(c) Which metal acted as the anode and which as the cathode? Justify your answer.
(d) On the diagram of the cell, indicate which way the electrons are flowing in the wire. Additionally, indicate any ionic movement occurring in the salt bridge.
(e) What would happen to the voltage of the reaction in the Cu/M cell if the concentration of the $CuSO_4$ increased while the concentration of the MNO_3 remained constant? Justify your answer.

6. Two electrodes are inserted into a solution of nickel (II) fluoride and a current of 2.20 A is run through them. A list of standard reduction potentials is as follows:

Half-reaction	$E°$
$O_2(g) + 4H^+ + 4e^- \rightarrow H_2O(l)$	1.23 V
$F_2(g) + 2e^- \rightarrow 2F^-$	2.87 V
$2H_2O(l) + 2e^- \rightarrow H_2(g) + 2OH^-$	−0.83 V
$Ni^{2+} + 2e^- \rightarrow Ni(s)$	−0.25 V

(a) Write the net ionic equation that takes place during this reaction.
(b) Qualitatively describe what an observer would see taking place at each electrode.
(c) Will the solution become acidic, basic, or remain neutral as the reaction progresses?
(d) How long would it take to create 1.2 g of Ni(s) at the cathode?

CHAPTER 5 ANSWERS AND EXPLANATIONS

Multiple-Choice

1. **B** Carbonates are insoluble when paired with iron. Iron has a charge of +3 and carbonate has a charge of –2. To cancel out, both charges need to have a magnitude of 6, requiring two iron atoms and three carbonate ions. The best representation of that is (B).

2. **D** When those solutions mix, a precipitate of AgCl will form, removing those ions from the solution. The remaining ions (K^+ and NO_3^-) do not react and remain the same as they were when they started.

3. **D** The precipitate that would form is $PbCl_2$, meaning that both NO_3^- and Na^+ are spectator ions. The precipitate would require twice as many chloride ions as lead ions. Therefore, if equal moles of both are used, the NaCl would be the limiting reagent, and almost all of the chloride ions would be present in the solid, with very few left in the solution. There would, however, still be significant Pb^{2+} remaining in solution, as it is in excess.

4. **C** Via the information given from the solubility rules, we can determine that the precipitate would be copper (I) carbonate, which has a formula of Cu_2CO_3. In the net ionic equation, spectator ions (in this case, K^+ and Cl^-) cancel out and do not appear in the final reaction.

5. **C** Z^{2+} was able to reduce to solid Z by taking electrons from metal X, so Z must have a higher reduction potential than X. Y^{2+} was unable to take electrons from metal X, and therefore Y must have a lower reduction potential than X.

6. **A** In an endothermic reaction, energy must be absorbed by the reactants for the reaction to occur. This is because the amount of energy necessary to break the bonds of the reactants is greater than the amount that is released by the formation of the bonds in the products.

7. **A** In (D), chromium is a pure element and has an oxidation number of 0. In (C), chromium's oxidation number is equal to its charge of +3, and in (B), it must balance the –2 on the oxygen, so it has a charge of +2. In (A), the total charge on the ion is –2, and each oxygen is –2. Solving the following, where X is the oxidation number on chromium: $X + -2(4) = -2$. So, the oxidation number is +6.

8. C For any reaction (exothermic or endothermic), the energy level of the activated complex is always above the energy level of both the products and reactants.

9. A The reaction will stop when no more precipitate can form, which will occur when either magnesium or phosphate ions run out. As there are no excess phosphate ions in solution after the reaction, the potassium phosphate must have run out first, leaving excess magnesium ions unreacted in solution.

10. D Moles = $\frac{\text{grams}}{\text{MW}}$

Moles of calcium hydroxide = $\frac{(148\text{ g})}{(74\text{ g/mol})}$ = 2 moles

Every mole of $Ca(OH)_2$ contains 2 moles of oxygen.

So there are (2)(2) = 4 moles of oxygen

Grams = (moles)(MW)

So grams of oxygen = (4 mol)(16 g/mol) = 64 grams

11. A Assume that we have 100 grams of the compound. That means that we have 31 grams of oxygen and 69 grams of chlorine.

Moles = $\frac{\text{grams}}{\text{MW}}$

Moles of oxygen = $\frac{(31\text{g})}{(16\text{ g/mol})}$ = slightly less than 2 mol

Moles of chlorine = $\frac{(69\text{ g})}{(35.5\text{ g/mol})}$ = slightly less than 2 mol

So the ratio of chlorine to oxygen is 1 to 1, and the empirical formula is ClO^-.

12. C $12.25\text{ g } KClO_3 \times \frac{1\text{ mol } KClO_3}{122.25\text{ g } KClO_3} = 0.1000\text{ mol } KClO_3$

$0.1000\text{ mol } KClO_3 \times \frac{3\text{ mol } O_2}{2\text{ mol } KClO_3} = 0.1500\text{ mol } O_2$

$0.1500\text{ mol } O_2 \times \frac{32.00\text{ g } O_2}{1\text{ mol } O_2} = 4.80\text{ g } O_2$

Note that even without a calculator, you are expected to be able to do simple calculations such as the ones above.

13. B $0.1500 \text{ mol } O_2 \times \frac{22.4 \text{ L}}{1 \text{ mol } O_2} = 3.36 \text{ L } O_2$

14. A Pressure is a measure of the amount of force with which the gas particles are hitting the container walls. An increase in temperature is indicative of increased molecular velocity. If the molecules are moving faster, they will not only have more energy when they hit the container walls, but they will also hit those walls more often. Both of those factors contribute to the increased pressure.

15. D A catalyst does not change the amount of bond energy present in either the reactants or the products, and catalysts always speed up the overall reaction. Many decomposition reactions occur more quickly with a catalyst, but even without one present, the reaction can progress, albeit at a slower rate.

16. C The molecular weight of $CuSO_4$ is 160 g/mol, so we have only 1 mole of the hydrate. The lost mass was due to water, so 1 mole of the hydrate must have contained 90 grams of H_2O.

$$\text{Moles} = \frac{\text{grams}}{\text{MW}}$$

$$\text{Moles of water} = \frac{(90 \text{ g})}{(18 \text{ g/mol})} = 5 \text{ moles}$$

So if 1 mole of hydrate contains 5 moles of H_2O, then the formula for the hydrate must be $CuSO_4 \bullet 5\,H_2O$.

17. C $\text{Moles} = \frac{\text{grams}}{\text{MW}}$

$$\text{Moles of } CaCO_3 = \frac{(150 \text{ g})}{(100 \text{ g/mol})} = 1.5 \text{ moles}$$

From the balanced equation, for every mole of $CaCO_3$ consumed, one mole of CO_2 is produced. So 1.5 moles of CO_2 are produced.

At STP, volume of gas = (moles)(22.4 L)

So the volume of CO_2 = (1.5)(22.4) = 34 L

18. B All of the reactants are consumed in the reaction and the temperature doesn't change, so the pressure will change only if the number of moles of gas changes over the course of the reaction. The number of moles of gas (3 moles) doesn't change in the balanced equation, so the pressure will remain the same (2 atm) at the end of the reaction as at the beginning.

19. **D** Let's say we have 100 grams of the compound.

$$\text{Moles} = \frac{\text{grams}}{\text{MW}}$$

$$\text{So moles of carbon} = \frac{(80\text{ g})}{(12\text{ g/mol})} = 6.7\text{ moles}$$

$$\text{and moles of hydrogen} = \frac{(20\text{ g})}{(1\text{ g/mol})} = 20\text{ moles}$$

According to our rough calculation, there are about three times as many moles of hydrogen in the compound as there are moles of carbon, so the empirical formula is CH_3.

The molar mass for the empirical formula is 15 g/mol, so we need to double the moles of each element to get a compound with a molar mass of 30 g/mol. That makes the molecular formula of the compound C_2H_6.

20. **C** The oxidation numbers of the reactants are Cr^{6+}, O^{2-}, I^-, and H^+, and the oxidation numbers of the products are Cr^{3+}, O^{2-}, I^0, and H^+.

Chromium gains electrons and is reduced, and iodine loses electrons and is oxidized; the oxidation states of oxygen and hydrogen are not changed.

21. **C** The oxidation state of hydrogen remains +1 throughout the reaction. We have O^- at the start. When water forms, oxygen has gained an electron and been reduced to O^{2-} (GER). When O_2 forms, oxygen has been oxidized to O^0 through the loss of an electron (LEO). As you can see, in this process, oxygen is both oxidized and reduced.

22. **A** We add the reduction potential for Fe^{2+} (–0.4 V) to the oxidation potential for Cu (–0.3 V, the reverse of the reduction potential) to get –0.7 V.

23. **D** To get the reaction potential, we add the reduction potential for the reduction half-reaction to the oxidation potential for the oxidation half-reaction.

Here are the reaction potentials for (A)–(D).

(A) (–1.2 V) + (–0.3 V) = –1.5 V

(B) (–1.2 V) + (+0.8 V) = –0.4 V

(C) (–0.8 V) + (–0.3 V) = –1.1 V

(D) (–0.8 V) + (+1.2 V) = +0.4 V

Choice (D) is the only reaction with a positive voltage, which means that it is the only reaction listed that is favored.

24. **C** First let's find out how many electrons are provided by the current.

Coulombs = (seconds)(amperes) = (600)(5.00)

$$\text{Moles of electrons} = \frac{(\text{coulombs})}{(96{,}500)} = \frac{(600)(5.00)}{(96{,}500)} \text{ moles}$$

In $AlCl_3$, we have Al^{3+}, so the half-reaction for the plating of aluminum is

$Al^{3+} + 3e^- \rightarrow Al(s)$

So we get $\frac{1}{3}$ as many moles of Al(*s*) as we have moles of electrons.

$$\text{Moles of Al}(s) = (\text{moles of electrons})\left(\frac{1}{3}\right) = \frac{(600)(5.00)}{(96{,}500)(3)} \text{ moles}$$

Now, grams = (moles)(MW)

$$\text{Grams of Al}(s) = \frac{(600)(5.00)}{(96{,}500)(3)}(27.0) = \frac{(600)(5.00)(27.0)}{(96{,}500)(3)} \text{ grams}$$

25. **B** The Pb half-cell has a greater reduction potential than the Cr half-cell. Pb^{2+} will be able to take electrons from the Cr(*s*) solution, and reduce to become Pb(*s*).

26. **C** The total net ionic equation for the cells is:

$$3Pb^{2+} + 2Cr(s) \rightarrow 3Pb(s) + 2Cr^{3+}$$

The Cr^{3+} ions are on the products side of the reaction, and the reaction would shift left, reducing the overall voltage for the cell.

27. **D** As the reaction progresses, Pb^{2+} is being reduced to create Pb(*s*). This decreases the amount of positive charge in the solution, which is replaced by the Na^+ from the salt bridge. At the anode, excess Cr^{3+} is being created, which is balanced by the Cl^- anions from the salt bridge.

28. **A** For a reaction to occur, the reduction potential of the ions in solution must exceed the reduction potential for the ions of the corresponding metal. If Cr(*s*) were placed in a solution containing Pb^{2+}, the Pb^{2+} would take electrons from the Cr(*s*) and create Pb(*s*) and Cr^{3+}.

29. **B** Oxidation occurs at the anode, and that entails the loss of electrons. As there is no hydrogen gas present in the solution to start, the only substance that can be reduced at the anode is the solid copper itself.

30. **A** The half-reaction occurring at the anode is $2H^+ + 2e^- \rightarrow H_2$ (g), as the only substance present at the beginning of the reaction can that can be reduced is the hydrogen ion (initially, there are no Zn^{2+} or Cu^{2+} ions in solution). With copper is being oxidized at the anode, the calculation then becomes (0.00 V) + (–0.34 V) = –0.34 V. That means a voltage differential of 0.34 V will be required for this reaction to occur.

31. **C** In the initial Cu/Zn cell, as the electrolysis proceeds Cu^{2+} ions are produced and become part of the solution. The SRP of those ions is greater than that the SRP of H^+ ions, so later in the reaction, the Cu^{2+} is reduced into solid copper at the zinc cathode—this is what causes the copper to plate out on the zinc. If nickel were used as the anode, Ni^{2+} would be in solution, however, H^+ has a higher reduction potential than Ni^{2+}, which means the H^+ reduction would continue to occur at the cathode, and no nickel would be plated out of solution.

Free-Response

1. (a) The beryllium ions only have one full energy level, while chloride ions have three and would thus be much larger. The positive beryllium ions would be attracted to the negative dipoles of the water molecules, and the negative chloride ions would be attracted to the positive dipoles of the water molecules.

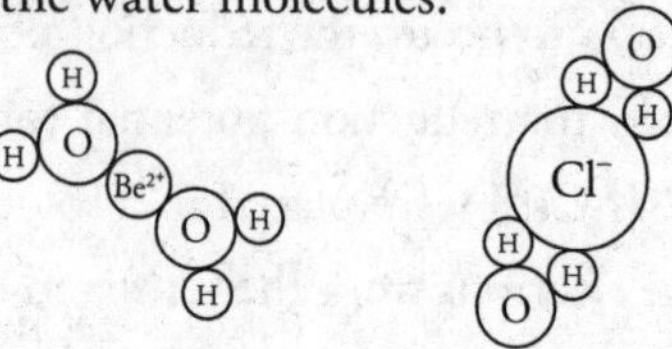

(b) First, we have to calculate the concentration of the $BeCl_2$.

$$2.54 \text{ g } BeCl_2 \times \frac{1 \text{ mol } BeCl_2}{79.91 \text{ g } BeCl_2} \times \frac{0.0317 \text{ mol}}{0.0500 \text{ L}} = 0.635\ M\ BeCl_2$$

For every mole of $BeCl_2$ that dissociates, one Be^{2+} and two Cl^- ions are produced. Thus, the concentration of the Be^{2+} will be the same as the base concentration of the $BeCl_2$, and the concentration of the Cl^- will be twice that. So, $[Be^{2+}] = 0.635\ M$ and $[Cl^-] = 1.27\ M$.

(c) $Pb^{2+} + 2Cl^- \rightarrow PbCl_2(s)$

(d) There must be one lead ion present for every two chloride ions in order to create the maximum amount of precipitate. We can figure out the number of moles of chloride from the concentration and volume, which we already have, and then convert:

$$1.27\ M\ Cl^- = \frac{n}{0.0500 \text{ L}}$$

$$n = 0.0635 \text{ mol } Cl^- \times \frac{1 \text{ mol } Pb^2}{2 \text{ mol } Cl^-} = 0.0318 \text{ mol } Pb^{2+}$$

From there, we can use the given concentration of the lead nitrate solution to determine how much of it is required.

$$0.850\ M = \frac{0.0318 \text{ mol}}{V} \qquad V = 0.0374 \text{ L} = 37.4 \text{ mL}$$

(e) If both reactants are mixed in equal amounts, either value can be used to calculate the amount of precipitate that will be created.

$$0.0318 \text{ mol } Pb^{2+} \times \frac{1 \text{ mol } PbCl_2}{1 \text{ mol } Pb^{2+}} \times \frac{278.1 \text{ g}}{1 \text{ mol } PbCl_2} = 8.84 \text{ g}$$

(f) (i) Boiling the water would evaporate all the water, but would also cause the spectator ions to no longer be dissolved in water. In addition to any $PbCl_2$ that was created, the excess beryllium and nitrate ions would adhere to the solid, giving a falsely high molar mass and am impure product.

(ii) While decanting the solution, smaller particles of precipitate may get poured out with the water. Decanting the solution into a filtered funnel setup would be a much better option, as the filter paper would then catch all of the precipitate while allowing the water (which would carry the spectator ions) through.

2. (a) In a full redox reaction, the electrons from each half-reaction must cancel. To do that, the reduction reaction must be multiplied by two, and the oxidation reaction multiplied by five. When combining the reactions after doing so, this yields:

$$6H^+ + 2MnO_4^- + 5H_2O_2(aq) \rightarrow 2Mn^{2+} + 8H_2O(l) + 5O_2(g)$$

Note that 10 of the hydrogen ions cancel out when the reactions combine.

(b) A gas will be evolved as the hydrogen peroxide oxidizes into oxygen gas. Additionally, there will be a color change as the purple permanganate solution reduces into colorless manganese ions.

(c) (i) 14.03 mL – 1.52 mL = 12.51 mL

(ii) First, we have to calculate the moles of permanganate that were titrated.

$$0.150\ M = \frac{n}{0.01251\ \text{L}} \qquad n = 0.00188\ \text{mol}\ MnO_4^-$$

We can then use the stoichiometric ratios from the balanced equation to convert to moles of hydrogen peroxide.

$$0.00188\ \text{mol}\ MnO_4^- \times \frac{5\ \text{mol}\ H_2O_2}{2\ \text{mol}\ MnO_4^-} = 0.00470\ \text{mol}\ H_2O_2$$

Finally, we can convert that to the concentration of H_2O_2. We will use a volume of 5.00 mL and not 50.0 mL in our calculations because we are interested in the concentration of the solution before it was diluted with water.

$$\frac{0.00470\ \text{mol}\ H_2O_2}{0.0050\ \text{L}\ H_2O_2} = 0.94\ M\ H_2O_2$$

(d) Beakers are precise to the ones places; you cannot accurately measure out a volume in a beaker to any number of decimal places. Graduated cylinders are accurate to one decimal place, so using a beaker would reduce the number of significant figures in your answer by one.

(e) (i) If the tip of the buret is not filled prior to dispensing the permanganate solution from it, the first milliliter or so that would be dispensed is air instead of solution. Thus, the experimental volume of permanganate would be artificially high (instead of 12.51 mL of permanganate, it would have really been about 11.5 mL of permanganate with the rest being air). If the recorded volume of the permanganate is too high, that will eventually lead to a calculated $[H_2O_2]$ that is also too high.

(ii) During a titration, it is best practice to regularly rinse down the sides of the flask you are titrating into, just in case any of the titrated solution splashes onto it and does not mix with the solution you are titrating into. If you do not do this, some of the permanganate that leaves the buret might stick to the sides of the flask and not react with the hydrogen peroxide. This would mean the recorded volume of permanganate is too high (not all of the measured permanganate actually reacted), which would again lead to an artificially high calculate concentration of hydrogen peroxide.

3. (a) The total number of valence electrons in acetylene is 10. In order to fill the octet for the carbon atoms, a triple bond will be required.

$H - C \equiv C - H$

(b) $C_2H_2 + \frac{5}{2}O_2 \rightarrow 2\ CO_2 + H_2O$

(c)

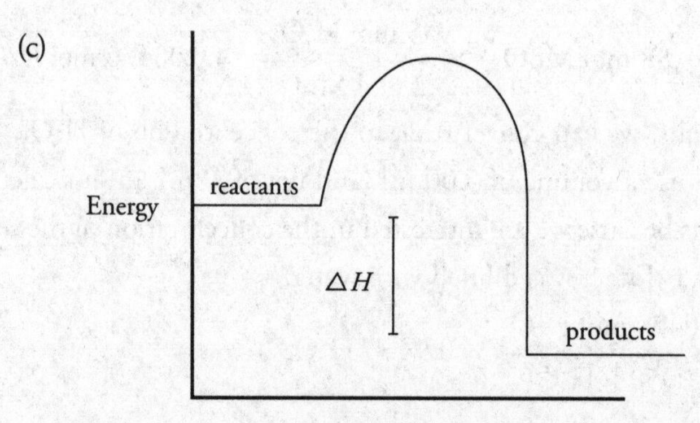

(d) The bonds in the products are more stable, as they are at a lower energy level.

4. (a) We need to find the limiting reagent. There's plenty of water, so it must be one of the other two reactants.

$$\text{Moles} = \frac{\text{grams}}{\text{MW}}$$

$$\text{Moles of Mg} = \frac{(1.46\ \text{g})}{(24.3\ \text{g/mol})} = 0.060\ \text{moles}$$

Moles = (molarity)(volume)

Moles of $CuSO_4$ = (0.200 M)(0.500 L) = 0.100 moles

From the balanced equation, Mg and $CuSO_4$ are consumed in a 1:1 ratio, so we'll run out of Mg first. Mg is the limiting reagent, and we'll use it to find the yield of H_2.

From the balanced equation, 1 mole of H_2 is produced for every 2 moles of Mg consumed, so the number of moles of H_2 produced will be half the number of moles of Mg consumed.

$$\text{Moles of } H_2 = \frac{1}{2}(0.060\ \text{mol}) = 0.030\ \text{moles}$$

(b) Mg is the limiting reagent, so some $CuSO_4$ will remain. From the balanced equation, Mg and $CuSO_4$ are consumed in a 1:1 ratio, so when 0.060 moles of Mg are consumed, 0.060 moles of $CuSO_4$ are also consumed.

Moles of $CuSO_4$ remaining = (0.100 mol) – (0.060 mol) = 0.040 moles

(c) From the balanced equation, 1 mole of Cu_2O is produced for every 2 moles of Mg consumed, so the number of moles of Cu_2O produced will be half the number of moles of Mg consumed.

$$\text{Moles of } Cu_2O = \frac{1}{2}(0.060\ \text{mol}) = 0.030\ \text{moles}$$

Grams = (moles)(MW)

Grams of Cu_2O = (0.030 mol)(143 g/mol) = 4.29 grams

(d) All of the Mg consumed ends up as Mg^{2+} ions in the solution.

$$\text{Molarity} = \frac{\text{moles}}{\text{liters}} \qquad [Mg^{2+}] = \frac{(0.060\ \text{mol})}{(0.500\ \text{L})} = 0.120\ M$$

5. (a) When metal M was placed in the copper solution, a reaction occurred. Therefore, the copper must have the higher reduction potential and thus is reduced in the Cu/M cell. The half-reactions are:

Reduction: $Cu^{2+} + 2e^- \rightarrow Cu(s)$
Oxidation: $M(s) \rightarrow M^+ + e^-$

The oxidation half-reaction must be multiplied by two to balance the electrons before combining the reactions to yield:

$Cu^{2+} + 2M(s) \rightarrow Cu(s) + 2M^+$

(b) $E_{red} + E_{ox} = +0.74$ V

The reduction potential for $Cu^{2+} + 2e^- \rightarrow Cu(s)$ is known:

$0.34 \text{ V} + E_{ox} = +0.74 \text{ V}$ $\qquad E_{ox} = +0.40 \text{ V}$

The reduction potential for metal M is the opposite of its oxidation potential.

$E_{red} = -0.40$ V

(c) Reduction occurs at the cathode, so copper is the cathode. Oxidation occurs at the anode, so M is the anode.

(d)

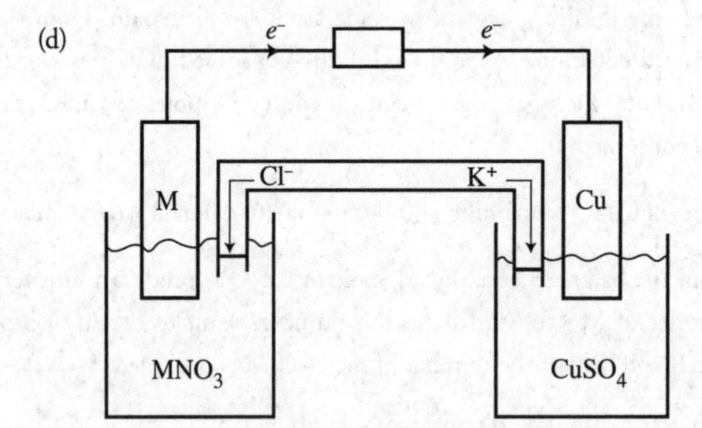

Electrons should be flowing from metal M to the copper bar. In the salt bridge, the K^+ ions will flow towards the copper solution to replace the Cu^{2+} being reduced into Cu(*s*). The Cl^- will flow towards the solution for metal M in order to balance out the charge of the extra M^+ ions being created via the oxidation of M(*s*).

(e) If the concentration of the Cu^{2+} increases, it will cause a shift to the right in accordance with Le Châtelier's principle. This will increase the cell's overall voltage.

6. (a) The two potential reduction reactions are the bottom two, as their reactants (H_2O and Ni^{2+}) are actually present at the start of the reaction. Of the two, the nickel reduction is more positive, and thus will take place. The top two reactions must be flipped to have the reactants (H_2O and F^-) on the reactants side, which also flips the sign. As a result, the water oxidation has the more positive value (–1.23 as opposed to –2.87) and will therefore occur.

Reduction: $Ni^{2+} + 2e^- \rightarrow Ni(s)$
Oxidation: $H_2O(l) \rightarrow O_2(g) + 4H^+ + 4e^-$

After multiplying the reduction half-reaction by 2 to balance the electrons and combining both half-reactions, the net cell reaction is:

$2Ni^{2+} + H_2O(l) \rightarrow 2Ni(s) + O_2(g) + 4H^+$

(b) At the cathode, solid nickel would begin to plate out of solution. At the anode, oxygen gas would form and bubble up to the surface.

(c) The solution will become more acidic due to the creation of hydrogen ions at the anode.

(d) $1.20 \text{ g Ni} \times \frac{1 \text{ mol Ni}}{58.69 \text{ g Ni}} \times \frac{2 \text{ mol } e^-}{1 \text{ mol Ni}} \times \frac{96500 \text{ C}}{1 \text{ mol } e^-} \times \frac{1.0 \text{ sec}}{2.20 \text{ C}} = 1790 \text{ seconds}$

Chapter 6
Big Idea #4: Chemical Reactions and Their Rates

Rates of chemical reactions are determined by details of the molecular collisions.

RATE LAW USING INITIAL CONCENTRATIONS

The rate law for a reaction describes the dependence of the initial rate of a reaction on the concentrations of its reactants. It includes the Arrhenius constant, *k*, which takes into account the activation energy for the reaction and the temperature at which the reaction occurs. The rate of a reaction is described in terms of the rate of appearance of a product or the rate of disappearance of a reactant. The rate law for a reaction cannot be determined from a balanced equation; it must be determined from experimental data, which is presented on the test in table form.

Here's How It's Done

The data below were collected for the following hypothetical reaction:

$$A + 2B + C \rightarrow D$$

Experiment	Initial Concentration of Reactants (*M*) [A]	[B]	[C]	Initial Rate of Formation of D (*M*/sec)
1	0.10	0.10	0.10	0.01
2	0.10	0.10	0.20	0.01
3	0.10	0.20	0.10	0.02
4	0.20	0.20	0.10	0.08

The rate law always takes the following form, using the concentrations of the reactants:

$$\text{Rate} = k[A]^x[B]^y[C]^z$$

The greater the value of a reactant's exponent, the more a change in the concentration of that reactant will affect the rate of the reaction. To find the values for the exponents *x*, *y*, and *z*, we need to examine how changes in the individual reactants affect the rate. The easiest way to find the exponents is to see what happens to the rate when the concentration of an individual reactant is doubled.

Let's Look at [A]

From experiment 3 to experiment 4, [A] doubles while the other reactant concentrations remain constant. For this reason, it is useful to use the rate values from these two experiments to calculate *x* (the order of the reaction with respect to reactant A).

As you can see from the table, the rate quadruples from experiment 3 to experiment 4, going from 0.02 *M*/sec to 0.08 *M*/sec.

We need to find a value for the exponent x that relates the doubling of the concentration to the quadrupling of the rate. The value of x can be calculated as follows:

$$(2)^x = 4, \text{ so } x = 2$$

Because the value of x is 2, the reaction is said to be second order with respect to A.

$$\text{Rate} = k[A]^2[B]^y[C]^z$$

Let's Look at [B]

From experiment 1 to experiment 3, [B] doubles while the other reactant concentrations remain constant. For this reason it is useful to use the rate values from these two experiments to calculate y (the order of the reaction with respect to reactant B).

As you can see from the table, the rate doubles from experiment 1 to experiment 3, going from 0.01 M/sec to 0.02 M/sec.

We need to find a value for the exponent y that relates the doubling of the concentration to the doubling of the rate. The value of y can be calculated as follows:

$$(2)^y = 2, \text{ so } y = 1$$

Because the value of y is 1, the reaction is said to be first order with respect to B.

$$\text{Rate} = k[A]^2[B][C]^z$$

Let's Look at [C]

From experiment 1 to experiment 2, [C] doubles while the other reactant concentrations remain constant.

The rate remains the same at 0.01 M.

The rate change is $(2)^z = 1$, so $z = 0$.

Because the value of z is 0, the reaction is said to be zero order with respect to C.

$$\text{Rate} = k[A]^2[B]$$

Because the sum of the exponents is 3, the reaction is said to be third order overall.

Once the rate law has been determined, the value of the rate constant can be calculated using any of the lines of data on the table. The units of the rate constant are dependent on the order of the reaction, so it's important to carry along units throughout all rate constant calculations.

Let's use experiment 3.

$$k = \frac{\text{Rate}}{[\text{A}]^2[\text{B}]} = \frac{(0.02\,M/\text{sec})}{(0.10\,M)^2(0.20\,M)} = 10\left(\frac{(M)}{(M)^3(\text{sec})}\right) = 10\ M^{-2}\cdot\text{sec}^{-1}$$

You should note that we can tell from the coefficients in the original balanced equation that the rate of appearance of D is equal to the rate of disappearance of A and C because the coefficients of all three are the same. The coefficient of D is half as large as the coefficient of B, however, so the rate at which D appears is half the rate at which B disappears.

RATE LAW USING CONCENTRATION AND TIME

It's also useful to have rate laws that relate the rate constant *k* to the way that concentrations change over time. The rate laws will be different depending on whether the reaction is first, second, or zero order, but each rate law can be expressed as a graph that relates the rate constant, the concentration of a reactant, and the elapsed time.

First-Order Rate Laws

The rate of a first-order reaction depends on the concentration of a single reactant raised to the first power.

$$\text{Rate} = k[\text{A}]$$

As the concentration of reactant A is depleted over time, the rate of reaction will decrease with a characteristic half-life. The **half-life** of a reaction describes the amount of time it takes for half of a sample to react. For the first order reaction graphed below, the half-life of A is 1 minute. In the first minute, the concentration of A drops by half, from 1.0 *M* to 0.5 *M*. In the second minute, it drops from 0.5 *M* to 0.25 *M*, and so forth. A concentration versus time graph for a first-order reactant will always display the typical half-life exponential decay curve as shown below.

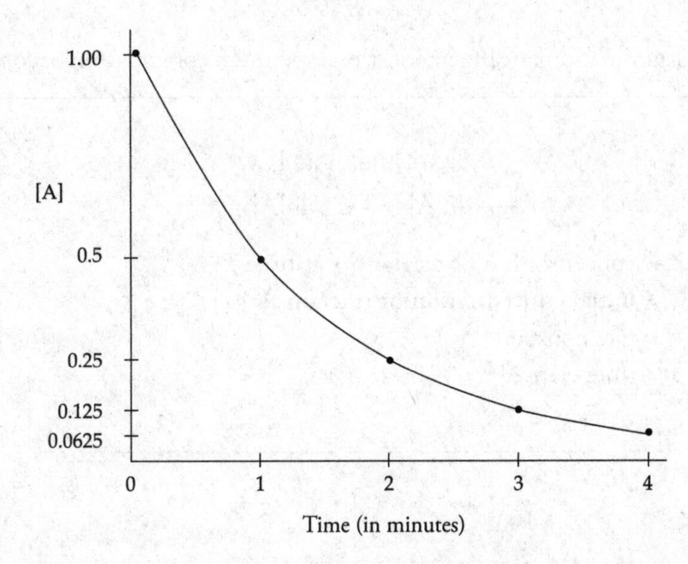

The rate law for a first-order reaction uses natural logarithms. This means that the use of natural logarithms in the rate law creates a linear graph comparing concentration and time. The use of natural logarithms in the rate law creates a linear graph comparing concentration and time. The slope of the line is given by $-k$ and the y-intercept is given by $\ln[A]_0$.

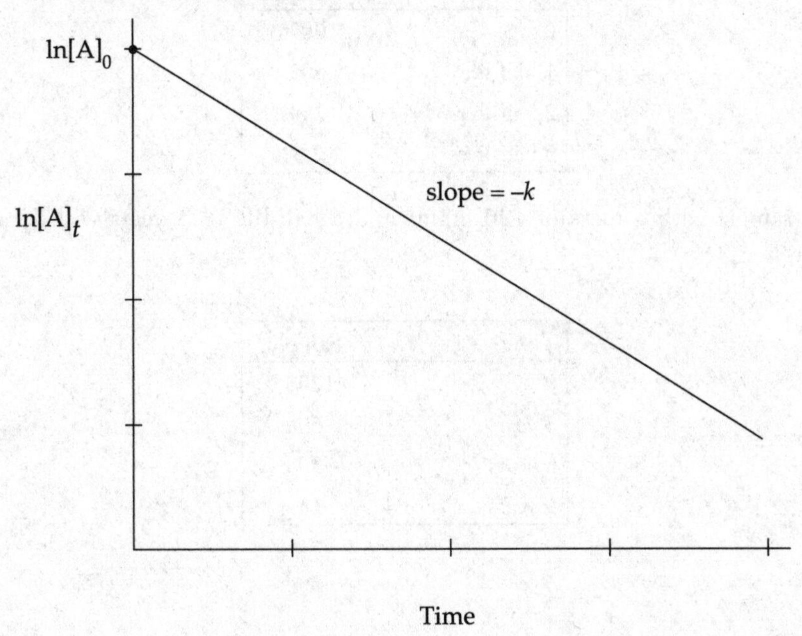

Using slope-intercept form, we can interpret this graph to come up with a useful equation.

$$y = mx + b$$

$y = \ln[A]_t$
$m = -k$
$x =$ time
$b = \ln [A]_0$

After substitution and rearrangement, the slope-intercept equation becomes:

First-Order Rate Law

$$\ln[A]_t = -kt + \ln[A]_0$$

$[A]_t$ = concentration of reactant A at time t
$[A]_0$ = initial concentration of reactant A
k = rate constant
t = time elapsed

Half-Life

Reminder
Don't forget that the chart should start with the time at zero.

The half-life of a reactant in a chemical reaction is the time it takes for half of the substance to react. Most half-life-problems can be solved by using a simple chart:

Time	Sample
0	100%
1 half-life	50%
2 half-lives	25%
3 half-lives	12.5%

So a sample with a mass of 120 grams and a half-life of 3 years will decay as follows:

Time	Sample
0	120 g
3 years	60 g
6 years	30 g
9 years	15 g

Additionally, the half-life of a first order reactant can be calculated from the graphical interpretation of the reaction using the following formula:

$$\text{Half-life} = \frac{\ln 2}{k} = \frac{0.693}{k}$$

Let's try an example based on the data below.

[A] (*M*)	Time (min)
2.0	0
1.6	5
1.2	10

(a) Let's find the value of k. We'll use the first two lines of the table.
$\ln[A]_t - \ln[A]_0 = -kt$
$\ln(1.6) - \ln(2.0) = -k(5 \text{ min})$
$-0.22 = -(5 \text{ min})k$
$k = 0.045 \text{ min}^{-1}$

(b) Now let's use k to find [A] when 20 minutes have elapsed.
$\ln[A]_t - \ln[A]_0 = -kt$
$\ln[A]_t - \ln(2.0) = -(0.045 \text{ min}^{-1})(20 \text{ min})$
$\ln[A]_t = -0.21$
$[A]_t = e^{-0.21} = 0.81\ M$

(c) Now let's find the half-life of the reaction.

We can look at the answer (b) to see that the concentration dropped by half (1.6 *M* to 0.8 *M*) from 5 minutes to 20 minutes. That makes the half-life about 15 minutes. We can confirm this using the half-life equation.

$$\text{Half-life} = \frac{0.693}{k} = \frac{0.693}{0.045 \text{ min}^{-1}} = 15.4 \text{ minutes}$$

Another use of half-life is to examine the rate of decay of a radioactive substance. A radioactive substance is one that will slowly decay into a more stable form as time goes on.

Second-Order Rate Laws

The rate of a second-order reaction depends on the concentration of a single reactant raised to the second power.

$$\text{Rate} = k[A]^2$$

The rate law for a second-order reaction uses the inverses of the concentrations.

Second-Order Rate Law

$$\frac{1}{[A]_t} = kt + \frac{1}{[A]_0}$$

$[A]_t$ = concentration of reactant A at time t
$[A]_0$ = initial concentration of reactant A
k = rate constant
t = time elapsed

The use of inverses in the rate law creates a linear graph comparing concentration and time. Notice that the line moves upward as the concentration decreases. The slope of the line is given by k and the y-intercept is given by $\frac{1}{[A]_0}$.

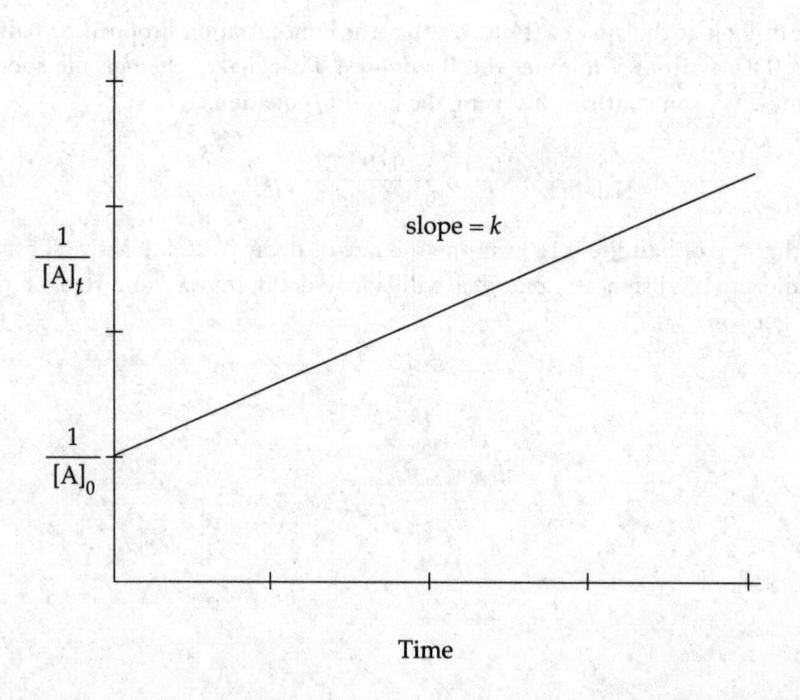

Zero-Order Rate Laws

The rate of a zero-order reaction does not depend on the concentration of reactants at all, so the rate of a zero-order reaction will always be the same at a given temperature.

$$\text{Rate} = k$$

The graph of the change in concentration of a reactant of a zero-order reaction versus time will be a straight line with a slope equal to $-k$.

COLLISION THEORY

According to collision theory, chemical reactions occur because reactants are constantly moving around and colliding with one another.

When reactants collide with sufficient energy (**activation energy, E_a**), a reaction occurs. These collisions are referred to as effective collisions, because they lead to a chemical reaction. Ineffective collisions do not produce a chemical reaction. At any given time during a reaction, a certain fraction of the reactant molecules will collide with sufficient energy to cause a reaction between them.

Reaction rate increases with increasing concentration of reactants because if there are more reactant molecules moving around in a given volume, then more collisions will occur.

Reaction rate increases with increasing temperature because increasing temperature means that the molecules are moving faster, which means that the molecules have greater average kinetic energy. The higher the temperature, the greater the number of reactant molecules colliding with each other with enough energy (E_a) to cause a reaction.

In addition, reactions only occur if the reactants collide with the correct orientation. For example, in the reaction $2NO_2F \rightarrow 2NO_2 + F_2$, there are many possible collision orientations. Two of them are drawn below:

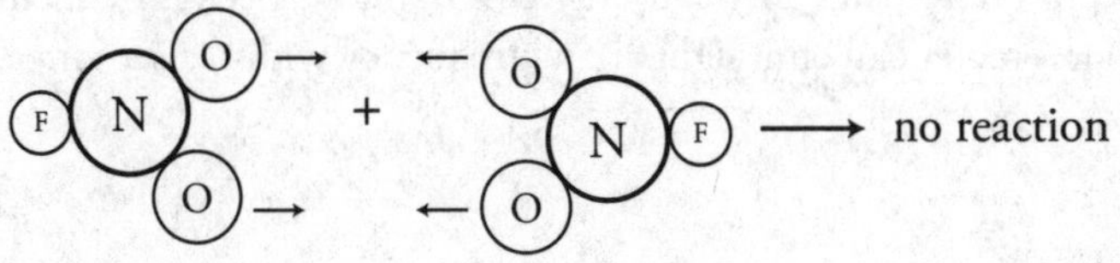

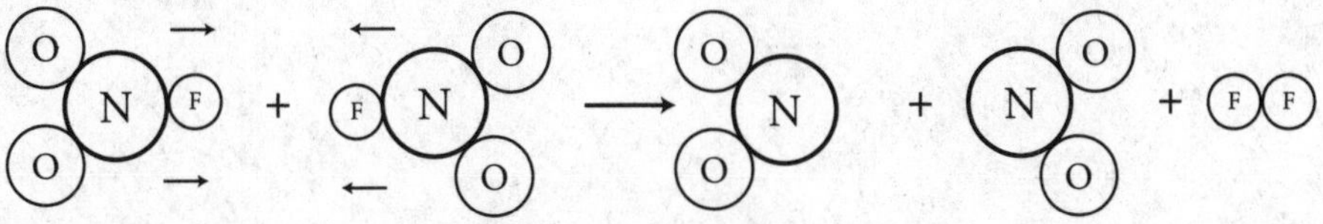

The reaction will only occur if the two NO_2F molecules collide in such a fashion that the N–F bonds can break and the F–F bonds can form. While there is no way to quantify collision orientation, it is an important part of collision theory, and you should be familiar with the basic concept underlying it.

BEER'S LAW

To measure the concentration of a solution over time, a device called a spectrophotometer can be used in some situations. A spectrophotometer measures the amount of light at a given wavelength that is absorbed by a solution. If a solution changes color as the reaction progresses, the amount of light that is absorbed will change. Absorbance can be calculated using Beer's law:

Beer's Law

$$A = abc$$

A = absorbance
a = molar absorptivity, a constant that depends on the solution
b = path length, the distance the light is traveling through the solution
c = concentration of the solution

As molar absorptivity and path length are constants when using a spectrophotometer, Beer's law is often interpreted as a direct relationship between absorbance and the concentration of the solution. Beer's law is most effective with solutions that visibly change color over the course of a reaction, but if a spectrophotometer that emits light in the ultraviolet region is used, Beer's law can be used to determine the concentrations of reactants in solutions that are invisible to the human eye.

You may also run into a device called a colorimeter while studying Beer's law. A colorimeter is simply a spectrophotometer that can only emit light at specific frequencies, whereas a spectrophotometer can emit light at any frequency within a set range.

REACTION MECHANISMS

Many chemical reactions are not one-step processes. Rather, the balanced equation is the sum of a series of individual steps, called elementary steps.

For instance, the hypothetical reaction

$$2A + 2B \rightarrow C + D \qquad \text{Rate} = k[A]^2[B]$$

could take place by the following three-step mechanism:

I.	$A + A \rightleftharpoons X$	(fast)
II.	$X + B \rightarrow C + Y$	(slow)
III.	$Y + B \rightarrow D$	(fast)

Species X and Y are called reaction **intermediates**, because they are produced but also fully consumed over the course of the reaction. Intermediates will always cancel out when adding up the various elementary steps in a reaction, as shown below.

I. $A + A \rightleftharpoons X$

II. $X + B \rightarrow C + Y$

III. $Y + B \rightarrow D$

$$A + A + X + B + Y + B \rightarrow X + C + Y + D$$

Cancel species that appear on both sides.

$$2A + 2B \rightarrow C + D$$

By adding up all the steps, we get the balanced equation for the overall reaction, so this mechanism is consistent with the balanced equation.

As in any process where many steps are involved, the speed of the whole process can't be faster than the speed of the slowest step in the process, so the slowest step of a reaction is called the **rate-determining step.** Since the slowest step is the most important step in determining the rate of a reaction, the slowest step and the steps leading up to it are used to see if the mechanism is consistent with the rate law for the overall reaction.

The rate for an elementary step can be determined using by taking the concentration of the reactants in that step and raising them to the power of any coefficient attached to that reactant. So, for the current reactions:

I. $A + A \rightleftharpoons X$ (so $2A \rightleftharpoons X$) Rate = $k[A]^2$ (fast)

II. $X + B \rightarrow Y$ Rate = $k[X][B]$ (slow)

III. $Y + B \rightarrow D$ Rate = $k[Y][B]$ (fast)

The rate law for the entire reaction is equal to that of the slowest elementary step, which is step II. However, step II has an intermediate (X) present in it. Looking at step 1, we can also see that [X] is equivalent to $[A]^2$ (as the sides are in equilibrium). If $[A]^2$ is substituted in for [X], the rate law for step II becomes rate = $k[A]^2[B]$, which is the overall rate law for this reaction.

It is important to emphasize that you can only use the coefficient method to determine the rate law of elementary steps. You cannot use the coefficients from the overall balanced equation in a similar fashion to determine the overall rate law. (For instance, the rate law for the above example is not rate = $k[A]^2[B]^2$). The only way to use the coefficient method to determine the rate law for a full reaction is by knowing which elementary step is the slowest and applying the above method to that step. If you do not know the relative speed of the elementary steps, you must use experimental data to determine the rate law.

As discussed previously, an energy diagram is a graphical representation which shows the energy level of the products and the reactants, as well as the required activation energy for a reaction to occur. As many reactions take place over several steps, an energy diagram can be broken down to examine the energy change in each individual step. Take the diagram below:

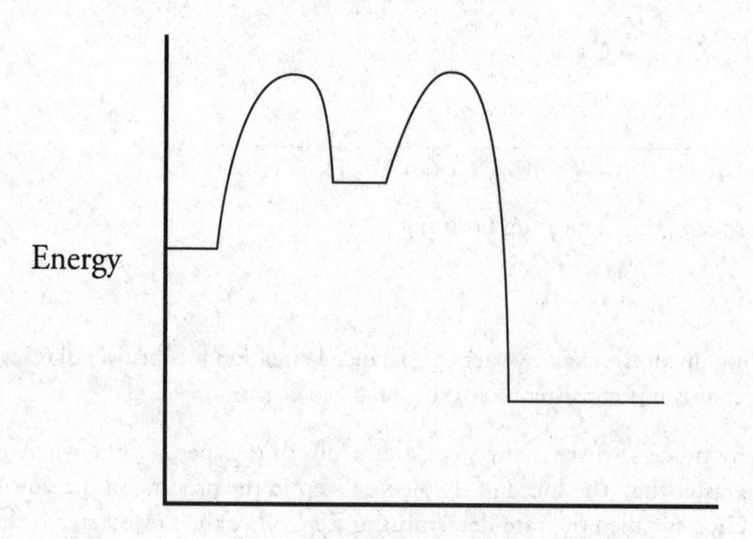

In this diagram, you can see the overall reaction is exothermic. However, the reaction takes place in two different steps: the first, which is endothermic, and the second, which is exothermic.

CATALYSTS

As we mentioned earlier in the book, a catalyst increases the rate of a chemical reaction without being consumed in the process; catalysts do not appear in the balanced equation. In some cases, a catalyst is a necessary part of a reaction because in its absence, the reaction would proceed at too slow of a rate to be at all useful.

When looking at elementary steps, a catalyst is present both before and after the overall reaction. Catalysts cancel out of the overall reaction but are present in elementary steps.

Example: $A + B \rightarrow C$

I. $A + X \rightarrow Y$

II. $B + Y \rightarrow C + X$

In the above example, X is a catalyst and Y is a reaction intermediate.

When a catalyst is introduced to a reaction, the ensuing reaction is said to undergo catalysis. There are many types of catalysis. One of the most common is surface catalysis, in which a reaction intermediate is formed as in the example above. Another is enzyme catalysis, in which the catalyst binds to the reactants in a way to reduce the overall activation energy of the reaction. Enzymes are very common in biological applications. Finally, in acid-base catalysis, reactants will gain or lose protons in order to change the rate of reaction. Acids and bases will be studied in more depth in Big Idea #6.

A catalyst increases the rate of a chemical reaction by providing an alternative reaction pathway with a lower activation energy.

CHAPTER 6 QUESTIONS

Multiple-Choice Questions

Use the following information to answer questions 1-4.

A multi-step reaction takes place with the following elementary steps:

Step I. $A + B \rightleftharpoons C$

Step II. $C + A \rightarrow D$

Step III. $C + D \rightarrow B + E$

1. What is the overall balanced equation for this reaction?

   (A) $2A + B + 2C + D \rightarrow C + D + B + E$
   (B) $A + B \rightarrow B + E$
   (C) $A + 2C \rightarrow D + E$
   (D) $2A + C \rightarrow E$

2. What is the function of species B in this reaction?

   (A) Without it, no reaction would take place.
   (B) It is a reaction intermediate which facilitates the progress of the reaction.
   (C) It is a catalyst which changes the overall order of the reaction.
   (D) It lowers the overall activation energy of the reaction.

3. If step II is the slow step for the reaction, what is the overall rate law?

   (A) Rate = $k[A]^2[B]$
   (B) Rate = $k[A][C]$
   (C) Rate = $k[A][B]$
   (D) Rate = $k[A]/[D]$

4. Why would increasing the temperature make the reaction rate go up?

   (A) It is an endothermic reaction that needs an outside energy source to function.
   (B) The various molecules in the reactions will move faster and collide more often.
   (C) The overall activation energy of the reaction will be lowered.
   (D) A higher fraction of molecules will have the same activation energy.

5. $A + B \rightarrow C$ rate = $k[A]$

Under which conditions would the order of the above reaction double?

(A) Doubling the initial concentration of A
(B) Doubling the initial concentration of B
(C) Reducing the concentration of A by half
(D) Changing the concentration of the reactants has no effect on reaction order

6. $A + B \rightarrow C + D$ rate = $k[A][B]^2$

What are the potential units for the rate constant for the above reaction?

(A) s^{-1}
(B) $s^{-1}M^{-1}$
(C) $s^{-1}M^{-2}$
(D) $s^{-1}M^{-3}$

7. The following mechanism is proposed for a reaction:

$2A \leftrightarrow B$ (fast equilibrium)
$C + B \rightarrow D$ (slow)
$D + A \rightarrow E$ (fast)

Which of the following is the correct rate law for compete reaction?

(A) Rate = $k[C]^2[B]$
(B) Rate = $k[C][A]^2$
(C) Rate = $k[C][A]^3$
(D) Rate = $k[D][A]$

8. $2NOCl \rightarrow 2NO + Cl_2$

The reaction above takes place with all of the reactants and products in the gaseous phase. Which of the following is true of the relative rates of disappearance of the reactants and appearance of the products?

(A) NO appears at twice the rate that NOCl disappears.
(B) NO appears at the same rate that NOCl disappears.
(C) NO appears at half the rate that NOCl disappears.
(D) Cl_2 appears at the same rate that NOCl disappears.

9. $$H_2(g) + I_2(g) \rightarrow 2HI(g)$$

When the reaction given above takes place in a sealed isothermal container, the rate law is

$$\text{Rate} = k[H_2][I_2]$$

If a mole of H_2 gas is added to the reaction chamber and the temperature remains constant, which of the following will be true?

(A) The rate of reaction and the rate constant will increase.
(B) The rate of reaction and the rate constant will not change.
(C) The rate of reaction will increase and the rate constant will decrease.
(D) The rate of reaction will increase and the rate constant will not change.

10. $$A + B \rightarrow C$$

When the reaction given above takes place, the rate law is

$$\text{Rate} = k[A]$$

If the temperature of the reaction chamber were increased, which of the following would be true?

(A) The rate of reaction and the rate constant will increase.
(B) The rate of reaction and the rate constant will not change.
(C) The rate of reaction will increase and the rate constant will decrease.
(D) The rate of reaction will increase and the rate constant will not change.

11. $$A + B \rightarrow C$$

Based on the following experimental data, what is the rate law for the hypothetical reaction given above?

Experiment	[A] (*M*)	[B] (*M*)	Initial Rate of Formation of C (mol/L·sec)
1	0.20	0.10	3×10^{-2}
2	0.20	0.20	6×10^{-2}
3	0.40	0.20	6×10^{-2}

(A) Rate = $k[A]$
(B) Rate = $k[A]^2$
(C) Rate = $k[B]$
(D) Rate = $k[A][B]$

12. A solution of Co^{2+} ions appears red when viewed under white light. Which of the following statements is true about the solution?

    (A) A spectrophotometer set to the wavelength of red light would read a high absorbance.
    (B) If the solution is diluted, the amount of light reflected by the solution will decrease.
    (C) All light with a frequency that is lower than that of red light will be absorbed by it.
    (D) Electronic transmissions within the solution match the wavelength of red light.

13.

$$A + B \rightarrow C$$

Based on the following experimental data, what is the rate law for the hypothetical reaction given above?

Experiment	[A] (*M*)	[B] (*M*)	Initial Rate of Formation of C (*M*/sec)
1	0.20	0.10	2.0×10^{-6}
2	0.20	0.20	4.0×10^{-6}
3	0.40	0.40	1.6×10^{-5}

(A) Rate = k[A]
(B) Rate = $k[A]^2$
(C) Rate = k[B]
(D) Rate = k[A][B]

14.

Time (Hours)	[A] *M*
0	0.40
1	0.20
2	0.10
3	0.05

Reactant A underwent a decomposition reaction. The concentration of A was measured periodically and recorded in the chart above. Based on the data in the chart, which of the following is the rate law for the reaction?

(A) Rate = k[A]

(B) Rate = $k[A]^2$

(C) Rate = $2k$[A]

(D) Rate = $\frac{1}{2}k$[A]

15. $A \rightarrow B + C$ rate $= k[A]^2$

Which of the following graphs may have been created using data gathered from the above reaction?

(A)

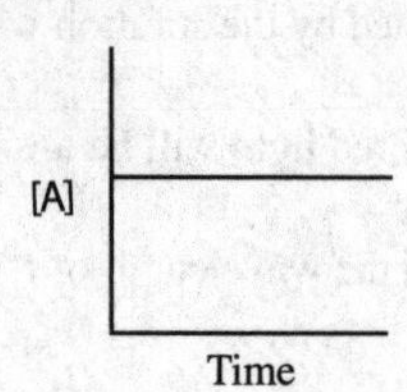

(B)

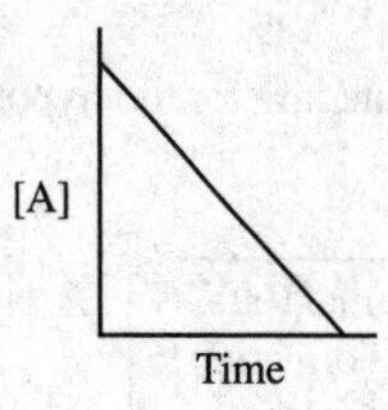

(C)

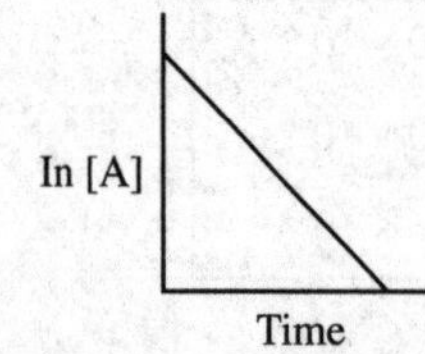

(D)

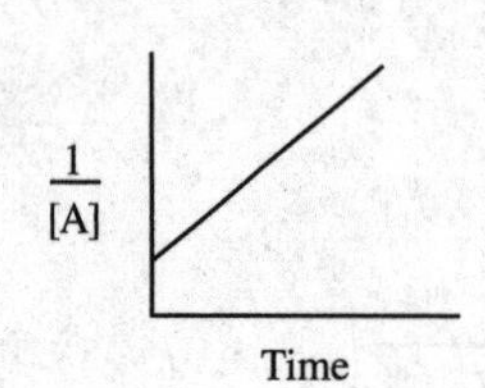

16. After 44 minutes, a sample of $^{44}_{19}K$ is found to have decayed to 25 percent of the original amount present. What is the half-life of $^{44}_{19}K$?

(A) 11 minutes
(B) 22 minutes
(C) 44 minutes
(D) 66 minutes

Free-Response Questions

1. $$A + 2B \rightarrow 2C$$

The following results were obtained in experiments designed to study the rate of the reaction above:

Experiment	Initial Concentration (mol/L) [A]	[B]	Initial Rate of Disappearance of A (M/sec)
1	0.05	0.05	3.0×10^{-3}
2	0.05	0.10	6.0×10^{-3}
3	0.10	0.10	1.2×10^{-2}
4	0.20	0.10	2.4×10^{-2}

(a) Determine the order of the reaction with respect to each of the reactants, and write the rate law for the reaction.

(b) Calculate the value of the rate constant, k, for the reaction. Include the units.

(c) If another experiment is attempted with [A] and [B], both 0.02-molar, what would be the initial rate of disappearance of A?

(d) The following reaction mechanism was proposed for the reaction above:

$$A + B \rightarrow C + D$$
$$D + B \rightarrow C$$

(i) Show that the mechanism is consistent with the balanced reaction.

(ii) Show which step is the rate-determining step, and explain your choice.

2. $$2NO(g) + Br_2(g) \rightarrow 2NOBr(g)$$

The following results were obtained in experiments designed to study the rate of the reaction above:

Experiment	Initial Concentration (mol/L) [NO]	[Br_2]	Initial Rate of Appearance of NOBr (M/sec)
1	0.02	0.02	9.6×10^{-2}
2	0.04	0.02	3.8×10^{-1}
3	0.02	0.04	1.9×10^{-1}

(a) Write the rate law for the reaction.
(b) Calculate the value of the rate constant, k, for the reaction. Include the units.
(c) In experiment 2, what was the concentration of NO remaining when half of the original amount of Br_2 was consumed?
(d) Which of the following reaction mechanisms is consistent with the rate law established in (a)? Explain your choice.

I. $NO + NO \rightleftharpoons N_2O_2$ (fast)

$N_2O_2 + Br_2 \rightarrow 2NOBr$ (slow)

II. $Br_2 \rightarrow Br + Br$ (slow)

$2(NO + Br \rightarrow NOBr)$ (fast)

3. $$2N_2O_5(g) \rightarrow 4NO_2(g) + O_2(g)$$

Dinitrogen pentoxide gas decomposes according to the equation above. The first-order reaction was allowed to proceed at 40°C and the data below were collected.

[N_2O_5] (M)	Time (min)
0.400	0.0
0.289	20.0
0.209	40.0
0.151	60.0
0.109	80.0

(a) Calculate the rate constant for the reaction using the values for concentration and time given in the table. Include units with your answer.
(b) After how many minutes will [N_2O_5] be equal to 0.350 M?
(c) What will be the concentration of N_2O_3 after 100 minutes have elapsed?
(d) Calculate the initial rate of the reaction. Include units with your answer.
(e) What is the half-life of the reaction?

4. The decomposition of phosphine occurs via the pathway below:

$$4PH_3(g) \rightarrow P_4(g) + 6H_2(g)$$

A scientist observing this reaction at 250 K plots the following data:

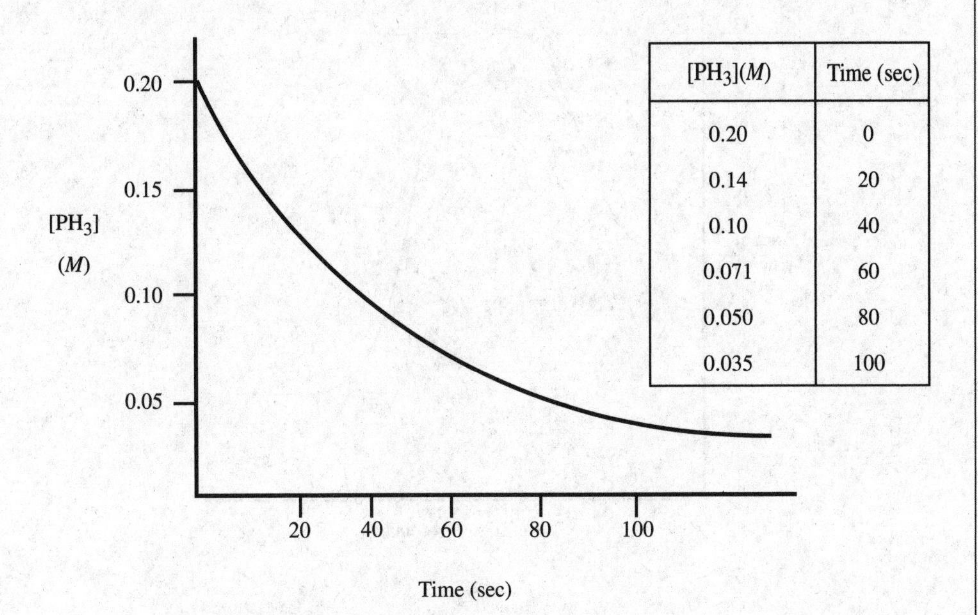

$[PH_3]$(M)	Time (sec)
0.20	0
0.14	20
0.10	40
0.071	60
0.050	80
0.035	100

(a) (i) What order is this reaction? Why?
(ii) What would the concentration of the PH_3 gas be after 120 sec?

(b) If the rate of disappearance of PH_3 is $2.5 \cdot 10^{-3}$ M/s at $t = 20$ s:
(i) What is the rate of appearance of P_4 at the same point in time?
(ii) How will the rate of appearance of P_4 change as the reaction progresses forward?

(c) The experiment is repeated with the same initial concentration of phosphine, but this time the temperature is set at 500 K. How and why would the following values change, if at all?
(i) The half-life of the phosphine
(ii) The rate law
(iii) The value of the rate constant

5.

$$A(g) + B(g) \rightarrow C(g)$$

The reaction above is second order with respect to A and zero order with respect to B. Reactants A and B are present in a closed container. Predict how each of the following changes to the reaction system will affect the rate and rate constant, and explain why.

(a) More gas A is added to the container.
(b) More gas B is added to the container.
(c) The temperature is increased.
(d) An inert gas D is added to the container.
(e) The volume of the container is decreased.

6. Use your knowledge of kinetics to answer the following questions. Justify your answers.

(a)

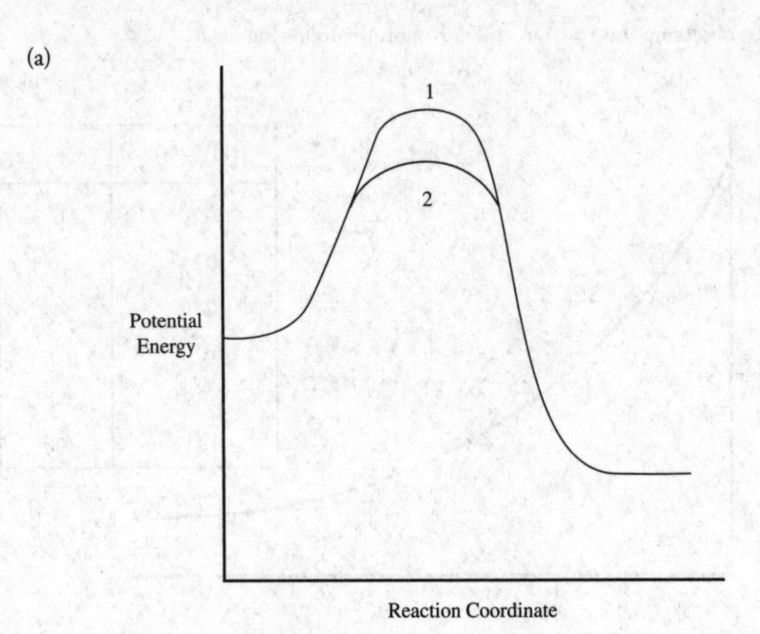

The two lines in the diagram above show different reaction pathways for the same reaction. Which of the two lines shows the reaction when a catalyst has been added?

(b)

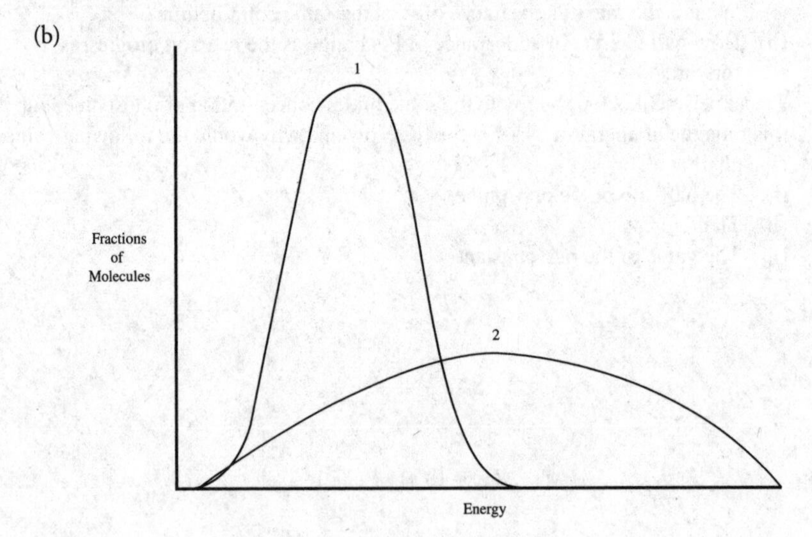

Which of the two lines in the energy distribution diagram shows the conditions at a higher temperature?

(c)

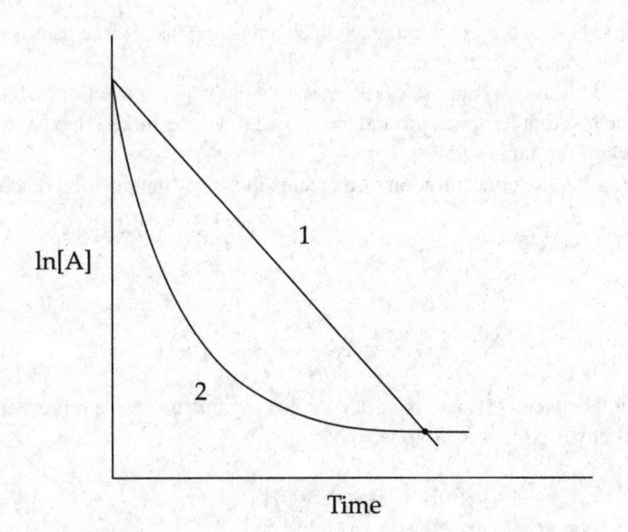

Which of the two lines in the diagram above shows the relationship of ln[A] to time for a first order reaction with the following rate law?

$$\text{Rate} = k[\text{A}]$$

(d)

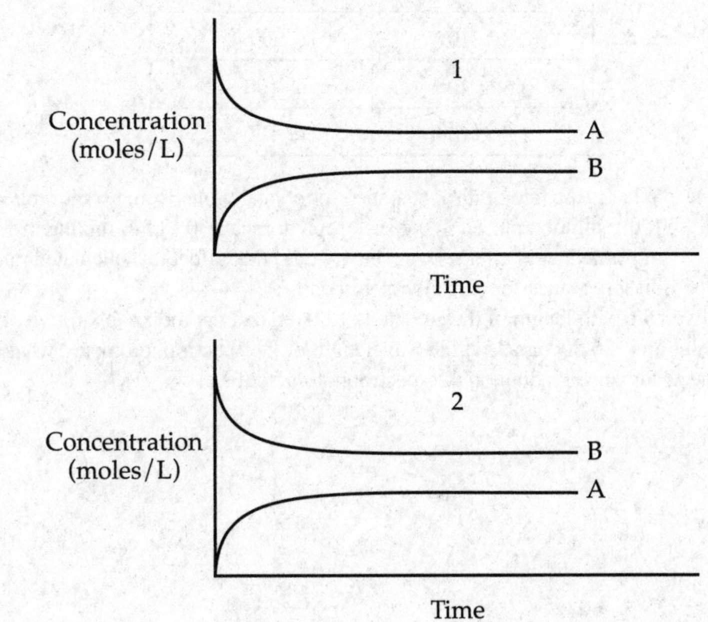

Which of the two graphs above shows the changes in concentration over time for the following reaction?

$$\text{A} \rightarrow \text{B}$$

7. Use your knowledge of kinetics to explain each of the following statements:

(a) An increase in the temperature at which a reaction takes place causes an increase in reaction rate.
(b) The addition of a catalyst increases the rate at which a reaction will take place.
(c) A catalyst that has been ground into powder will be more effective than a solid block of the same catalyst.
(d) Increasing the concentration of reactants increases the rate of a reaction.

8. The reaction between crystal violet (a complex organic molecule represented by CV^+) and sodium hydroxide is as follows:

$$\underset{\text{(violet)}}{CV^+} + \underset{\text{(colorless)}}{OH^-} \rightarrow \underset{\text{(colorless)}}{CVOH}$$

As the crystal violet is the only colored species in the reaction, a spectrophotometer calibrated to a specific wavelength can be used to determine its concentration over time. The following data was gathered:

$[CV^+]$ (*M*)	Time (s)
5.5×10^{-5}	0
3.8×10^{-5}	60
2.6×10^{-5}	120
1.8×10^{-5}	180

(a) (i) What is the rate of disappearance for crystal violet from $t = 60$ s to $t = 120$ s?
(ii) If the solution placed in the spectrophotometer 30 s after mixing instead of immediately after mixing, how would that affect the calculated rate of disappearance for crystal violet in part (i)?
(b) Given the path length of the cuvette is 1.00 cm and the molar absorptivity of the solution is 26,000 cm/*M* at the wavelength of the spectrophotometer, what would the absorbance reading on the spectrophotometer be at $t = 60$ s?

(c) This reaction is known to be first order with respect to crystal violet. On the provided axes, graph a function of $[CV^+]$ vs. time that will provide you with a straight line graph.

The following data was also gathered over the course of three experiments:

Experiment	$[CV^+]_{init}$ (M)	$[OH^-]_{init}$ (M)	Initial rate of formation of CVOH (M/s)
1	5.5×10^{-5}	0.12	3.60×10^{-7}
2	5.5×10^{-5}	0.24	7.20×10^{-7}
3	4.1×10^{-5}	0.18	?

(d) Write the rate law for this reaction.

(e) What is the rate constant, k, for this reaction? Include units in your answer.

(f) Determine the initial rate of formation of CVOH for experiment 3.

CHAPTER 6 ANSWERS AND EXPLANATIONS

Multiple-Choice

1. **D** The answer to (A) shows all of the reactants and products that are part of this reaction. However, species that appear on both sides need to cancel. That eliminates B, one C, and D, leaving $2A + C \rightarrow E$

2. **D** Species B is a catalyst as it is present before and after the reaction but cancels out of the overall equation. However, catalysts do not change the order of the reaction; they simply lower the activation energy so the reaction can proceed more quickly. The reaction would still take place without the catalyst, but it would proceed very slowly.

3. **A** Only reactants can appear in rate laws. In Step II, you can replace C with A + B, as these are shown to be equivalent in step I. Step II now reads:

$$A + B + A \rightarrow D \quad \text{or} \quad 2A + B \rightarrow D$$

As step II is the slow step, the coefficients then become exponents in the overall rate law, leading us to (A).

4. **B** Increasing the temperature increases the speed of the molecules. This will increase how often those molecules collide, meaning that a successful collision is more likely to happen. Additionally, there will also be more energy within each collision, making it more likely that the collision will exceed the activation energy barrier.

5. **D** While the rate of reaction may be increased by changing the concentration of A, the rate order, which describes how changing the concentration of A affects the speed of the reaction, will not change.

6. **C** To answer this we can do a unitless dimensional analysis:

$$M/\text{s} = k(M)(M)^2$$
$$M/\text{s} = k(M)^3$$
$$k = \text{s}^{-1}M^{-2}$$

7. **B** The coefficients on the reactants in the slowest elementary step match up with the order of those reactants in the overall rate law. However, B is an intermediate that is not present at the start of the reaction and as such cannot be part of the rate law. B is in equilibrium with 2A in the first step, though, so the 2A can be substituted for B in the slow step, which then yields (B) when the rate law is determined.

8. **B** For every two NO molecules that form, two NOCl molecules must disappear, so NO is appearing at the same rate that NOCl is disappearing. Choice (D) is wrong because for every mole of Cl_2 that forms, two moles of NOCl are disappearing, so Cl_2 is appearing at *half* the rate that NOCl is disappearing.

9. **D** From the rate law given in the question (Rate = $k[H_2][I_2]$), we can see that increasing the concentration of H_2 will increase the rate of reaction. The rate constant, *k*, is not affected by changes in the concentration of the reactants.

10. **A** When temperature increases, the rate constant increases to reflect the fact that more reactant molecules are likely to have enough energy to react at any given time. So both the rate constant and the rate of reaction will increase.

11. **C** From a comparison of experiments 1 and 2, when [B] is doubled while [A] is held constant, the rate doubles. That means that the reaction is first order with respect to B.

    From a comparison of experiments 2 and 3, when [A] is doubled while [B] is held constant, the rate doesn't change. That means that the reaction is zero order with respect to A and that A will not appear in the rate law.

    So the rate law is Rate = $k[B]$.

12. **D** A spectrophotometer works by reading light absorbance. A red solution appears to be red because it transmits red light. The amount of light reflected by the solution does not change as it dilutes; the amount of light that is transmitted and absorbed does. The solution has a red color because the distance between the energy levels within the cobalt ions corresponds with the wavelength of red light.

13. **D** From a comparison of experiments 1 and 2, when [B] is doubled while [A] is held constant, the rate doubles. That means that the reaction is first order with respect to B.

From a comparison of experiments 2 and 3, when both [A] and [B] are doubled, the rate increases by a factor of 4. We would expect the rate to double based on the change in B; because the rate is in fact multiplied by 4, the doubling of A must also change the rate by a factor of 2, so the reaction is also first order with respect to A.

So the rate law is Rate = k[A][B].

14. **A** The key to this question is to recognize that reactant A is disappearing with a characteristic half-life. This is a signal that the reaction is first order with respect to A. So the rate law must be Rate = k[A].

15. **D** The reaction is a second order reaction, of which a graph of the inverse of concentration always produces a straight line with a slope equal to k, the rate constant.

16. **B** Make a chart. Always start at time = 0.

Half-Lives	Time	Stuff
0	0	100%
1	X	50%
2	44 min.	25%

It takes two half-lives for the amount of $^{44}_{19}K$ to decrease to 25 percent. If two half-lives takes 44 minutes, one half-life must be 22 minutes.

Free-Response

1. (a) When we compare the results of experiments 3 and 4, we see that when [A] doubles, the rate doubles, so the reaction is first order with respect to A.

When we compare the results of experiments 1 and 2, we see that when [B] doubles, the rate doubles, so the reaction is first order with respect to B.

Rate = k[A][B]

(b) Use the values from experiment 3, just because they look the simplest.

$$k = \frac{\text{Rate}}{[A][B]} = \frac{(1.2 \times 10^{-2}\ M/\text{sec})}{(0.10\ M)(0.10\ M)} = 1.2\ M^{-1}\text{sec}^{-1}$$

(c) Use the rate law.

Rate = k[A][B]

Rate = $(1.2\ M^{-1}\text{sec}^{-1})(0.02\ M)(0.02\ M) = 4.8 \times 10^{-4}\ M/\text{sec}$

(d) (i) $A + B \rightarrow C + D$

$D + B \rightarrow C$

The two reactions add up to

$A + 2B + D \rightarrow 2C + D$

D's cancel, and we're left with the balanced equation.

$A + 2B \rightarrow 2C$

(ii) $A + B \rightarrow C + D$ (slow)

$D + B \rightarrow C$ (fast)

The first part of the mechanism is the slow, rate-determining step because its rate law is the same as the experimentally determined rate law.

2. (a) When we compare the results of experiments 1 and 2, we see that when [NO] doubles, the rate quadruples, so the reaction is second order with respect to NO.

When we compare the results of experiments 1 and 3, we see that when $[Br_2]$ doubles, the rate doubles, so the reaction is first order with respect to Br_2.

Rate = $k[NO]^2[Br_2]$

(b) Use the values from experiment 1, just because they look the simplest.

$$k = \frac{\text{Rate}}{[NO]^2[Br_2]} = \frac{(9.6\times10^{-2}\ M/\text{sec})}{(0.02\ M)^2(0.02\ M)} =$$

$1.2 \times 10^4\ M^{-2}\text{sec}^{-1}$

(c) In experiment 2, we started with $[Br_2]$ = 0.02 *M*, so 0.01 *M* was consumed.

From the balanced equation, 2 moles of NO are consumed for every mole of Br_2 consumed. So 0.02 *M* of NO are consumed.

[NO] remaining = 0.04 *M* – 0.02 *M* = 0.02 *M*

(d) Choice (I) agrees with the rate law.

$NO + NO \leftrightarrow N_2O_2$ (fast)

$N_2O_2 + Br_2 \rightarrow 2NOBr$ (slow)

The slow step is the rate-determining step, with the following rate law:

Rate = $k[N_2O_2][Br_2]$

Looking at the first step in the proposed mechanism, we can replace the N_2O_2 with 2NO. Doing so gives us:

$2NO + Br_2 \rightarrow 2NOBr$

The rate law for that elementary step matches the overall rate law determined previously.

3. (a) Use the first-order rate law and insert the first two lines from the table.

$$\ln[N_2O_5]_t - \ln[N_2O_5]_0 = -kt$$

$$\ln(0.289) - \ln(0.400) = -k(20.0 \text{ min})$$

$$-.325 = -k(20.0 \text{ min})$$

$$k = 0.0163 \text{ min}^{-1}$$

(b) Use the first-order rate law.

$$\ln[N_2O_5]_t - \ln[N_2O_5]_0 = -kt$$

$$\ln(0.350) - \ln(0.400) = -(0.0163 \text{ min}^{-1})t$$

$$-0.134 = -(0.0163 \text{ min}^{-1})t$$

$$t = 8.22 \text{ min}$$

(c) Use the first-order rate law.

$$\ln[N_2O_5] - \ln[N_2O_5]_0 = -kt$$

$$\ln[N_2O_5] - \ln(0.400) = -(0.0163 \text{ min}^{-1})(100 \text{ min})$$

$$\ln[N_2O_5] + 0.916 = 1.63$$

$$\ln[N_2O_5] = -2.55$$

$$[N_2O_5] = e^{-2.55}\, M = 0.078\, M$$

(d) For a first-order reaction, Rate = $k[N_2O_5] = (0.0163 \text{ min}^{-1})(0.400\, M) = 0.00652\, M/\text{min}$.

(e) You can see from the numbers on the table that the half-life is slightly over 40 min. To calculate it exactly, use the formula $\text{Half-life} = \frac{0.693}{k} = \frac{0.693}{0.0163 \text{ min}^{-1}} = 42.5 \text{ min}$.

4. (a) (i) The reaction is first order. The half-life is constant; it takes 40 seconds for the concentration to drop from 0.20 *M* to 0.10 *M*, and another 40 seconds for it to drop from 0.10 *M* to 0.050 *M*. This type of exponential decay is typical of a first order reaction.

(ii) After t = 80 s, another 40 seconds is an additional half-life, and it would bring the concentration down to 0.025 *M* (half of 0.050 *M*).

(b) (i) We can use the stoichiometric ratios to determine the rate of appearance of P_4:

$$2.5 \times 10^{-3}\ M/\text{s PH}_3 \times \frac{1 \text{ mol P}_4}{4 \text{ mol PH}_3} = 6.3 \times 10^{-4}\ M/\text{s P}_4$$

(ii) As the reaction progresses, the rate at which the phosphine disappears will decrease, as indicated by the decreasing slope of the line. Thus, the rate of appearance of P_4 will also decrease.

(c) (i) The half-life of the phosphine will decrease. This is because at a higher temperature, the reaction will proceed faster. It will thus take less time for the phosphine to decay.

(ii) The rate law is unaffected by outside conditions and will remain unchanged.

(iii) The rate constant will increase. There is a directly proportional relationship between the rate constant and temperature. As the temperature increases, so will the rate constant.

5. (a) The rate of the reaction will increase because the rate depends on the concentration of A as given in the rate law: Rate = $k[A]^2$.

The rate constant is independent of the concentration of the reactants and will not change.

(b) The rate of the reaction will not change. If the reaction is zero order with respect to B, then the rate is independent of the concentration of B.

The rate constant is independent of the concentration of the reactants and will not change.

(c) The rate of the reaction will increase with increasing temperature because the rate constant increases with increasing temperature.

The rate constant increases with increasing temperature because at a higher temperature more gas molecules will collide with enough energy to overcome the activation energy for the reaction.

(d) Neither the rate nor the rate constant will be affected by the addition of an inert gas.

(e) The rate of the reaction will increase because decreasing the volume of the container will increase the concentration of A: Rate = $k[A]^2$.

The rate constant is independent of the concentration of the reactants and will not change.

6. (a) Line 2 is the catalyzed reaction. Adding a catalyst lowers the activation energy of the reaction, making it easier for the reaction to occur.

(b) Line 2 shows the higher temperature distribution. At a higher temperature, more of the molecules will be at higher energies, causing the distribution to flatten out and shift to the right.

(c) Line 1 is correct. The ln[reactant] for a first order reaction changes in a linear fashion over time, as shown in the following equation.

$$\ln[A]_t = -kt + \ln[A]_0$$

$$y = mx + b$$

Notice the similarity to the slope-intercept form for a linear equation.

(d) Graph 1 is correct, showing a decrease in the concentration of A as it is consumed in the reaction and a corresponding increase in the concentration of B as it is produced.

7. (a) An increase in temperature means an increase in the energy of the molecules present. If the molecules have more energy, then more of them will collide more often with enough energy to overcome the activation energy required for a reaction to take place, causing the reaction to proceed more quickly.

(b) A catalyst offers a reaction an alternative pathway with a lower activation energy. If the activation energy is lowered, then more molecular collisions will occur with enough energy to overcome the activation energy, causing the reaction to proceed more quickly.

(c) The effectiveness of a solid catalyst depends on the surface area of the catalyst that is exposed to the reactants. Grinding a solid into powder greatly increases its surface area.

(d) Increasing the concentration of reactants results in crowding the reactants more closely together, making it more likely that they will collide with one another. The more collisions that occur, the more likely that collisions that will result in reactions will occur.

8. (a) (i) $$\frac{(3.8\times10^{-5}\ M) - (2.6\times10^{-5}\ M)}{60\ s - 120\ s} = 2.0\times10^{-7}\ M/s$$

(ii) The solution starts reacting (and thus, fading) immediately after it is mixed. If the student waited 30 seconds before putting the cuvette in, the calculated rate of disappearance would thus decrease.

(b) $A = abc$

$A = (26{,}000\ \text{cm}/M)(1.00\ \text{cm})(3.8 \times 10^{-5}\ M)$

$A = 0.98$

(c) A reaction that is first order with respect to $[CV^+]$ will create a straight line in a graph of ln $[CV^+]$ vs. time.

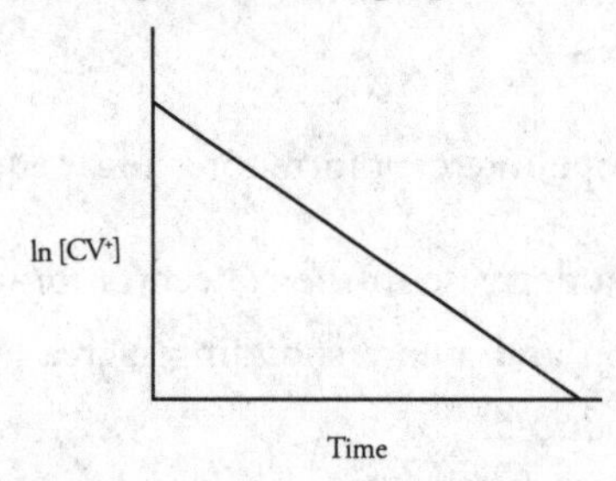

(d) Between trials 1 and 2, the value of the $[CV^+]$ remained constant while the $[OH^-]$ doubled. At the same time, the rate of reaction also doubled, meaning there is a direct relationship between $[OH^-]$ and the rate, so the reaction is first order with respect to $[OH^-]$. The rate with respect to $[CV^+]$ has already been established as first order, so:

rate = $k[CV^+][OH^-]$

(e) Either of the trials can be used for this. Taking trial 2:

$7.20 \times 10^{-7}\ M/s = k(5.5 \times 10^{-5}\ M)(0.24\ M)$

$k = 0.055\ M^{-1}s^{-1}$

(f) Rate = $k[CV^+][OH^-]$

Rate = $(0.055\ M^{-1}s^{-1})(4.1 \times 10^{-5}\ M)(0.18\ M)$

Rate = $4.1 \times 10^{-7}s^{-1}$

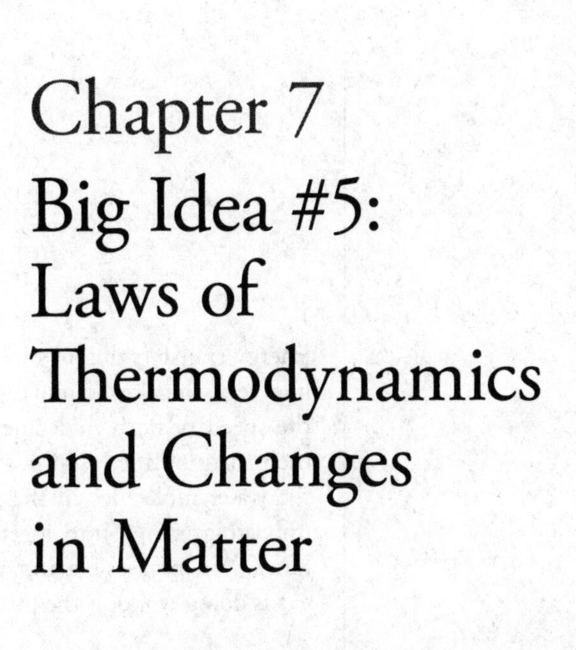

Chapter 7
Big Idea #5: Laws of Thermodynamics and Changes in Matter

The laws of thermodynamics describe the essential role of energy and explain and predict the direction of changes in matter.

ENERGY TRANSFER

Suppose we had two objects of different temperatures, one at 100°C and the other at 0°C, and we put the two in contact with one another. Spontaneously (that is, it happens on its own), heat will transfer from the object with the higher temperature to the one with the lower temperature. Over time, the colder object will exhibit a temperature rise, indicating that energy has transferred from the warmer object. This is an example of heat transfer.

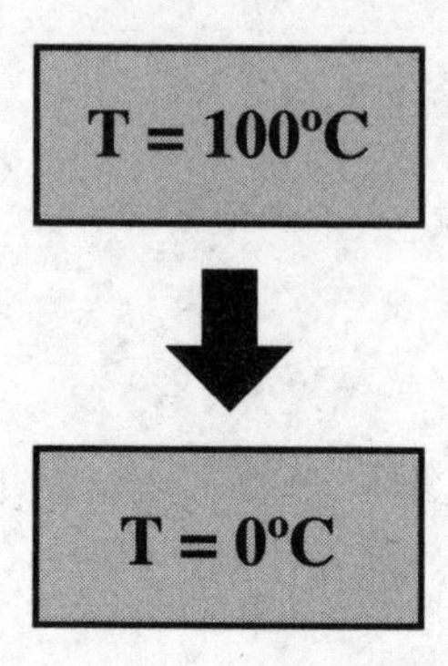

Energy transfers due to molecular collisions. As the faster moving molecules collide with the slower moving ones, they transfer some of their energy, changing the speed of both molecules. Energy can also be transferred from molecules to their surroundings. Think about a steam engine; as water is converted into steam, the water molecules in the steam will collide with the piston inside the engine, imparting energy into it. If enough energy is imparted, the piston will start to move. When gases expand and cause things like pistons to move, we say that the gas is doing work on the piston.

FIRST AND SECOND LAWS OF THERMODYNAMICS

The first law of thermodynamics says that the energy of the universe is constant. Energy can be neither created nor destroyed, so while energy can be *converted* in a chemical process, the total energy remains constant.

The second law of thermodynamics says that if a process is favored in one direction, then it can't be favored in the reverse direction, and an increase in entropy promotes favored reactions.

STATE FUNCTIONS

Enthalpy change (ΔH), entropy change (ΔS), and free-energy change (ΔG) are **state functions.** That means they all depend only on the change between the initial and final states of a system, not on the process by which the change occurs. For a chemical reaction, this means that the thermodynamic state functions are independent of the reaction pathway; for instance, the addition of a catalyst to a reaction will have no effect on the overall energy or entropy change of the reaction.

Standard State Conditions

When the values of thermodynamic quantities are given on the test, they are almost always given for standard state conditions. A thermodynamic quantity under standard state conditions is indicated by the little superscript circle, so the following is true under standard state conditions:

$$\Delta H = \Delta H^\circ$$
$$\Delta S = \Delta S^\circ$$
$$\Delta G = \Delta G^\circ$$

Standard State Conditions

- All gases are at 1 atmosphere pressure.
- All liquids are pure.
- All solids are pure.
- All solutions are at 1-molar (1 *M*) concentration.
- The energy of formation of an element in its normal state is defined as zero.
- The temperature used for standard state values is almost invariably room temperature: 25°C (298 K). Standard state values can be calculated for other temperatures, however.

HEAT OF FORMATION, ΔH°_f

Heat of formation is the change in energy that takes place when one mole of a compound is formed from its component pure elements under standard state conditions. Heat of formation is almost always calculated at a temperature of 25°C (298 K).

Remember: ΔH°_f for a pure element is defined as zero. This is even true for elements that, in their pure state, appear as diatomic molecules (such as oxygen and hydrogen).

- If ΔH°_f for a compound is negative, energy is released when the compound is formed from pure elements, and the product is *more* stable than its constituent elements. That is, the process is exothermic.
- If ΔH°_f for a compound is positive, energy is absorbed when the compound is formed from pure elements, and the product is *less* stable than its constituent elements. That is, the process is endothermic.

If the ΔH°_f values of the products and reactants are known, ΔH for a reaction can be calculated.

$$\Delta H^\circ = \Sigma\, \Delta H^\circ_f \text{ products} - \Sigma\, \Delta H^\circ_f \text{ reactants}$$

The ΔH for a reaction describes the energy change in that reaction when it goes to completion as written. On the AP Exam, the units for reaction enthalpy are kJ/mol_{rxn}. These units are often simply just written as kJ/mol, but either way they indicate the same quantity—the amount of heat released (negative ΔH) or absorbed (positive ΔH) when the reaction occurs with the mole quantities represented in the balanced equation. Let's find ΔH° for the reaction below.

$$2CH_3OH(g) + 3O_2(g) \rightarrow 2CO_2(g) + 4H_2O(g)$$

Compound	ΔH°_f (kJ/mol)
$CH_3OH(g)$	–201
$O_2(g)$	0
$CO_2(g)$	–394
$H_2O(g)$	–242

$$\Delta H^\circ = \Sigma\, \Delta H^\circ_f \text{ products} - \Sigma\, \Delta H^\circ_f \text{ reactants}$$
$$\Delta H^\circ = [(2)(\Delta H^\circ_f CO_2) + (4)(\Delta H^\circ_f H_2O)] - [(2)(\Delta H^\circ_f CH_3OH) + (3)(\Delta H^\circ_f O_2)]$$
$$\Delta H^\circ = [(2)(-394 \text{ kJ}) + (4)(-242 \text{ kJ})] - [(2)(-201 \text{ kJ}) + (3)(0 \text{ kJ})]$$
$$\Delta H^\circ = (-1{,}756 \text{ kJ}) - (-402 \text{ kJ})$$
$$\Delta H^\circ = -1{,}354 \text{ kJ/mol}_{rxn}$$

In the example above, when 2 moles of CH_3OH react with 3 moles of oxygen, exactly 1354 kJ of energy is released. However, when reactions occur, they are rarely in exact quantities like that. Instead, you have to combine stoichiometry concepts with thermodynamics to determine the energy change.

Example:

How much heat is released when 5.00 g of CH_3OH is combusted in excess oxygen?

In this case, the prompt tells us that CH_3OH is the limiting reactant. Otherwise, we would have to determine which reactant was limiting. The limiting reactant not only limits the amount of products formed, but it also limits the amount of heat change that occurs during the reaction.

Since we know the CH_3OH is limiting, we can set up some stoichiometry to get to our answer:

$$5.00 \text{ g } CH_3OH \times \frac{1 \text{ mol } CH_3OH}{32.05 \text{ g } CH_3OH} \times \frac{1 \text{ mol}_{rxn}}{2 \text{ mol } CH_3OH} \times \frac{-1354 \text{ kJ}}{1 \text{ mol}_{rxn}} = -106 \text{ kJ}$$

So, 106 kJ of energy is released (indicated by the negative sign). Don't forget to take mole ratios into account when doing thermodynamics problems that involve stoichiometry. As a general rule, whenever you are given specific amounts of reactants (either in grams or moles) and asked about the energy change for a reaction, you will have to use stoichiometry to get the correct answer.

BOND ENERGY

Bond energy is the energy required to break a bond. Because the breaking of a bond is an endothermic process, bond energy is always a positive number. When a bond is formed, energy equal to the bond energy is released.

$$\Delta H^\circ = \Sigma \text{ Bond energies of bonds broken} - \Sigma \text{ Bond energies of bonds formed}$$

The bonds broken will be the reactant bonds, and the bonds formed will be the product bonds.

The number of bonds broken and formed is affected by both the number of that particular type of bond within a molecule, as well as how many of those molecules there are in a balanced reaction. For instance, in the equation below, there are four O–H bonds being formed because each water molecule has two O–H bonds, and there are two water molecules present.

Let's find ΔH° for the following reaction.

$$2H_2(g) + O_2(g) \rightarrow 2H_2O(g)$$

Bond	Bond Energy (kJ/mol)
H–H	436
O=O	499
O–H	463

$\Delta H° = \Sigma$ Bond energies of bonds broken – Σ Bond energies of bonds formed

$$\Delta H° = [(2)(\text{H–H}) + (1)(\text{O=O})] - [(4)(\text{O–H})]$$
$$\Delta H° = [(2)(436 \text{ kJ}) + (1)(499 \text{ kJ})] - [(4)(463 \text{ kJ})]$$
$$\Delta H° = (1{,}371 \text{ kJ}) - (1{,}852 \text{ kJ})$$
$$\Delta H° = -481 \text{ kJ/mol}$$

HESS'S LAW

Hess's law states that if a reaction can be described as a series of steps, then ΔH for the overall reaction is simply the sum of the ΔH values for all the steps.

When manipulating equations for use in enthalpy calculations, there are three rules:

1. If you flip the equation, flip the sign on the enthalpy value.
2. If you multiply or divide an equation by an integer, also multiply/divide the enthalpy value by that same integer.
3. If several equations, when summed up, create a new equation, you can also add the enthalpy values of those component equations to get the enthalpy value for the new equation.

For example, let's say you want to calculate the enthalpy change for the following reaction:

$$4NH_3(g) + 5O_2(g) \rightarrow 4NO(g) + 6H_2O(g)$$

Given:

Equation A: $N_2(g) + O_2(g) \rightarrow 2NO(g)$ $\quad \Delta H = 180.6$ kJ/mol
Equation B: $N_2(g) + 3H_2(g) \rightarrow 2NH_3(g)$ $\quad \Delta H = -91.8$ kJ/mol
Equation C: $2H_2(g) + O_2(g) \rightarrow 2H_2O(g)$ $\quad \Delta H = -483.7$ kJ/mol

First, we want to make sure all of the equations have the products and reactants on the correct side. A quick glance shows us the NH_3 should end up on the left but is given to us on the right, so equation B needs to be flipped, which will change the ΔH value to 91.8 kJ/mol. All other species appear to be on the side they need to be on, so we'll leave the other two reactions alone.

Next, we're going to see what coefficients we need to get to. The NO needs to have a 4 but is only 2 in equation A, so we'll multiply that equation by 2, which will change the ΔH value to 180.6 × 2 = 361.2 kJ/mol. The NH_3 also needs to have a 4 but is only 2 in equation B, so we'll also multiply that equation by 2, changing ΔH to 91.8 * 2 = 183.6 kJ/mol. Finally, the H_2O needs to have a 6 but only has a 2 in equation C. We now have:

$2N_2(g) + 2O_2(g) \rightarrow 4NO(g)$	$\Delta H = 361.2$ kJ/mol
$4NH_3(g) \rightarrow 2N_2(g) + 6H_2(g)$	$\Delta H = 183.6$ kJ/mol
$6H_2(g) + 3O_2(g) \rightarrow 6H_2O(g)$	$\Delta H = -1451$ kJ/mol

Fortunately, the two O_2 on the products side add up to 5, and both the N_2 and H_2 cancel out completely, giving us the final equation that we wanted. So, our final value for the enthalpy will be 361.2 + 183.6 – 1451 = –906 kJ/mol.

ENTHALPY OF SOLUTION

When an ionic substance dissolves in water, a certain amount of heat is emitted or absorbed. The bond between the cation and anion is breaking, which requires energy, but energy is released when those ions form new attractions to the dipoles of the water molecules. Typically, this process can be broken down into three steps. We will use NaCl dissolving in water for this example.

Step 1: Breaking the Solute Bonds

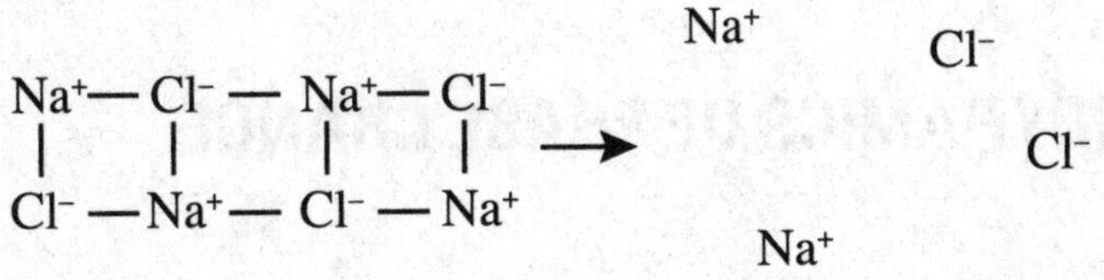

$Na^+ — Cl^- — Na^+ — Cl^-$
$Cl^- — Na^+ — Cl^- — Na^+$
→ Na^+ Cl^- Cl^- Na^+

The bonds between the Na^+ and Cl^- ions must be broken. The amount of energy needed to break this bond is equal to the lattice energy (see **page 106** for a more detailed explanation on lattice energy). As energy is required to break the bonds, this step always has a positive ΔH.

Step 2: Separating the Solvent Molecules

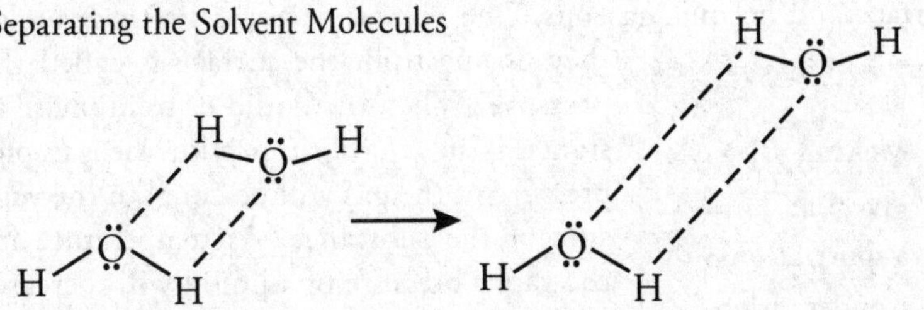

The water molecules must spread out to make room for the Na^+ and Cl^- ions. This requires energy to weaken (but not break) the intermolecular forces between them, so this step always has a positive ΔH.

Step 3: Creating New Attractions

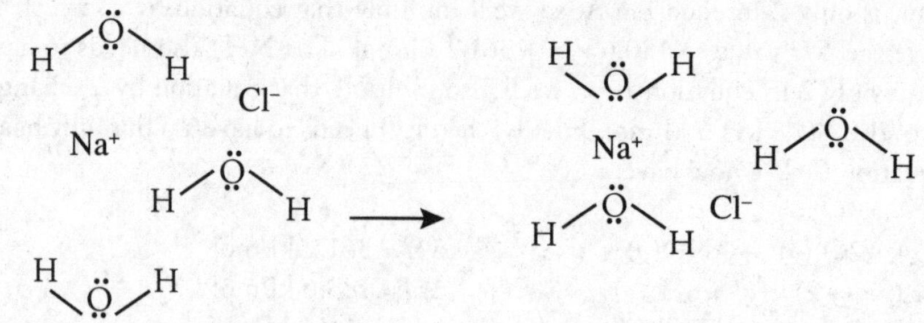

The last step involves the free-floating ions being attracted to the dipoles of the water molecules. Note that while bonds are not being formed (no electrons are transferring or being shared), energy is still released in this process. As such, this step always has a negative ΔH.

The energy values from step 2 and step 3 combined are often called the **hydration energy.** This value is always negative, as the magnitude of the energy change in step 3 exceeds the magnitude of the energy change in step 2. As with lattice energy, hydration energy is a Coulombic energy and thus increases as the ions either increase in charge (i.e. $Mg^{2+} > Na^{+}$) or decrease in size (i.e. $Na^{+} > K^{+}$).

If you add the energy values for all three steps together, you can determine the enthalpy of solution for that compound. Thus, if the hydration energy exceeds the lattice energy, the enthalpy of solution is negative. However, if the lattice energy exceeds the hydration energy, the enthalpy of solution is positive.

THERMODYNAMICS OF PHASE CHANGE

Naming the Phase Changes

Phase changes occur because of changes in temperature and/or pressure. Phase changes are always physical changes, as they do not involve breaking any bonds or the creation of new substances.

Some particles in a liquid or solid will have enough energy to break away from the surface and become gaseous. The pressure exerted by these molecules as they escape from the surface is called the **vapor pressure**. When the liquid or solid phase of a substance is in equilibrium with the gas phase, the pressure of the gas will be equal to the vapor pressure of the substance. As temperature increases, the vapor pressure of a liquid will increase. When the vapor pressure of a liquid increases to the point where it is equal to the surrounding atmospheric pressure, the liquid boils.

Solid to *liquid*	—	Melting
Liquid to *solid*	—	Freezing
Liquid to *gas*	—	Vaporization
Gas to *liquid*	—	Condensation
Solid to *gas*	—	Sublimation
Gas to *solid*	—	Deposition

Heat of Fusion

The **heat of fusion** is the energy that must be put into a solid to melt it. This energy is needed to overcome the forces holding the solid together. Alternatively, the heat of fusion is the heat given off by a substance when it freezes. The intermolecular forces within a solid are more stable and, therefore, have lower energy than the forces within a liquid, so energy is released in the freezing process.

Heat of Vaporization

The **heat of vaporization** is the energy that must be put into a liquid to turn it into a gas. This energy is needed to overcome the forces holding the liquid together. Alternatively, the heat of vaporization is the heat given off by a substance when it condenses. Intermolecular forces become stronger when a gas condenses; the gas becomes a liquid, which is more stable, and energy is released.

As heat is added to a substance, the temperature of the substance can increase or the substance can change phases, but both changes cannot occur simultaneously. Therefore, when a substance is changing phases, the temperature of that substance remains constant.

Phase Diagram for Water

A phase diagram tells you the state of a substance at various pressure/temperature combinations. The phase diagram for water is below.

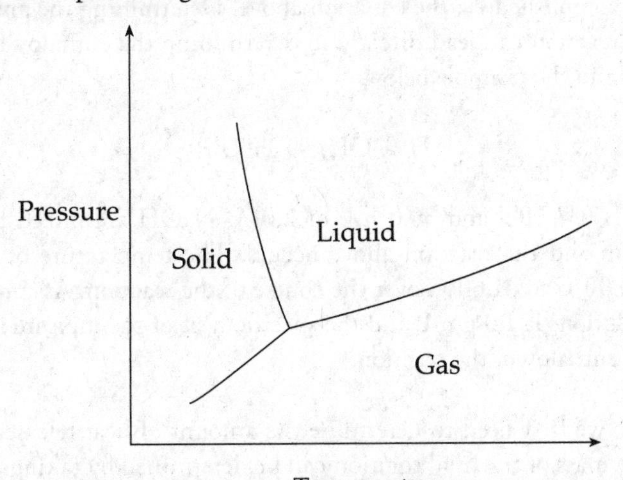

In the phase diagram for substances other than water, the solid-liquid equilibrium line slopes upward. In the phase diagram for water, the solid-liquid equilibrium line slopes *downward.* What this means is that when pressure is increased, a normal substance will change from liquid to solid, but water will change from solid to liquid.

Water has this odd property because its hydrogen bonds form a lattice structure when it freezes. This forces the molecules to remain farther apart in ice than in water, making the solid phase less dense than the liquid phase. That's why ice floats on water, and this natural phenomenon preserves most marine life.

CALORIMETRY

The specific heat of a substance is the amount of heat required to raise the temperature of one gram (or kilogram) of that substance by one degree Celsius (or one Kelvin). An object with a large **specific heat** can absorb a lot of heat without undergoing much of a temperature change, while a substance with a low specific heat will experience much greater temperature changes.

The easiest way to think about this is by considering the temperature changes you experience when wearing a black shirt as opposed to a white shirt on a hot summer day. You are absorbing the same amount of heat from the Sun regardless of which shirt you are wearing. However, the black shirt has a considerably lower specific heat than the white shirt, and thus will experience a greater temperature change—making you a lot hotter!

Specific Heat

$$q = mc\Delta T$$

q = heat added (J or cal)
m = mass of the substance (g or kg)
c = specific heat
ΔT = temperature change (K or °C)

Calorimetry is the measurement of heat changes during chemical reactions, and is frequently accomplished via the equation above. Determining the amount of heat transfer in a reaction can lead directly to determining the enthalpy for that reaction, as shown in the example below.

$$H^+ + OH^- \rightarrow H_2O(l)$$

25.0 mL of 1.5 *M* HCl and 30.0 mL of 2.0 *M* NaOH are mixed together in a styrofoam cup and the reaction above occurs. The temperature of the reaction rises from 23.00°C to 31.60°C over the course of the reaction. Assuming the density of the solutions is 1.0 g/mL and the specific heat of the mixture is 4.18 J/g°C, calculate the enthalpy of the reaction.

To solve this, we first need to determine the amount of heat released during the reaction. The mass of the final solution can be determined by taking the total volume of the solution (55.0 mL) and multiplying that by the density, yielding 55.0g. The temperature change is 31.60 – 23.00 = 8.60°C. So:

$$q = mc\Delta T$$
$$q = (55.0\text{g})(4.18\ \text{J/g°·C})(8.60\text{°C})$$
$$q = 1970\ \text{J}$$

From here, we need to determine how many moles of product are formed in this reaction. Looking at the two reactants, we can see that there is going to be less of the HCl, which has a lower molarity and volume. As everything here is in a 1:1

ratio, that means the HCl will be limiting and we can use it to determine how many moles of product will form.

$$\text{Molarity} = \text{mol/volume}$$

$$1.5\ M = n/0.025\ \text{L}$$

$$n = 0.038\ \text{mol HCl} = 0.038\ \text{mol}\ H_2O\ \text{formed}$$

The calculated heat gain from earlier (1970 J) was the heat gained by the water. Due to conservation of energy, this is also the heat lost by the reaction itself. To calculate the enthalpy of reaction, we have to flip that sign to indicate heat is lost, and then divide that value by the number of moles.

$$\Delta H = -1970\ \text{J}/0.038\ \text{mol} = -52{,}000\ \text{J/mol} = -52\ \text{kJ/mol}$$

It is important to emphasize that ΔH is always measured in joules per mole (or kilojoules per mole). Enthalpy is not just heat; it's the amount of heat given per mole of product that is created. Limiting reagent calculations can be done in situations where it is not easy to determine which reactant is limiting by inspection, as we did above.

Also, bear in mind that if you completely dissolve a solid into water, you must add the mass of that solid to the mass of the water when doing your calculations. So, if you were to dissolve 5.0 g of NaCl into 200.0 mL of water (remember, the density of water is 1.0 g/mL), the final mass of the solution would be 205.0 g.

HEATING CURVES

A heating (or cooling) curve shows what happens to the temperature of a substance as heat is added. If the substance is in a single phase, the temperature of the substance will increase. The amount of the temperature increase can be calculated using calorimetry. If the substance is in the process of a phase change, the temperature will remain constant and the amount of heat required to fully cause the substance to change phases can be calculated using the heat of fusion or the heat of vaporization.

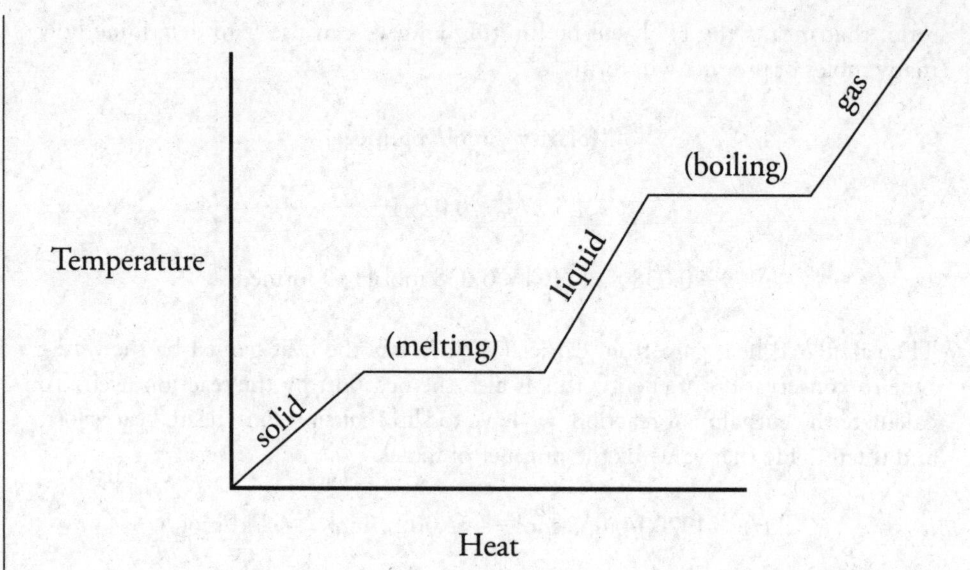

There are a lot of different units that can be used when talking about these heat versus temperature curves, so it is very important to pay extremely close attention to what your units are when doing calculations.

Example:

A 1.53 g piece of ice is in a freezer and initially at a temperature of –15.1°C. The ice is removed from the freezer and melts completely after reaching a temperature of 0.0°C. If the specific heat of ice is 2.03 J/g°C and its molar heat of fusion is 6.01 kJ/mol, how much heat is required for the entire process to occur?

There are two parts to this problem. The first part involves the temperature change of the ice.

$$q = mc\Delta T$$
$$q = (1.53\text{g})(2.03\ \text{J/g°C})(15.1\text{°C})$$
$$q = 46.9\ \text{J}$$

The second part involves the ice melting. As this is a molar heat capacity, the number of moles of ice must first be determined.

$$1.53\ \text{g H}_2\text{O} \times \frac{1\ \text{mol H}_2\text{O}}{18.01\ \text{g}} = 0.0850\ \text{mol H}_2\text{O}$$

$$6.01\ \text{kJ/mol} \times 0.0850\ \text{mol} = 0.511\ \text{kJ} = 511\text{J}$$

Finally, the two values can be added together.

$$46.9\ \text{J} + 511\ \text{J} = 558\ \text{J}$$

ENTROPY

The entropy, S, of a system is a measure of the randomness or disorder of the system; the greater the disorder of a system, the greater its entropy. Because zero entropy is defined as a solid crystal at 0 K, and because 0 K has never been reached experimentally, all substances that we encounter will have some positive value for entropy. Standard entropies, $S°$, are calculated at 25°C (298 K).

The standard entropy change, $\Delta S°$, that has taken place at the completion of a reaction is the difference between the standard entropies of the products and the standard entropies of the reactants.

$$\Delta S° = \Sigma S°_{\text{products}} - \Sigma S°_{\text{reactants}}$$

You should be familiar with several simple rules concerning entropies.

- Liquids have higher entropy values than solids.
- Gases have higher entropy values than liquids.
- Particles in solution have higher entropy values than solids.
- Two moles of a gaseous substance have a higher entropy value than one mole of a gaseous substance.

GIBBS FREE ENERGY

The **Gibbs free energy**, or simply free energy, G, of a process is a measure of whether or not the process will proceed without the input of outside energy. A process that occurs without outside energy input is said to be thermodynamically favored, while one that does not is said to be thermodynamically unfavored. Prior to the 2014 exam, the terms *spontaneous* and *nonspontaneous* were used to described thermodynamically favored and unfavored reactions, respectively. It would be good for you to be familiar with both sets of terms, although this text will use the updated terms.

FREE ENERGY CHANGE, ΔG

The standard free energy change, ΔG, for a reaction can be calculated from the standard free energies of formation, $G°_f$, of its products and reactants in the same way that $\Delta S°$ was calculated.

$$\Delta G° = \Sigma \Delta G°_{f\ \text{products}} - \Sigma \Delta G°_{f\ \text{reactants}}$$

For a given reaction:

- if ΔG is negative, the reaction is thermodynamically favored
- if ΔG is positive, the reaction is thermodynamically unfavored
- if $\Delta G = 0$, the reaction is at equilibrium

ΔG, ΔH, AND ΔS

In general, nature likes to move toward two different and seemingly contradictory states—low energy and high disorder, so thermodynamically favored processes must result in decreasing enthalpy or increasing entropy or both.

There is an important equation that relates favorability (ΔG), enthalpy (ΔH), and entropy (ΔS) to one another.

Nothing To See Here
Note that for all phase changes, no chemical reaction is occurring; therefore, the ΔG for any phase change is 0.

$$\Delta G° = \Delta H° - T\Delta S°$$
$$T = \text{absolute temperature (K)}$$

Note that you should make sure your units always match up here. Frequently, ΔS is given in J/mol•K and ΔH is given in kJ/mol. The convention is to convert the $T\Delta S$ term to kilojoules (kJ), as that is what Gibbs free energy is usually measured in. However, you can also convert the ΔH term to joules instead, so long as you are keeping track and labeling all units appropriately.

The chart below shows how different values of enthalpy and entropy affect spontaneity.

ΔH	ΔS	T	ΔG	
–	+	Low	–	Always favored
		High	–	
+	–	Low	+	Never favored
		High	+	
+	+	Low	+	Not favored at low temperature
		High	–	Favored at high temperature
–	–	Low	–	Favored at low temperature
		High	+	Not favored at high temperature

You should note that at low temperature, enthalpy is dominant, while at high temperature, entropy is dominant.

VOLTAGE AND FAVORABILITY

A redox reaction will be favored if its potential has a positive value. We also know from thermodynamics that a reaction that is favored has a negative value for free-energy change. The relationship between reaction potential and free energy for a redox reaction is given by the equation below, which serves as a bridge between thermodynamics and electrochemistry.

$$\Delta G° = -nFE°$$

$\Delta G°$ = Standard Gibbs free energy change (J/mol)

n = the number of moles of electrons exchanged in the reaction (mol)

F = Faraday's constant, 96,500 coulombs/mole (that is, 1 mole of electrons has a charge of 96,500 coulombs)

$E°$ = Standard reaction potential (V, which is equivalent to J/C)

From this equation we can see a few important things. If $E°$ is positive, $\Delta G°$ is negative and the reaction is thermodynamically favored, and if $E°$ is negative, $\Delta G°$ is positive and the reaction is thermodynamically unfavored.

Let's take a look at an example.

Calculate the ΔG value for the below reaction under standard conditions:

$$Zn(s) + 2Ag^+ \rightarrow Zn^{2+} + 2Ag(s) \qquad E° = +1.56 \text{ V}$$

For this reaction, two moles of electrons are being transferred as the silver is reduced and the zinc is oxidized. So, $n = 2$. With that in mind:

$$\Delta G° = -(2)(96500)(1.56)$$

$$\Delta G° = -301{,}000 \text{ J/mol}$$

As the $\Delta G°$ value is negative, we would expect this reaction to be favored under standard conditions, which is in line with the positive value for the cell.

Note that your units on the free energy are in J/mol and not kJ/mol; this will always be the case with this equation as one volt is equivalent to one joule/coulomb, so the answer comes out in joules instead of kilojoules.

CHAPTER 7 QUESTIONS

Multiple-Choice Questions

Use the following information to answer questions 1-5.

Reaction 1: $N_2H_4(l) + H_2(g) \rightarrow 2NH_3(g)$	$\Delta H = ?$
Reaction 2: $N_2H_4(l) + CH_4O(l) \rightarrow CH_2O(g) + N_2(g) + 3H_2(g)$	$\Delta H = -37\ kJ/mol_{rxn}$
Reaction 3: $N_2(g) + 3H_2(g) \rightarrow 2NH_3(g)$	$\Delta H = -46\ kJ/mol_{rxn}$
Reaction 4: $CH_4O(l) \rightarrow CH_2O(g) + H_2(g)$	$\Delta H = -65\ kJ./mol_{rxn}$

1. What is the enthalpy change for reaction 1?

   (A) $-148\ kJ/mol_{rxn}$
   (B) $-56\ kJ/mol_{rxn}$
   (C) $-18\ kJ/mol_{rxn}$
   (D) $+148\ kJ/mol_{rxn}$

2. If reaction 2 were repeated at a higher temperature, how would the reaction's value for ΔG be affected?

   (A) It would become more negative because entropy is a driving force behind this reaction.
   (B) It would become more positive because the reactant molecules would collide more often.
   (C) It would become more negative because the gases will be at a higher pressure.
   (D) It will stay the same; temperature does not affect the value for ΔG.

3. Under what conditions would reaction 3 be thermodynamically favored?

   (A) It is always favored.
   (B) It is never favored.
   (C) It is only favored at low temperatures.
   (D) It is only favored at high temperatures.

4. If 64 g of CH_4O were to decompose via reaction 4, approximately how much energy would be released or absorbed?

   (A) 65 kJ of energy will be absorbed.
   (B) 65 kJ of energy will be released.
   (C) 130 kJ of energy will be absorbed.
   (D) 130 kJ of energy will be released.

5. Which of the following is true for all of the reactions?

   (A) The entropy increases.
   (B) If completed inside a sealed container, the pressure will decrease as the reactions go to completion.
   (C) The activation energy for all four reactions will decrease with increasing temperature.
   (D) More energy is released when the bonds in the products form than is necessary to break down the bonds of the reactants.

6. $$2Al(s) + 3Cl_2(g) \rightarrow 2AlCl_3(s)$$

The reaction above is not thermodynamically favored under standard conditions, but it becomes thermodynamically favored as the temperature decreases toward absolute zero. Which of the following is true at standard conditions?

(A) ΔS and ΔH are both negative.
(B) ΔS and ΔH are both positive.
(C) ΔS is negative, and ΔH is positive.
(D) ΔS is positive, and ΔH is negative.

7. 1.50 g of $NaNO_3$ is dissolved into 25.0 mL of water, causing the temperature to increase by 2.2°C. The density of the final solution is found to be 1.02 g/mL. Which of the following expressions will correctly calculate the heat gained by the water as the $NaNO_3$ dissolves? Assume the volume of the solution remains unchanged.

(A) (25.0)(4.18)(2.2)

(B) $\frac{(26.5)(4.18)(2.2)}{1.02}$

(C) $\frac{(1.02)(4.18)(2.2)}{1.50}$

(D) (25.0)(1.02)(4.18)(2.2)

8. Water can be broken down into hydrogen and oxygen gas if sufficient current is run through it via an inert platinum electrode. What must be true about the values of E°_{cell} and ΔG for the electrolysis of water?

	E°_{cell}	ΔG
(A)	Positive	Positive
(B)	Positive	Negative
(C)	Negative	Positive
(D)	Negative	Negative

9. $C(s) + 2H_2(g) \rightarrow CH_4(g)$ $\Delta H^\circ = x$

$C(s) + O_2(g) \rightarrow CO_2(g)$ $\Delta H^\circ = y$

$H_2(g) + \frac{1}{2}O_2(g) \rightarrow H_2O(l)$ $\Delta H^\circ = z$

Based on the information given above, what is ΔH° for the following reaction?

$CH_4(g) + 2O_2(g) \rightarrow CO_2(g) + 2H_2O(l)$

(A) $x + y + z$
(B) $x + y - z$
(C) $z + y - 2x$
(D) $2z + y - x$

10. A student studies two solutions. Solution A has a volume of 100 mL and is at a temperature of 25.0°C. Solution B has a volume of 1000 mL and is at a temperature of 22.0°C. Which of the following statements must be true regarding both solutions?

(A) Solution A has more heat than solution B.
(B) The specific heat capacity of solution A is greater than that of solution B.
(C) If the solutions were to be mixed, heat would transfer from B to A.
(D) Solution B has more thermal energy than solution A.

11. $$2H_2(g) + O_2(g) \rightarrow 2H_2O(g)$$

Based on the information given in the table below, what is $\Delta H°$ for the above reaction?

Bond	Average bond energy (kJ/mol)
H–H	500
O=O	500
O–H	500

(A) –2,000 kJ
(B) –500 kJ
(C) +1,000 kJ
(D) +2,000 kJ

12. $$H_2O(l) \rightarrow H_2O(s)$$

Which of the following is true for the above reaction?

(A) The value for ΔS is positive.
(B) The value for ΔG is zero.
(C) The value for ΔH is negative.
(D) The reaction is favored at 1.0 atm and 298 K.

Use the following diagram to answer questions 13-15.

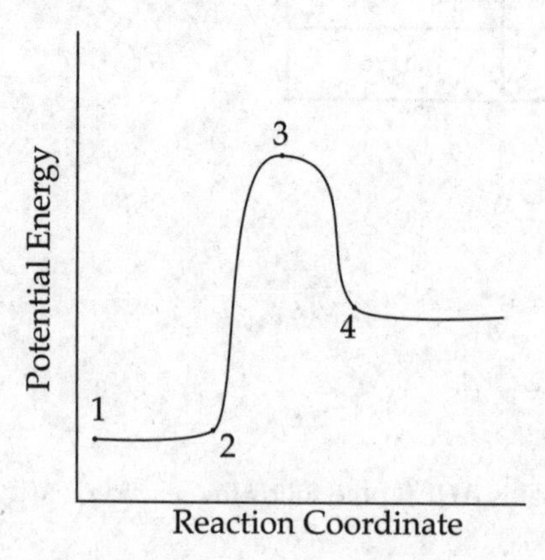

13. Which point on the graph shown above corresponds to activated complex or transition state?

(A) 1
(B) 2
(C) 3
(D) 4

14. The distance between which two points is equal to the enthalpy change for this reaction?

(A) Points 1 and 2
(B) Points 1 and 3
(C) Points 1 and 4
(D) Points 2 and 3

15. What would happen to this graph if a catalyst were to be added?

(A) Point 3 would be lower.
(B) The distance between points 2 and 4 would be decreased.
(C) The slope of the line between points 2 and 3 would increase.
(D) Point 4 would be higher.

16.

$$C(s) + O_2(g) \rightarrow CO_2(g) \qquad \Delta H° = -390 \text{ kJ/mol}$$

$$H_2(g) + \frac{1}{2} O_2(g) \rightarrow H_2O(l) \qquad \Delta H° = -290 \text{ kJ/mol}$$

$$2C(s) + H_2(g) \rightarrow C_2H_2(g) \qquad \Delta H° = +230 \text{ kJ/mol}$$

Based on the information given above, what is $\Delta H°$ for the following reaction?

$$C_2H_2(g) + \frac{5}{2} O_2(g) \rightarrow 2CO_2(g) + H_2O(l)$$

(A) –1,300 kJ
(B) –1,070 kJ
(C) –840 kJ
(D) –780 kJ

17. In which of the following reactions is entropy increasing?

(A) $2SO_2(g) + O_2(g) \rightarrow 2SO_3(g)$
(B) $CO(g) + H_2O(g) \rightarrow H_2(g) + CO_2(g)$
(C) $H_2(g) + Cl_2(g) \rightarrow 2HCl(g)$
(D) $2NO_2(g) \rightarrow 2NO(g) + O_2(g)$

18. When pure sodium is placed in an atmosphere of chlorine gas, the following reaction occurs without the addition of additional energy:

$$Na(s) + Cl_2(g) \rightarrow 2NaCl(s)$$

Which of the following correctly identifies the values for ΔG, ΔS, and ΔH?

(A) $G > 0, S > 0, H < 0$
(B) $G < 0, S < 0, H < 0$
(C) $G < 0, S < 0, H > 0$
(D) $G > 0, S < 0, H < 0$

19.

$$H_2(g) + F_2(g) \rightarrow 2HF(g)$$

Gaseous hydrogen and fluorine combine in the reaction above to form hydrogen fluoride with an enthalpy change of –540 kJ. What is the value of the heat of formation of HF(*g*)?

(A) –1,080 kJ/mol
(B) –270 kJ/mol
(C) 270 kJ/mol
(D) 540 kJ/mol

Use the following information to answer questions 20-23.

When calcium chloride ($CaCl_2$) dissolves in water, the temperature of the water increases dramatically.

20. Which of the following must be true regarding the enthalpy of solution?

(A) The lattice energy in $CaCl_2$ exceeds the bond energy within the water molecules.
(B) The hydration energy between the water molecules and the solute ions exceeds the lattice energy within $CaCl_2$.
(C) The strength of the intermolecular forces between the solute ions and the dipoles on the water molecules must exceed the hydration energy.
(D) The hydration energy must exceed the strength of the intermolecular forces between the water molecules.

21. During this reaction, heat transfers from

(A) the reactants to the products
(B) the reactants to the system
(C) the system to the surroundings
(D) the products to the surroundings

22. Which is the primary driving factor behind this reaction?

(A) Entropy
(B) Enthalpy
(C) Both enthalpy and entropy
(D) Neither enthalpy nor entropy

23. Compared to $CaCl_2$, what must be true regarding the hydration energy of CaF_2?

(A) It would be greater because fluoride is smaller than chloride.
(B) It would be the same because the charges of fluoride and chloride are identical.
(C) It would be the same because hydration energy is only dependent on the IMFs present in water.
(D) It would be smaller because the molar mass of CaF_2 is smaller than that of $CaCl_2$.

24.

$$2S(s) + 3O_2(g) \rightarrow 2SO_3(g)$$
$$\Delta H = +800 \text{ kJ/mol}$$

$$2SO_3(g) \rightarrow 2SO_2(g) + O_2(g)$$
$$\Delta H = -200 \text{ kJ/mol}$$

Based on the information given above, what is ΔH for the following reaction?

$$S(s) + O_2(g) \rightarrow SO_2(g)$$

(A) 300 kJ
(B) 500 kJ
(C) 600 kJ
(D) 1000 kJ

Free-Response Questions

1.

Substance	Absolute Entropy, $S°$ (J/mol•K)	Molar Mass (g/mol)
$C_6H_{12}O_6(s)$	212.13	180
$O_2(g)$	205	32
$CO_2(g)$	213.6	44
$H_2O(l)$	69.9	18

Energy is released when glucose is oxidized in the following reaction, which is a metabolism reaction that takes place in the body.

$$C_6H_{12}O_6(s) + 6O_2(g) \rightarrow 6CO_2(g) + 6H_2O(l)$$

The standard enthalpy change, $\Delta H°$, for the reaction is –2,801 kJ/mol_{rxn} at 298 K.

(a) Calculate the standard entropy change, $\Delta S°$, for the oxidation of glucose.
(b) Calculate the standard free energy change, $\Delta G°$, for the reaction at 298 K.
(c) Using the axis below, draw an energy profile for the reaction and indicate the magnitude of ΔH.

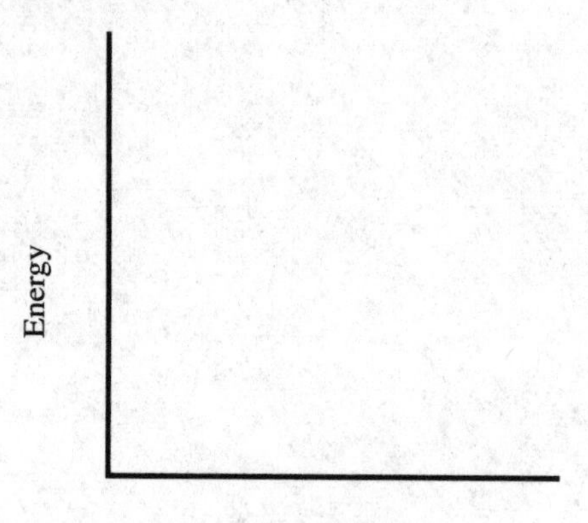

Reaction Progress

(d) How much energy is given off by the oxidation of 1.00 gram of glucose?

2.

Bond	Average Bond Dissociation Energy (kJ/mol)
C–H	415
O=O	495
C=O	799
O–H	463

$$CH_4(g) + 2O_2(g) \rightarrow CO_2(g) + 2H_2O(g)$$

The standard free energy change, $\Delta G°$, for the reaction above is –801 kJ/mol$_{rxn}$ at 298 K.

(a) Use the table of bond dissociation energies to find $\Delta H°$ for the reaction above.
(b) How many grams of methane must react with excess oxygen in order to release 1500 kJ of heat?
(c) What is the value of $\Delta S°$ for the reaction at 298 K?
(d) Give an explanation for the size of the entropy change found in (c).

3.

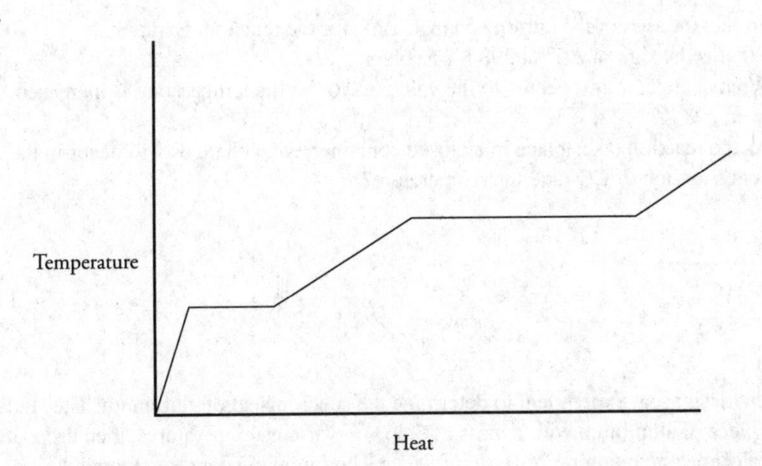

The diagram above shows how the temperature of a certain covalent substance changes as heat is added to it.

(a) Which is greater for the substance: the heat of fusion or the heat of vaporization? How do you know?
(b) If additional heat is added to the substance, the line would continue at its current slope and never become horizontal again. Why is this?
(c) Reading the graph above, a student theorizes that the specific heat capacity of the substance is greatest in the solid phase. Do you agree? Why or why not?
(d) What would the signs be for the enthalpy and entropy changes as heat continues to be added? Justify your answer.

4. $$CH_3OH\ (l) \rightarrow CH_3OH\ (g)$$

For the boiling of methanol, CH_3OH, $\Delta H = {}^{+}37.6$ kJ/mol and $\Delta S = +111$ J/mol•K.

(a) (i) Why is the ΔH value positive for this process?
(ii) Why is the ΔS value positive for this process?
(b) What is the boiling point of methanol in degrees Celsius?
(c) How much heat is required to boil 50.0 mL of ethanol if the density of ethanol is 0.789 g/mL?
(d) What will happen to the temperature of the methanol as it boils? Explain.
(e) Would methanol be soluble with water? Why or why not?
(f) Would you expect the boiling point of ethanol, CH_3CH_2OH, to be less than, greater than, or the same as methanol? Justify your answer.

5. $$CaO(s) + CO_2(g) \rightarrow CaCO_3(s)$$

The reaction above is thermodynamically favored at 298 K, and the heat of reaction, $\Delta H°$, is -178 kJ/mol$_{rxn}$.

(a) Predict the sign of the entropy change, $\Delta S°$, for the reaction. Explain.
(b) What is the sign of $\Delta G°$ at 298 K? Explain.
(c) What change, if any, occurs to the value of $\Delta G°$ as the temperature is increased from 298 K?
(d) As the reaction takes place in a closed container, what changes will occur in the concentration of CO_2 and the temperature?

6. A student designs an experiment to determine the specific heat of aluminum. The student heats a piece of aluminum with a mass of 5.86 g to various temperatures, then drops it into a calorimeter containing 25.0 mL of water. The following data is gathered during one of the trials:

Initial Temperature of Al (°C)	Initial Temperature of H_2O (°C)	Final Temperature of Al + H_2O (°C)
109.1	23.2	26.8

(a) Given that the specific heat of water is 4.18 J/g·°C and assuming its density is exactly 1.00 g/mL, calculate the heat gained by the water.
(b) Calculate the specific heat of aluminum from the experimental data given.
(c) Calculate the enthalpy change for the cooling of aluminum in water in kJ/mol.
(d) If the accepted specific heat of aluminum is 0.900 J/g·°C, calculate the percent error.
(e) Suggest two potential sources of error that would lead the student's experimental value to be different from the actual value. Be specific in your reasoning and make sure any identified error can be quantitatively tied to the student's results.

CHAPTER 7 ANSWERS AND EXPLANATIONS

Multiple-Choice

1. C To get reaction 1, all that is needed is to flip reaction 4 and then add it to reactions 2 and 3.

$N_2H_4(l) + CH_4O(l) \rightarrow CH_2O(g) + N_2(g) + 3H_2(g)$
$N_2(g) + 3H_2(g) \rightarrow 2NH_3(g)$
$CH_2O(g) + H_2(g) \rightarrow CH_4O(l)$

$(-37\ \text{kJ/mol}_{rxn}) + (-46\ \text{kJ/mol}_{rxn}) + (65\ \text{kJ/mol}_{rxn}) = -18\ \text{kJ/mol}_{rxn}$

2. A Entropy is positive during reaction 2, as creating gas molecules out of liquid molecules demonstrates an increase in disorder. Using $\Delta G = \Delta H - T\Delta S$, if the temperature increases, the $T\Delta S$ term will also increase. As the overall value for that term is negative, increasing $T\Delta S$ makes ΔG more negative.

3. C Use the $\Delta G = \Delta H - T\Delta S$ equation. To do that, determine the signs of both ΔH and the $T\Delta S$ term.

We know that ΔH is negative. ΔS is negative as well, because the reaction is going from four molecules to two molecules, meaning it is becoming more ordered. That, in turn, means the $T\Delta S$ term is positive.

Reactions are only favored when ΔG is negative. As temperature increases, the $T\Delta S$ term becomes larger and more likely to be greater in magnitude than the ΔH term. If the temperature remains low, the $T\Delta S$ value is much more likely to be smaller in magnitude than the ΔH value, meaning ΔG is more likely to be negative and the reaction will be favored.

4. D The molar mass of CH_4O is 32 g/mol, so 64 g of it represents two moles. For every one mole of CH_4O that decomposes, 65 kJ of energy is released—this is indicated by the negative sign on the H value. If two moles decompose, twice that amount—130 kJ—will be released.

5. D All four reactions have a negative value for enthalpy. Energy is absorbed (positive ΔH) when bonds break and released (negative ΔH) when bonds form, so in order for the overall ΔH value to be negative, more energy must be released than is absorbed during each reaction.

6. **A** Remember that $\Delta G = \Delta H - T\Delta S$.

If the reaction is thermodynamically favored only when the temperature is very low, then ΔG is negative only when T is very small. This can happen only when ΔH is negative and ΔS is negative. A very small value for T will eliminate the influence of ΔS.

7. **D** Use the formula $q = mc\Delta T$. The mass is the solution's mass, which can be obtained by multiplying 25.0 mL by 1.02 g/mL.

8. **C** If an outside current is required for the reaction to occur, the reaction is not favored, which means $E°_{cell}$ is negative and ΔG is greater than 0.

9. **D** The equations given above the question give the heats of formation of all the reactants and products (remember: the heat of formation of O_2, an element in its most stable form, is zero).

$\Delta H°$ for a reaction = ($\Delta H°$ for the products) – ($\Delta H°$ for the reactants).

First, the products:

From $2H_2O$, we get $2z$
From CO_2, we get y
So $\Delta H°$ for the products = $2z + y$

Now the reactants:

From CH_4, we get x. The heat of formation of O_2 is defined to be zero, so that's it for the reactants.
$\Delta H°$ for the reaction = $(2z + y) - (x) = 2z + y - x$.

10. **D** Heat is a measure of energy transfer, so a solution cannot "have" heat. Solutions can have thermal energy, though, and even though solution B is at a lower temperature, there is a lot more of it, and the amount of energy contained within it will thus be greater.

11. **B** The bond energy is the energy that must be put into a bond to break it. First let's figure out how much energy must be put into the reactants to break their bonds.

To break 2 moles of H–H bonds, it takes (2)(500) kJ = 1,000 kJ.

To break 1 mole of O=O bonds, it takes 500 kJ.

So to break up the reactants, it takes +1,500 kJ.

Energy is given off when a bond is formed; that's the negative of the bond energy. Now let's see how much energy is given off when 2 moles of H_2O are formed.

2 moles of H_2O molecules contain 4 moles of O–H bonds, so (4)(–500) kJ = –2,000 kJ are given off.

So the value of $\Delta H°$ for the reaction is

(–2,000 kJ, the energy given off) + (1,500 kJ, the energy put in) = –500 kJ.

12. **B** During a phase change, the products and reactants are at equilibrium with each other, meaning ΔG is zero. When a liquid turns into a solid, disorder decreases (negative ΔS) and energy is absorbed (positive ΔH). Choice (D) is incorrect because 298 K = 25°C, and water does not freeze at that temperature at 1.0 atm.

13. **C** Point 3 represents the activated complex, which is the point of highest energy. This point is the transition state between the reactants and the products.

14. **C** The enthalpy change is the distance between the energy level of the reactants (on which points 1 and 2 lie) and the energy level of the products (on which point 4 lies).

15. **A** A catalyst lowers the activation energy of the reaction without changing the energy level of either the reactants or products.

16. **A** The equations given on top give the heats of formation of all the reactants and products (remember: the heat of formation of O_2, an element in its most stable form, is zero).

$\Delta H°$ for a reaction = ($\Delta H°$ for the products) – ($\Delta H°$ for the reactants).

First, the products:

From CO_2, we get (2)(–390 kJ) = –780 kJ.

From H_2O, we get –290 kJ.

So $\Delta H°$ for the products = (–780 kJ) + (–290 kJ) = –1,070 kJ.

Now the reactants:

From C_2H_2, we get +230 kJ. The heat of formation of O_2 is defined as zero, so that's it for the reactants.

$\Delta H°$ for the reaction = (–1,070 kJ) – (+230 kJ) = –1,300 kJ.

17. **D** Choice (D) is the only reaction where the number of moles of gas is increasing, going from 2 moles of gas on the reactant side to 3 moles of gas on the product side. In all the other choices, the number of moles of gas either decreases or remains constant.

18. **B** For the reaction to occur without the input of additional energy, it must be thermally favored, meaning $\Delta G < 0$. The product is a solid, which is more ordered than the gas which is part of the reactants, meaning $\Delta S < 0$.

Via $\Delta G = \Delta H - T\Delta S$, if ΔS is negative, the $T\Delta S$ term is positive. Thus, for ΔG to be negative as it is in this reaction, the value for ΔH must be negative as well.

19. **B** The reaction that forms 2 moles of HF(*g*) from its constituent elements has an enthalpy change of –540 kJ. The heat of formation is given by the reaction that forms 1 mole from these elements, so you can just divide –540 kJ by 2 to get –270 kJ.

20. **B** For the temperature of the water to increase, the reaction must be exothermic. If this is the case, the hydration energy of the solution exceeds the lattice energy within the solute.

21. **C** The reaction must be exothermic, as the water heats up. In an exothermic reaction, heat is transferred from the system (the reaction itself) to the surroundings (the water).

22. **C** The reaction is exothermic, meaning it has a negative ΔH, which indicates that it is favored in terms of enthalpy. It is also favored in terms of entropy, as the ions start organized into the ionic lattice, and end up in a more disordered state as free-floating ions in solution.

23. **A** Hydration energy (like lattice energy) is based on Coulomb's law, which has factors of both charge and size. The charges are the same in both substances, but fluoride is smaller than chloride. As size is in the denominator of Coulomb's law, a smaller size means there will be more hydration energy.

24. **A** You can use Hess's law. Add the two reactions together, and cancel things that appear on both sides.

$2S(s) + 3O_2(g) \rightarrow 2SO_3(g)$ $\Delta H = +800$ kJ

$2SO_3(g) \rightarrow 2SO_2(g) + O_2(g)$ $\Delta H = -200$ kJ

$2S(s) + 3O_2(g) + 2SO_3(g) \rightarrow 2SO_2(g) + 2SO_3(g) + O_2(g)$

This reduces to:

$2S(s) + 2O_2(g) \rightarrow 2SO_2(g)$ $\Delta H = +600$ kJ

Now we can cut everything in half to get the equation we want.

$S(s) + O_2(g) \rightarrow SO_2(g)$ $\Delta H = +300$ kJ

Free-Response

1. (a) Use the entropy values in the table.

$$\Delta S° = \Sigma S°_{products} - \Sigma S°_{reactants}$$

$$\Delta S° = [(6)(213.6) + (6)(69.9)] - [(212.13) + (6)(205)] \text{ J/mol·K}$$

$$\Delta S° = 259 \text{ J/mol·K}$$

(b) Use the equation below. Remember that enthalpy values are given in kJ and entropy values are given in J.

$$\Delta G° = \Delta H° - T\Delta S°$$

$$\Delta G° = (-2{,}801 \text{ kJ/mol}_{rxn}) - (298 \text{ K})(0.259 \text{ kJ/mol}_{rxn}\text{·K}) = -2{,}880 \text{ kJ/mol}_{rxn}$$

(c)

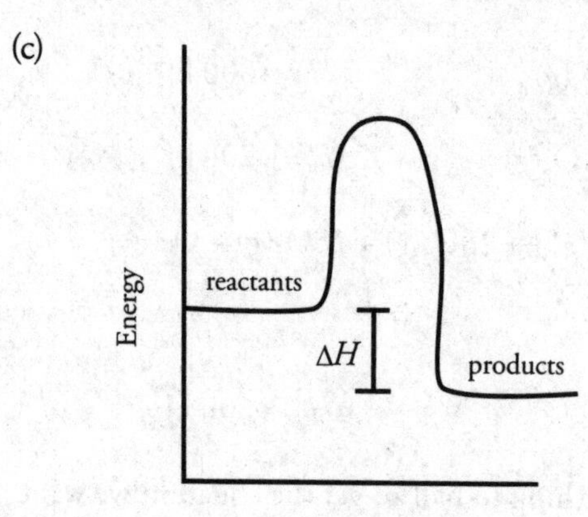

The reaction is exothermic; therefore the reactants must have more energy than the products, as indicated. The difference in energy is equal to the ΔH for this reaction.

(d)

$$1.00 \text{ g glucose} \times \frac{1 \text{ mol glucose}}{180 \text{ g glucose}} \times \frac{1 \text{ mol}_{rxn}}{1 \text{ mol glucose}} \times \frac{-2801 \text{ kJ}}{1 \text{ mol}_{rxn}} = -15.5 \text{ kJ}$$

2. (a) Use the relationship below.

$$\Delta H° = \Sigma \text{ Energies of the bonds broken } - \Sigma \text{ Energies of the bonds formed}$$

$$\Delta H° = [(4)(415) + (2)(495)] - [(2)(799) + (4)(463)] \text{ kJ/mol}$$

$$\Delta H° = -800 \text{ kJ/mol}$$

(b) $$1{,}500\,\text{kJ} \times \frac{1\,\text{mol}\,CH_4}{800\text{ kJ}} = 1.9 \text{ mol } CH_4 \times \frac{16.04\,\text{g}\,CH_4}{1\,\text{mol}\,CH_4} = 30 \text{ g } CH_4$$

(c) Use $\Delta G° = \Delta H° - T\Delta S°$

Remember that enthalpy values are given in kJ and entropy values are given in J.

$$\Delta S° = \frac{\Delta H - \Delta G}{T} = \frac{(-800\text{ kJ/mol}) - (-801\text{ kJ/mol})}{(298\text{ K})}$$

$$\Delta S° = 0.003 \text{ kJ/K} = 3 \text{ J/K}$$

(d) $\Delta S°$ is very small, which means that the entropy change for the process is very small. This makes sense because the number of moles remains constant, the number of moles of gas remains constant, and the complexity of the molecules remains about the same.

3. (a) The heat of fusion is represented by the lower horizontal line, and the heat of vaporization by the higher one. As the higher line is longer, that means more heat is required to vaporize the substance than to melt it.

(b) The substance is a gas at its highest temperature, and would not change phase again.

(c) The student is incorrect. The greater the specific heat of a substance, the more heat that is required to change the temperature of that substance. On the graph, that is represented by a shallower slope. As the slope of the solid (leftmost) line is greater than the liquid (central) or gas (rightmost) lines, the solid would actually have a lower specific heat than either the liquid or the gas.

(d) Heat is being added to the system, so enthalpy change is positive. As the substance first melts, and then boils, it is becoming more disordered. Even after it becomes a gas, the gas molecules will continue to spread out as the temperature increases. So, the entropy change is also positive.

4. (a) (i) In order to boil a liquid, the intermolecular forces between the various molecules of the liquid must be broken. This requires the input of energy, thus ΔH is positive.

(ii) A liquid turning into a gas leads to an increase in disorder, which yields a positive ΔS.

(b) When a phase change is occurring, the value for ΔG is always zero, as no real chemical reaction is occurring. Using that knowledge and making sure the units are matching (by converting the entropy to kJ/mol × K), we get:

$$\Delta G = \Delta H^\circ - T\Delta S^\circ$$

$$0 = 37.6 \text{ kJ/mol} - T(0.111 \text{ kJ/mol} \times \text{K})$$

$$T = 339 \text{ K}$$

Converting to Celsius: 339 K – 273.15 = 65.9°C

(c) We first need to figure out how many moles of ethanol are present.

$$D = \frac{m}{V} \quad 0.789 \text{ g/mL} = \frac{m}{50.0 \text{ mL}} \quad m = 39.5 \text{ g } CH_3OH$$

$$39.5 \text{ g} \times \frac{1 \text{ mol } CH_3OH}{32.05 \text{ g}} = 1.23 \text{ mol } CH_3OH$$

Then, we have to use the ΔH° value to calculate the amount of heat needed.

$$1.23 \text{ mol} \times \frac{76.3 \text{ kJ}}{\text{mol}} = 93.5 \text{ kJ}$$

(d) As methanol (or any other liquid) boils, the temperature will remain constant as the added heat is going into breaking intermolecular forces instead of adding extra kinetic energy to the liquid.

(e) Yes, methanol would be soluble with water. Methanol is a polar molecule, as is water, and in terms of molecular solubility, like dissolves like.

(f) It would be greater than that of methanol. Both molecules are polar with H-bonds, but ethanol has more electrons, and thus its electron cloud is more polarizable. This causes the London dispersion forces in ethanol to be stronger than those in methanol, leading to a higher boiling point.

5. (a) $\Delta S°$ is negative because the products are less random than the reactants. That's because two moles of reactants are converted to one mole of products and gas is converted into solid in the reaction.

(b) $\Delta G°$ is negative because the reaction is favored.

(c) Use $\Delta G° = \Delta H° - T\Delta S°$.

$\Delta G°$ will become less negative because as temperature is increased, the entropy change of a reaction becomes more important in determining its favorability. The entropy change for this reaction is negative, which discourages favorability, so increasing temperature will make the reaction less thermodynamically favored, thus making $\Delta G°$ less negative.

(d) The concentration of CO_2 will decrease as the reaction proceeds in the forward direction and the reactants are consumed. The temperature will increase as heat is given off by the exothermic reaction.

6. (a) $q = mc\Delta T$

$q = (25.0\text{ g})(4.18\text{ J/g}\cdot°\text{C})(3.6°\text{C})$

$q = 376\text{ J}$

(b) The heat gained by the water is the same as the heat lost by the aluminum.

$q = mc\Delta T$

$-376\text{ J} = (5.86\text{ g})(c)(-82.3°\text{C})$

$c = 0.780\text{ J/g}\cdot°\text{C}$

(c) $5.86\text{ g Al} \quad \dfrac{1\text{ mol Al}}{26.98\text{ g Al}} = 0.217\text{ mol Al}$

376 J/0.217 mol = 1730 J/mol = 1.73 kJ/mol

(d) $\%\text{ error} = \dfrac{|\text{experimental} - \text{accepted}|}{\text{accepted}} \times 100\%$

$\dfrac{|0.780 - 0.900|}{0.900} \times 100\% = 13.3\%\text{ error}$

(e) Error 1: If some of the heat that was lost by the aluminum was not absorbed by the water, that would cause the calculated heat gained by the water in part (a) to be artificially low. This, in turn, would reduce the value of the specific heat of aluminum as calculated in part (b).

Error 2: If there was more than 25.0 mL of water in the calorimeter, that would mean the mass in part (a) was artificially low, which would make the calculation for the heat gained by the water also artificially low. This, in turn, would reduce the value of the specific heat of aluminum calculated in part (b).

There are many potential errors here, but as long as you can quantitatively follow them to the conclusion that the experimental value would be too low, any error (within reason) can be acceptable.

Chapter 8
Big Idea #6: Equilibrium, Acids and Bases, Titrations, and Solubility

Any bond or intermolecular attraction that can be formed can be broken. These two processes are in dynamic competition, sensitive to initial conditions and external perturbations.

THE EQUILIBRIUM CONSTANT, K_{eq}

Most chemical processes are reversible. That is, reactants react to form products, but those products can also react to form reactants. A reaction is at equilibrium when the rate of the forward reaction is equal to the rate of the reverse reaction. The relationship between the concentrations of reactants and products in a reaction at equilibrium is given by the equilibrium expression, also called the **law of mass action.**

The Equilibrium Expression

For the reaction

$$aA + bB \rightleftharpoons cC + dD$$

$$K_{eq} = \frac{[C]^c[D]^d}{[A]^a[B]^b}$$

1. [A], [B], [C], and [D] are molar concentrations or partial pressures at equilibrium.
2. Products are in the numerator, and reactants are in the denominator.
3. Coefficients in the balanced equation become exponents in the equilibrium expression.
4. Solids and pure liquids are not included in the equilibrium expression because they cannot change their concentration. Only gaseous and aqueous species are included in the expression.
5. Units are not given for K_{eq}.

Let's look at a few examples:

1. $HC_2H_3O_2(aq) \rightleftharpoons H^+(aq) + C_2H_3O_2^-(aq)$

$$K_{eq} = K_a = \frac{[H^+][C_2H_3O_2^-]}{[HC_2H_3O_2]}$$

This reaction shows the dissociation of acetic acid in water. All of the reactants and products are aqueous particles, so they are all included in the equilibrium expression. None of the reactants or products have coefficients, so there are no exponents in the equilibrium expression. This is the standard form of K_a, the acid dissociation constant.

2. $2H_2S(g) + 3O_2(g) \rightleftharpoons 2H_2O(g) + 2SO_2(g)$

$$K_{eq} = K_c = \frac{[H_2O]^2[SO_2]^2}{[H_2S]^2[O_2]^3}$$

$$K_{eq} = K_p = \frac{(P_{H_2O})^2 (P_{SO_2})^2}{(P_{H_2S})^2 (P_{O_2})^3}$$

All of the reactants and products in this reaction are gases, so K_{eq} can be expressed in terms of concentration (K_c, moles/liter or molarity) or in terms of partial pressure (K_p, atmospheres). All of the reactants and products are included here, and the coefficients in the reaction become exponents in the equilibrium expression.

3. $CaF_2(s) \rightleftharpoons Ca^{2+}(aq) + 2F^-(aq)$
$K_{eq} = K_{sp} = [Ca^{2+}][F^-]^2$

This reaction shows the dissociation of a slightly soluble salt. There is no denominator in this equilibrium expression because the reactant is a solid. Solids are left out of the equilibrium expression because the concentration of a solid is constant. There must be some solid present for equilibrium to exist, but you do not need to include it in your calculations. This form of K_{eq} is called the solubility product, K_{sp}.

4. $NH_3(aq) + H_2O(l) \rightleftharpoons NH_4^+(aq) + OH^-(aq)$

$$K_{eq} = K_b = \frac{[NH_4^+][OH^-]}{[NH_3]}$$

This is the acid-base reaction between ammonia and water. We can leave water out of the equilibrium expression because it is a pure liquid. This is the standard form for K_b, the base dissociation constant.

Here is a roundup of the equilibrium constants you need to be familiar with for the test.

- K_c is the constant for molar concentrations.
- K_p is the constant for partial pressures.
- K_{sp} is the solubility product, which has no denominator because the reactants are solids.
- K_a is the acid dissociation constant for weak acids.
- K_b is the base dissociation constant for weak bases.
- K_w describes the ionization of water ($K_w = 1 \times 10^{-14}$).

A large value for K_{eq} means that products are favored over reactants at equilibrium, while a small value for K_{eq} means that reactants are favored over products at equilibrium.

The equilibrium constant has a lot of aliases, but they all take the same form and tell you the same thing. The **equilibrium constant** tells you the relative amounts of products and reactants at equilibrium.

Manipulating K_{eq}

Much as Hess's law allows for the determination of the enthalpy for a reaction given the enthalpy values for similar reactions, you can determine the equilibrium constant of a reaction by manipulating similar reactions with known equilibrium constants. However, the rules for doing so are different than the rules for shifting enthalpy values:

1. If you flip a reaction, you take the reciprocal of the equilibrium constant to get the new equilibrium constant.

2. If you multiply a reaction by a coefficient, you take the equilibrium constant to that power to get the new constant.

3. If you add two reactions together, you multiply the equilibrium constants of those reactions to get the new constant.

Example:

$2SO_2(g) + O_2(g) \leftrightarrow 2SO_3(g) \quad K = 4.0 \times 10^{24}$

$2NO(g) + O_2(g) \leftrightarrow 2NO_2(g) \quad K = 0.064$

According to the information above, what is the equilibrium constant for each of the following reactions?

a) $2SO_3(g) \leftrightarrow 2SO_2(g) + O_2(g)$

This reaction is the opposite of the first reaction, so $K = \frac{1}{4.0 \times 10^{24}} = 2.5 \times 10^{-25}$

b) $SO_2(g) + \frac{1}{2}O_2(g) \leftrightarrow SO_3(g)$

This reaction is the first reaction multiplied by one half. Thus, the new equilibrium constant is:

$K = (4.0 \times 10^{24})^{1/2} = 2.0 \times 10^{12}$

c) $SO_2(g) + NO_2(g) \leftrightarrow SO_3(g) + NO(g)$

To get this reaction, we first have to flip the second reaction:

$2NO_2 \leftrightarrow 2NO + O_2$

$$K = \frac{1}{0.065} = 16$$

We can then add the flipped second reaction to the original first reaction. The O_2 will cancel out, yielding the following:

$2SO_2(g) + 2NO_2(g) \leftrightarrow 2SO_3(g) + 2NO(g)$

$K = 4.0 \times 10^{24}(16) = 6.4 \times 10^{25}$

Finally, we have to multiply that equation by ½ to get our desired reaction.

$SO_2(g) + NO_2(g) \leftrightarrow SO_3(g) + NO(g)$

$K = (6.4 \times 10^{24})^{1/2} = 8.0 \times 10^{12}$

LE CHÂTELIER'S PRINCIPLE

At equilibrium, the rates of the forward and reverse reactions are equal. A "shift" in a certain direction means the rate of the forward or reverse reaction increases. Le Châtelier's Principle states that whenever a stress is placed on a system at equilibrium, the system will shift in response to that stress. If the forward rate increases, we say the reaction has shifted right, which will create more products. If the reverse rate increases, we say the reaction has shifted left, which will create more reactants.

Let's use the **Haber process,** which is used in the industrial preparation of ammonia, as an example.

$N_2(g) + 3H_2(g) \rightleftharpoons 2NH_3(g) \qquad \Delta H° = -92.6 \text{ kJ/mol}_{rxn}$

Concentration

When the concentration of a reactant or product is increased, the reaction will shift in the direction that allows it to use up the added substance. If N_2 or H_2 is added, the reaction shifts right. If NH_3 is added, the reaction shifts left.

When the concentration of a species is decreased, the reaction will shift in the direction that allows it to create the substance that has been removed. If N_2 or H_2 is removed, the reaction shifts left. If NH_3 is removed, the reaction shifts right.

Pressure

When the external pressure is increased, the reaction will shift to the side with fewer gas molecules. When the external pressure is decreased, the reaction will shift to the side with more gas molecules. In this case, an increase in pressure would cause a shift to the right, and a decrease in pressure would cause a shift to the left.

A common way to cause a pressure shift on a reaction is to change the volume of the container that the reaction is occurring in. A decrease in volume will lead to an increase in pressure (provided temperature remains constant), and thus cause a shift to the side with fewer gas molecules. If there is no gas involved in the reaction, or if both sides of the equilibrium have the same number of moles of gas, then changing the pressure and/or volume has no effect on the reaction.

Temperature

There's a trick to figure out what happens when the temperature changes. First, rewrite the equation to include the heat energy on the side that it would be present on. The Haber process is exothermic, so heat is generated:

$$N_2(g) + 3H_2(g) \leftrightarrow 2NH_3(g) + \text{energy}$$

Then, if the temperature goes up, the reaction will proceed in the reverse direction (shifting away from the added energy). If the temperature goes down, the reaction will proceed in the forward direction (creating more energy). The reverse would be true in an endothermic reaction, as the energy would be part of the reactants.

CHANGES IN THE EQUILIBRIUM CONSTANT

It is important to understand that shifts caused by concentration or pressure changes are temporary shifts, and do not change the value of the equilibrium constant itself. Eventually, the concentrations of the products and reactants will re-establish the same ratio as they originally had at equilibrium.

However, shifts caused by temperature changes are different. Because changing temperature also affects reaction kinetics by adding (or removing) energy from the equilibrium system, a change in temperature will also affect the equilibrium constant for the reaction itself, in addition to causing a shift. This means the ratio of the products to reactants at equilibrium will change as the temperature changes.

In the Haber process, an increase in temperature causes a shift to the left, and it would permanently affect the value of the equilibrium constant in a way that is consistent with the shift. A shift to the left causes an increase in the concentrations of the reactants (the denominator in the mass action expression) while simultaneously causing a decrease in the concentration of the product (the numerator in the mass action expression). Thus, increasing the temperature decreases the value of the equilibrium constant—that is, it causes there to be a larger amount of reactants present at equilibrium compared to the amount of products present. The reverse would be true for a temperature decrease; the shift to the right would cause an increase in the equilibrium constant.

THE REACTION QUOTIENT, *Q*

The reaction quotient is essentially the quantitative application of Le Châtelier's Principle. It is determined using the law of mass action. However, you can use the concentration or pressure values at any point in the reaction to calculate Q. (Remember, you can only use concentrations or pressures of reactions at equilibrium to calculate K). The value for the reaction quotient can be compared to the value for the equilibrium constant to predict in which direction a reaction will shift from the given set of initial conditions.

The Reaction Quotient

For the reaction

$$aA + bB \rightleftharpoons cC + dD$$

$$Q = \frac{[C]^c[D]^d}{[A]^a[B]^b}$$

[A], [B], [C], and [D] are initial molar concentrations or partial pressures.

- If Q is less than K, the reaction shifts right.
- If Q is greater than K, the reaction shifts left.
- If $Q = K$, the reaction is already at equilibrium.

Let's take a look at an example. The following reaction takes place in a sealed flask:

$$2CH_4(g) \leftrightarrow C_2H_2(g) + 3H_2(g)$$

(a) Determine the equilibrium constant if the following concentrations are found at equilibrium.

$[CH_4] = 0.0032\ M$ $\quad$ $[C_2H_2] = 0.025\ M$ $\quad$ $[H_2] = 0.040\ M$

To solve this, we first need to create the equilibrium constant expression:

$$K_c = \frac{[C_2H_2][H_2]^3}{[CH_4]^2}$$

By plugging in the numbers, we get:

$$K_c = \frac{(0.025)(0.040)^3}{(0.0032)^2} \qquad K_c = \frac{1.6 \times 10^{-6}}{1.0 \times 10^{-5}} \qquad K_c = 0.16$$

(b) Upon testing this reaction at another point, the following concentrations are found:

$[CH_4]$ = 0.0055 *M* $[C_2H_2]$ = 0.026 *M* $[H_2]$ = 0.029 *M*

Use the reaction quotient to determine which way the reaction needs to shift to reach equilibrium.

Q uses the exact same ratios as the equilibrium constant expression, so:

$$Q = \frac{[C_2H_2][H_2]^3}{[CH_4]^2} = \frac{(0.026)(0.029)^3}{(0.0055)^2} = \frac{6.3 \times 10^{-7}}{3.0 \times 10^{-5}} = 0.021$$

0.021 is less than the equilibrium constant value of 0.16, so the reaction must proceed to the right, creating more products and reducing the amount of reactant in order to come to equilibrium.

Voltaic Cells

When examining voltaic cells, the reduction potentials are always given at standard conditions; that is 25°C, 1.0 atm, and with all species having a concentration of 1.0 *M*. If any of those conditions deviate, it will also cause the cell potential to deviate.

Voltaic cells are all very favored, having equilibrium constants significantly greater than 1. If the reaction quotient for a voltaic cell were to ever become equal to the equilibrium constant (so, at $Q = K$), the voltage of the cell would drop to zero. Given that knowledge, the best way to determine how a cell's potential will change if standard conditions are deviated from is to use the reaction quotient.

As all concentrations are equal to 1.0 *M* at standard conditions, we can infer that at standard conditions, the reaction quotient will also be equal to 1. Any change to the cell that would cause the reaction quotient to increase (say, an increase in the concentration of a product or the decrease in the concentration of a reactant) would cause the reaction quotient to become closer to the equilibrium constant and would thus decrease the cell potential (remember, if *Q* ever reaches *K*, the cell potential becomes zero!). Any change that would cause the reaction quotient to decrease would move it further from the equilibrium constant, and thus increase the potential of the cell.

If a cell has a gas at either the cathode or the anode, a change in pressure can also affect the reaction quotient, and a change in temperature may as well. However, concentration changes, which usually occur constantly as the reaction in a cell is progressing, are the most common reason that a voltaic cell will deviate from its standard potential.

Example:

A voltaic cell using silver and zinc is connected, and the below reaction occurs.

$2\ Ag^{+} + Zn(s) \rightarrow Zn^{2+} + 2\ Ag(s)$

What would happen to the voltage of the above cell as the reaction progresses?

As the reaction progresses, $[Ag^{+}]$ decreases as it is reduced into $Ag(s)$, and $[Zn^{2+}]$ increases as it is oxidized from $Zn(s)$. Both of these changes would have the effect of increasing the reaction quotient, which would bring the cell closer to equilibrium and thus decrease the overall potential.

SOLUBILITY

Roughly speaking, a salt can be considered "soluble" if more than 1 gram of the salt can be dissolved in 100 milliliters of water. Soluble salts are usually assumed to dissociate completely in aqueous solution. Most, but not all, solids become more soluble in a liquid as the temperature is increased.

Solubility Product (K_{sp})

Salts that are "slightly soluble" and "insoluble" still dissociate in solution to some extent. The solubility product (K_{sp}) is a measure of the extent of a salt's dissociation in solution. The K_{sp} is one of the forms of the equilibrium expression. The greater the value of the solubility product for a salt, the more soluble the salt.

Solubility Product

For the reaction

$$A_aB_b(s) \rightleftharpoons a\,A^{b+}(aq) + b\,B^{a-}(aq)$$

The solubility expression is

$$K_{sp} = [A^{b+}]^a[B^{a-}]^b$$

Here are some examples:

$CaF_2(s) \rightleftharpoons Ca^{2+}(aq) + 2F^-(aq)$ $\qquad K_{sp} = [Ca^{2+}][F^-]^2$

$Ag_2CrO_4(s) \rightleftharpoons 2Ag^+(aq) + CrO_4^{2-}(aq)$ $\qquad K_{sp} = [Ag^+]^2[CrO_4^{2-}]$

$CuI(s) \rightleftharpoons Cu^+(aq) + I^-(aq)$ $\qquad K_{sp} = [Cu^+][I^-]$

The solubility of salts can be described by the K_{sp} or by the molar solubility. The molar solubility of the salt describes the number of moles of salt that can be dissolved per liter of solution. The K_{sp} value for Ag_2CrO_4 is 8.0×10^{-12}. We can determine the molar solubility using the following calculations:

$K_{sp} = [Ag^+]^2[CrO_4^{2-}]$

$8.0 \times 10^{-12} = (2x)^2(x)$

$8.0 \times 10^{-12} = 4x^3$

$x = [CrO_4^{2-}] = 1.3 \times 10^{-4}\ M$

$2x = [Ag^+] = 2.6 \times 10^{-4}\ M$

The molar solubility of the salt will also be equal to the concentration of any ion that occurs in a 1:1 ratio with the salt. In the example above, the molar solubility of the salt is $1.3 \times 10^{-4}\ M$, the same as the concentration of the CrO_4^{2-} ions.

Note that the concentration of the silver ions is both doubled (because twice as many of them will be in solution) AND squared in the K_{sp} expression. Thus, the 2 coefficient on the silver is represented twice in the solubility product calculations. Forgetting to account for that coefficient in the ion concentration (like using just x instead of $2x$) is a very common mistake made by students; take care to avoid it!

Typically, the molar solubility of most salts will increase with rising temperatures. This is because a higher temperature has more energy available to force the water molecules apart and make room for the solute ions. There are some salts that see their solubility decrease with increasing temperature, but there is no easy way to predict when that will occur.

The Common Ion Effect

Let's look at the solubility expression for AgCl.

$$K_{sp} = [Ag^+][Cl^-] = 1.6 \times 10^{-10}$$

If we throw a block of solid AgCl into a beaker of water, we can tell from the K_{sp} what the concentrations of Ag^+ and Cl^- will be at equilibrium. For every unit of AgCl that dissociates, we get one Ag^+ and one Cl^-, so we can solve the equation above as follows:

$$[Ag^+][Cl^-] = 1.6 \times 10^{-10}$$

$$(x)(x) = 1.6 \times 10^{-10}$$

$$x^2 = 1.6 \times 10^{-10}$$

$$x = [Ag^+] = [Cl^-] = 1.3 \times 10^{-5}\ M$$

So there are very small amounts of Ag^+ and Cl^- in the solution.

Let's say we add 0.10 mole of NaCl to 1 liter of the AgCl solution. NaCl dissociates completely, so that's the same thing as adding 0.1 mole of Na^+ ions and 0.1 mole of Cl^- ions to the solution. The Na^+ ions will not affect the AgCl equilibrium, so we can ignore them; but the Cl^- ions must be taken into account. That's because of the **common ion effect.**

The common ion effect says that the newly added Cl^- ions will affect the AgCl equilibrium, although the newly added Cl^- ions did not come from AgCl.

Let's look at the solubility expression again. Now we have 0.10 mole of Cl^- ions in 1 liter of the solution, so $[Cl^-] = 0.10\ M$.

$$[Ag^+][Cl^-] = 1.6 \times 10^{-10}$$

$$[Ag^+](0.10\ M) = 1.6 \times 10^{-10}$$

$$[Ag^+] = \frac{(1.6 \times 10^{-10})}{(0.10)} M$$

$$[Ag^+] = 1.6 \times 10^{-9}\ M$$

Now the number of Ag^+ ions in the solution has decreased drastically because of the Cl^- ions introduced to the solution by NaCl. So when solutions of AgCl and NaCl, which share a common Cl^- ion, are mixed, the more soluble salt (NaCl) can cause the less soluble salt (AgCl) to precipitate. In general, when two salt solutions that share a common ion are mixed, the salt with the lower value for K_{sp} will precipitate first.

Standard Free Energy Change and the Equilibrium Constant

The amount of Gibbs free energy in any given reaction can also be calculated if you know the equilibrium constant expression for that reaction.

$$\Delta G° = -RT \ln K$$

R = the gas constant, 8.31 J/mol·K
T = absolute temperature (K)
K = the equilibrium constant

Notice that if $\Delta G°$ is negative, K must be greater than 1, and products will be favored at equilibrium. Alternatively, if $\Delta G°$ is positive, K must be less than 1, and reactants will be favored at equilibrium.

ACIDS AND BASES DEFINITIONS

Arrhenius

S. A. Arrhenius defined an acid as a substance that ionizes in water and produces hydrogen ions (H^+ ions). For instance, HCl is an acid.

$$HCl \rightarrow H^+ + Cl^-$$

He defined a base as a substance that ionizes in water and produces hydroxide ions (OH^- ions). For instance, NaOH is a base.

$$NaOH \rightarrow Na^+ + OH^-$$

Brønsted-Lowry

J. N. Brønsted and T. M. Lowry defined an acid as a substance that is capable of donating a proton, which is the same as donating an H^+ ion, and they defined a base as a substance that is capable of accepting a proton. This definition is the one that will be used most frequently on the exam.

Look at the reversible reaction below.

$$HC_2H_3O_2 + H_2O \leftrightarrow C_2H_3O_2^- + H_3O^+$$

According to Brønsted-Lowry

$HC_2H_3O_2$ and H_3O^+ are acids.

$C_2H_3O_2^-$ and H_2O are bases.

Now look at this reversible reaction.

$$NH_3 + H_2O \leftrightarrow NH_4^+ + OH^-$$

According to Brønsted-Lowry

NH_3 and OH^- are bases.

H_2O and NH_4^+ are acids.

So in each case, the species with the H^+ ion is the acid, and the same species without the H^+ ion is the base; the two species are called a **conjugate pair.** The following are the acid-base conjugate pairs in the reactions above:

$HC_2H_3O_2$ and $C_2H_3O_2^-$

NH_4^+ and NH_3

H_3O^+ and H_2O

H_2O and OH^-

Notice that water can act either as an acid or base. Any substance which has that ability is called **amphoteric.**

pH

Many of the concentration measurements in acid-base problems are given to us in terms of pH and pOH.

p (anything) = –log (anything)

$pH = -\log [H^+]$
$pOH = -\log [OH^-]$
$pK_a = -\log K_a$
$pK_b = -\log K_b$

In a solution

- when $[H^+] = [OH^-]$, the solution is neutral, and pH = 7
- when $[H^+]$ is greater than $[OH^-]$, the solution is acidic, and pH is less than 7
- when $[H^+]$ is less than $[OH^-]$, the solution is basic, and pH is greater than 7

It is important to remember that *increasing* pH means *decreasing* $[H^+]$, which means that there are fewer H^+ ions floating around and the solution is *less acidic.* Alternatively, *decreasing* pH means *increasing* $[H^+]$, which means that there are more H^+ ions floating around and the solution is *more acidic.*

ACID STRENGTHS

Strong Acids

Strong acids dissociate completely in water, so the reaction goes to completion and they never reach equilibrium with their conjugate bases. Because there is no equilibrium, there is no equilibrium constant, so there is no dissociation constant for strong acids or bases.

> Important Strong Acids
>
> HCl, HBr, HI, HNO_3, $HClO_4$, H_2SO_4
>
> Important Strong Bases
>
> $LiOH$, $NaOH$, KOH, $Ba(OH)_2$, $Sr(OH)_2$

Because the dissociation of a strong acid goes to completion, there is no tendency for the reverse reaction to occur, which means that the conjugate base of a strong acid must be extremely weak.

It's much easier to find the pH of a strong acid solution than it is to find the pH of a weak acid solution. That's because strong acids dissociate completely, so the final concentration of H^+ ions will be the same as the initial concentration of the strong acid.

Let's look at a 0.010-molar solution of HCl.

HCl dissociates completely, so $[H^+] = 0.010\ M$
$pH = -\log [H^+] = -\log (0.010) = -\log (10^{-2}) = 2$

So you can always find the pH of a strong acid solution directly from its concentration.

Weak Acids

When a weak acid (often symbolized with HA) is placed in water, a small fraction of its molecules will dissociate into hydrogen ions (H^+) and conjugate base ions (A^-). Most of the acid molecules will remain in solution as undissociated aqueous particles.

The dissociation constants, K_a and K_b, are measures of the strengths of weak acids and bases. K_a and K_b are just the equilibrium constants specific to acids and bases.

Acid Dissociation Constant

$$K_a = \frac{[H^+][A^-]}{[HA]}$$

$[H^+]$ = concentration of hydrogen ions (*M*)
$[A^-]$ = concentration of conjugate base ions (*M*)
$[HA]$ = concentration of undissociated acid molecules (*M*)

Base Dissociation Constant

$$K_b = \frac{[HB^+][OH^-]}{[B]}$$

$[HB^+]$ = concentration of conjugate acid ions (*M*)
$[OH^-]$ = concentration of hydroxide ions (*M*)
$[B]$ = concentration of unprotonated base molecules (*M*)

The greater the value of K_a, the greater the extent of the dissociation of the acid and the stronger the acid. The same thing goes for K_b, but in the case of K_b, the base is not dissociating. Instead, it is accepting a hydrogen ion (proton) from an ion. So, a base does not dissociate—it protonates (or ionizes).

If you know the K_a for an acid and the concentration of the acid, you can find the pH. For instance, let's look at 0.20-molar solution of $HC_2H_3O_2$, with $K_a = 1.8 \times 10^{-5}$.

First we set up the K_a equation, plugging in values.

$$HC_2H_3O_2 \rightarrow H^+ + C_2H_3O_2^-$$

$$K_a = \frac{[H^+][C_2H_3O_2^-]}{[HC_2H_3O_2]}$$

Therefore, the ICE (Initial, Change, Equilibrium) table for the above problem is as follows:

	$[HC_2H_3O_2]$	$[H^+]$	$[C_2H_3O_2^-]$
Initial	0.20	0.0	0.0
Change	$-x$	$+x$	$+x$
Equilibrium	$0.20 - x$	x	x

Because every acid molecule that dissociates produces one H^+ and one $C_2H_3O_2^-$,

$$[H^+] = [C_2H_3O_2^-] = x$$

and because, strictly speaking, the molecules that dissociate should be subtracted from the initial concentration of $HC_2H_3O_2$, $[HC_2H_3O_2]$ should be $(0.20\ M - x)$. In practice, however, x is almost always insignificant compared with the initial concentration of acid, so we just use the initial concentration in the calculation.

$$[HC_2H_3O_2] = 0.20\ M$$

Now we can plug our values and variable into the K_a expression.

$$1.8 \times 10^{-5} = \frac{x^2}{0.20}$$

Solve for x.

$$x = [H^+] = 1.9 \times 10^{-3}$$

Now that we know $[H^+]$, we can calculate the pH.

$$\text{pH} = -\log [H^+] = -\log (1.9 \times 10^{-3}) = 2.7$$

This is the basic approach to solving many of the weak acid/base problems that will be on the test.

Another way to write out the dissociation of a weak acid is:

$$HA\ (aq) + H_2O\ (l) \rightleftharpoons A^- + H_3O^+$$

The above includes water molecules in the reaction, and is technically more accurate overall, as an acid will not dissociate unless it has a base to give protons to. The H_3O^+ ion is called the hydronium ion, and you can substitute it for H^+ in any acid/base reaction. So, $-\log [H_3O^+] = \text{pH}$.

Percent Dissociation

The primary factor when it comes to determining acid strength is that the more H^+ ions that an acid can donate, the stronger that acid will be. How easily an acid dissociates is often determined in part by its structure.

Consider the binary acids composed of hydrogen and a halogen. Of those, HI, HBr, and HCl are all strong acids, meaning they dissociate completely. HF, however, is not a strong acid. The reason for this is that fluorine is extremely electronegative, and thus fluorine will "hold on" to the hydrogen more effectively.

If you consider oxoacids, though, the reverse trend is true. Consider the Lewis diagrams for HOF and HOBr:

$$\mathrm{H-\ddot{\underset{\cdot\cdot}{O}}-\ddot{\underset{\cdot\cdot}{F}}\!:}\qquad \mathrm{H-\ddot{\underset{\cdot\cdot}{O}}-\ddot{\underset{\cdot\cdot}{Br}}\!:}$$

In this case, fluorine's very high electronegativity affects the O–F bond, drawing the shared electrons towards the fluorine. However, fluorine is so electronegative that it also attracts the shared electrons in the H–O bond as well, which weakens the overall H–O bond and makes the H more likely to dissociate. Thus, HOF is a stronger acid than HOBr.

The strength of an acid is very case-specific, but the one rule that is true no matter what the situation is that the easier it is for the H^+ ion to break free, the stronger the acid will be.

The other factor that affects how easily an acid can dissociate is the concentration of the acid. The lower the concentration is for an acid, the higher the percent dissociation will be. This is because the forward reaction involves the acid donating a proton to a water molecule. In an acid, there is an overabundance of water molecules, so it is very easy for the acid to find a water molecule to donate to. However, the reverse reaction is the hydronium ion donating a proton to the conjugate base. There are a much smaller number of both hydronium and conjugate base ions in a dilute solution, so the reverse reaction is kinetically hindered from happening. This means that more H_3O^+ ions will stay dissociated.

The greater the concentration of the acid, the more conjugate base there will be, and the easier it will be for that reverse reaction to take place. The easier it is for the reverse reaction to take place, the more HA there will be present in solution, and the less H_3O^+, leading to a lower overall percent dissociation.

HAt is also considered to be a strong acid, however, given the extremely high reactivity as well as the radioactive nature of astatine, HAt is extremely unstable and of no practical laboratory use.

Polyprotic Acids

Some acids, such as H_2SO_4 and H_3PO_4, can give up more than one hydrogen ion in solution. These are called **polyprotic** acids.

Polyprotic acids are always more willing to give up their first protons than later protons. For example, H_3PO_4 gives up an H^+ ion (proton) more easily than does $H_2PO_4^-$, so H_3PO_4 is a stronger acid. In the same way, $H_2PO_4^-$ is a stronger acid than HPO_4^{2-}.

This also means that, in a solution of a polyprotic acid, the percent dissociation decreases with each ensuing dissociation. So, in a solution of H_3PO_4, $[H_3PO_4] > [H_2PO_4^-] > [HPO_4^{2-}] > [PO_4^{3-}]$.

The Equilibrium Constant of Water (K_w)

Water comes to equilibrium with its ions according to the following reaction:

$$H_2O(l) \rightleftharpoons H^+(aq) + OH^-(aq) \qquad K_w = 1 \times 10^{-14} \text{ at } 25°C.$$

$$K_w = 1 \times 10^{-14} = [H^+][OH^-]$$

$$pH + pOH = 14$$

The common ion effect tells us that the hydrogen ion and hydroxide ion concentrations for any acid or base solution must be consistent with the equilibrium for the ionization of water. That is, no matter where the H^+ and OH^- ions came from, when you multiply $[H^+]$ and $[OH^-]$, you must get 1×10^{-14}. So for any aqueous solution, if you know the value of $[H^+]$, you can find out the value of $[OH^-]$ and vice versa.

The acid and base dissociation constants for conjugates must also be consistent with the equilibrium for the ionization of water.

$$K_w = 1 \times 10^{-14} = K_aK_b$$

$$pK_a + pK_b = 14$$

So if you know K_a as a weak acid, you can find K_b for its conjugate base and vice versa.

It is worth noting that pH is not limited to a 0 to 14 range. While most substances will fall into that range, substances that are very acidic can have negative pHs, and substances that are very basic can have pHs that are greater than 14. While you won't find these extreme pHs in most everyday substances, they do exist.

Most acids that you would find in an acids cabinet of a chemistry storeroom are very concentrated and may have negative pH values. For instance, nitric acid is typically manufactured at a concentration of 16 *M*. Doing the math:

$$-\log(16) = -1.2$$

Needless to say, you should exercise extreme caution when working with concentrated acids or bases, and it should be done only while wearing the proper personal protective equipment.

Something that is important to note is that the K_w for water is 1.0×10^{-14} at a temperature of 25°C only. Like any equilibrium constant, it will change if the temperature does. In this case, the dissociation of water is an endothermic process, as bonds are being broken without any new bonds being formed. So, as temperature increases, the reaction shifts to the right, which increases the value for K_w.

This will have an effect on the pH of water. For instance, at 50°C, $K_w = 5.48 \times 10^{-14}$. Thus, $[H^+] = [OH^-] = 2.34 \times 10^{-7}$ M and pH = pOH = 6.63. So, the pH of pure water at 50°C is NOT 7. The K_w of water is often measured by looking at the pK_w value ($-\log K_w$). As temperature and K_w increases, pK_w decreases, as illustrated in the table below.

Temperature (°C)	K_w	pK_w	pH
0	1.14×10^{-15}	14.94	7.47
10	6.81×10^{-15}	14.17	7.27
25	1.00×10^{-14}	14.00	7.00
50	5.48×10^{-14}	13.26	6.63
100	5.13×10^{-13}	12.29	6.14

NEUTRALIZATION REACTIONS

When an acid and a base mix, the acid will donate protons to the base in what is called a neutralization reaction. There are four different mechanisms for this, depending on the strengths of the acids and bases.

1. Strong acid + strong base

When a strong acid mixes with a strong base, both substances are dissociated completely. The only important ions in this type of reaction are the hydrogen and hydroxide ions (Even though not all bases have hydroxides, all strong bases do!).

Ex: $HCl + NaOH$

Net ionic: $H^+ + OH^- \leftrightarrow H_2O(l)$

The net ionic equation for all strong acid/strong base reactions is identical—it is always the creation of water. The other ions involved in the reaction (in the example above, Cl^- and Na^+) act as spectator ions and do not take part in the reaction.

2. Strong acid + weak base

In this reaction, the strong acid (which dissociates completely), will donate a proton to the weak base. The product will be the conjugate acid of the weak base.

Ex: $HCl + NH_3$

Net Ionic: $H^+ + NH_3(aq) \leftrightarrow NH_4^+$

3. Weak acid + strong base

In this reaction, the strong base will accept protons from the weak acid. The products are the conjugate base of the weak acid and water.

Ex: $HC_2H_3O_2 + NaOH$

Net ionic: $HC_2H_3O_2(aq) + OH^- \leftrightarrow C_2H_3O_2^- + H_2O(l)$

4. Weak acid + weak base

This is a simple proton transfer reaction, in which the acid gives protons to the base.

Ex: $HC_2H_3O_2 + NH_3$

Net ionic: $HC_2H_3O_2(aq) + NH_3(aq) \leftrightarrow C_2H_3O_2^- + NH_4^+$

During neutralization reactions, the final pH of the solution is dependent on whether the excess ions at equilibrium are due to the strong acid/base or due to a weak acid/base. When a strong acid/base is in excess, the pH calculation is fairly straightforward.

Example 1: 35.0 mL of 1.5 *M* HCN, a weak acid ($K_a = 6.2 \times 10^{-10}$) is mixed with 25.0 mL of 2.5 *M* KOH. Calculate the pH of the final solution.

To do this, we can modify our ICE chart to determine what species are present at equilibrium. First, we must determine the number of moles of the reactants:

HCN = (1.5 *M*)(0.035 L) = 0.052 mol   KOH = (2.5 *M*)(0.025 L) = 0.062 mol

Then we set up our ICE chart (leaving values for water out, as it is a pure liquid)

	HCN	OH^-	CN^+	H_2O
Initial	0.052	0.062	0	X
Change	–0.052	–0.052	+0.052	X
Equilibrium	0	0.010	0.052	X

The reaction will continue until the HCN runs out of protons to donate to the hydroxide. While there is some weak conjugate base left in solution, it is a weak base and its contribution to the pH of the solution will be irrelevant compared to the strength of the pure hydroxide ions. We now need to determine the concentration of the hydroxide ions at equilibrium. The total volume of the solution is the sum of both the acid and the base. So:

$$(35.0 \text{ mL} + 25.0 \text{ mL} = 60.0 \text{ mL})$$
$$[OH^-] = 0.010 \text{ mol}/0.060 \text{ L} = 0.17\ M$$
$$pOH = -\log(0.17) = 0.76$$
$$pH = 14 - 0.76 = 13.24$$

As you can see, it does not take much excess H^+ or OH^- to drive the pH of a solution to a fairly high or low value. Note that the K_a of the acid did not matter in this case, as the strong acid/base was in excess.

When a weak acid or base is in excess, is it easiest to use the Henderson-Hasselbalch equation:

The Henderson-Hasselbalch Equation

$$pH = pK_a + \log\frac{[A^-]}{[HA]}$$

[HA] = molar concentration of undissociated weak acid (*M*)

[A⁻] = molar concentration of conjugate base (*M*)

$$pOH = pK_b + \log\frac{[HB^+]}{[B]}$$

[B] = molar concentration of weak base (*M*)

[HB⁺] = molar concentration of conjugate acid (*M*)

Example 2: 25.0 mL of 1.0 *M* HCl is mixed with 60.0 mL of 0.50 *M* pyridine (C_5H_5N), a weak base ($K_b = 1.5 \times 10^{-9}$). Determine the pH of the solution.

Starting out, the number of moles of each reactant must be determined:

$H^+ = (1.0\ M)(0.025\ L) = 0.025$ mol H^+

$C_5H_5N = (0.50\ M)(0.060\ L) = 0.030$ mol C_5H_5N

Then we set up the ICE chart:

	H^+	C_5H_5N	HCH_5N^+
Initial	0.025	0.030	0.000
Change	–0.025	–0.025	+0.025
Equilibrium	0	0.005	0.025

The reaction continues until the H^+ runs out, leaving pyridine and its conjugate acid at equilibrium. Next, the new concentrations must be determined for both the weak base and its conjugate acid, as both will contribute to the final pH value. The total volume of the new solution is the sum of both the acid and base added, in this case, 25.0 + 60.0 = 85.0 mL. So:

$[C_5H_5N] = 0.005\ \text{mol}/0.085\ \text{L} = 0.059\ M$

$[HCH_5N^+] = 0.025\ \text{mol}/0.085\ \text{L} = 0.29\ M$

Using Henderson-Hasselbalch:

$$pOH = pK_b + \log \frac{[HCH_5N^+]}{[C_5H_5N]}$$

$$pOH = -\log (1.5 \times 10^{-9}) + \log \frac{(0.29)}{(0.059)}$$

$$pOH = 8.8 + \log (4.9)$$

$$pOH = 8.8 + 0.69$$

$$pOH = 9.5$$

$$pH = 14 - 9.5 = 4.5$$

When a weak acid or base is in excess, the pH of the solution does not change as quickly as when a strong acid or base is in excess.

BUFFERS

A **buffer** is a solution with a very stable pH. You can add acid or base to a buffer solution without greatly affecting the pH of the solution. The pH of a buffer will also remain unchanged if the solution is diluted with water or if water is lost through evaporation.

A buffer is created by placing a large amount of a weak acid or base into a solution along with its conjugate, in the form of salt. A weak acid and its conjugate base can remain in solution together without neutralizing each other. This is called the **common ion effect.**

When both the acid and the conjugate base are together in the solution, any hydrogen ions that are added will be neutralized by the base, while any hydroxide ions that are added will be neutralized by the acid, without this having much of an effect on the solution's pH.

Let's say we have a buffer solution with concentrations of 0.20 M $HC_2H_3O_2$ and 0.50 M $C_2H_3O_2^-$. The acid dissociation constant for $HC_2H_3O_2$ is 1.8×10^{-5}. Let's find the pH of the solution.

We can just plug the values we have into the Henderson-Hasselbalch equation for acids.

$$pH = pK_a + \log \frac{[C_2H_3O_2^-]}{[HC_2H_3O]}$$

$$pH = -\log (1.8 \times 10^{-5}) + \log \frac{(0.50M)}{(0.20M)}$$

$$pH = -\log (1.8 \times 10^{-5}) + \log (2.5)$$

$$pH = (4.7) + (0.40) = 5.1$$

Now let's see what happens when $[HC_2H_3O_2]$ and $[C_2H_3O_2^-]$ are both equal to 0.20 *M*.

$$pH = pK_a + \log \frac{[C_2H_3O_2^-]}{[HC_2H_3O]}$$

For this titration, the pH at the equivalence point is exactly 7 because the titration of a strong acid by a strong base produces a neutral salt solution.

$$pH = -\log(1.8 \times 10^{-5}) + \log \frac{(0.20M)}{(0.20M)}$$

$$pH = -\log(1.8 \times 10^{-5}) + \log(1)$$

$$pH = (4.7) + (0) = 4.7$$

Notice that when the concentrations of acid and conjugate base in a solution are the same, pH = pK_a (and pOH = pK_b). When you choose an acid for a buffer solution, it is best to pick an acid with a pK_a that is close to the desired pH. That way you can have almost equal amounts of acid and conjugate base in the solution, which will make the buffer as flexible as possible in neutralizing both added H^+ and OH^-.

You cannot create a buffer solution from a strong acid and its conjugate, because the conjugate base of a strong acid will be very weak. Taking HCl as an example; the Cl^- ion that is left after the acid dissociates completely is a very weak base and will not readily accept protons. The same is true for strong bases; you cannot form a buffer from a strong base and its conjugate for similar reasons.

TITRATION

Neutralization reactions are generally performed by titration, where a base of known concentration is slowly added to an acid (or vice versa). The progress of a neutralization reaction can be shown in a titration curve. The diagram below shows the titration of a strong acid by a strong base.

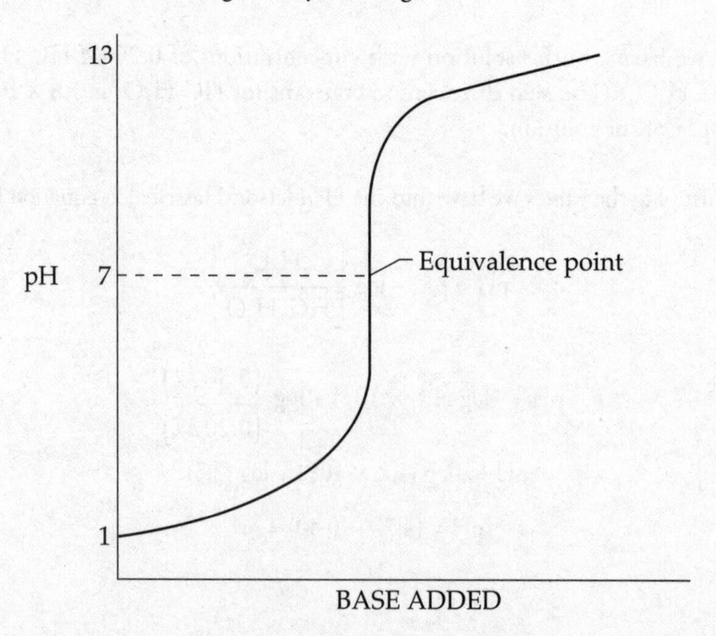

In the diagram above, the pH increases slowly but steadily from the beginning of the titration until just before the equivalence point. The **equivalence point** is the point in the titration when exactly enough base has been added to neutralize all the acid that was initially present. Just before the equivalence point, the pH increases sharply as the last of the acid is neutralized. The equivalence point of a titration can be recognized through the use of an indicator. An indicator is a substance that changes color over a specific pH range. When choosing an indicator, it's important to make sure the equivalence point falls within the pH range for the color change.

The following diagram shows the titration of a weak acid by a strong base:

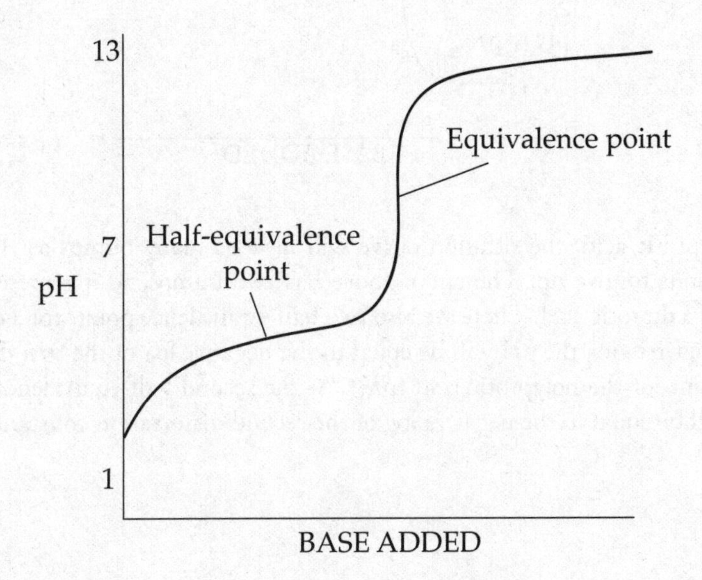

In this diagram, the pH increases more quickly at first, then levels out into a buffer region. At the center of the buffer region is the **half-equivalence point.** At this point, enough base has been added to convert exactly half of the acid into conjugate base; here the concentration of acid is equal to the concentration of conjugate base $[HA] = [A^-]$). Putting that into our weak acid dissociation expression:

$$K_a = \frac{[H^+][A^-]}{[HA]}$$

If $[A^-]$ and $[HA]$ are equal and cancel out, that leaves us with $K_a = [H^+]$, which is often represented by $pK_a = pH$. This is a good way to determine the K_a of a weak acid from a titration curve such as the one above.

The curve remains fairly flat until just before the equivalence point, when the pH increases sharply. For this titration, the pH at the equivalence point is greater than 7 because the only ion present in significant amounts at equilibrium is the conjugate base of the weak acid. Likewise, if you titrate a weak base with a strong acid, the solution will be acidic at equilibrium because the only ion present in significant amounts will be the conjugate acid of the weak base.

The following diagram shows the titration curve of a polyprotic acid:

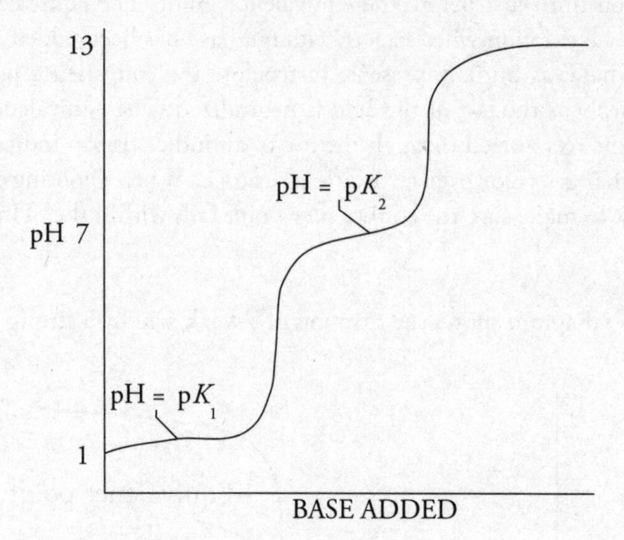

For a polyprotic acid, the titration curve will have as many bumps as there are hydrogen ions to give up. The curve above has two bumps, so it represents the titration of a diprotic acid. There are also two half-equivalence points for a diprotic acid. At the first one, the pH will be equal to the negative log of the first dissociation constant for the polyprotic acid (pK_1). At the second half-equivalence point, the pH will be equal to the negative log of the second dissociation constant (pK_2).

CHAPTER 8 QUESTIONS

Multiple-Choice

Use the following information to answer questions 1-5.

A student titrates 20.0 mL of 1.0 *M* NaOH with 2.0 *M* formic acid, HCO_2H ($K_a = 1.8 \times 10^{-4}$). Formic acid is a monoprotic acid.

1. How much formic acid is necessary to reach the equivalence point?

   (A) 10.0 mL
   (B) 20.0 mL
   (C) 30.0 mL
   (D) 40.0 mL

2. At the equivalence point, is the solution acidic, basic, or neutral? Why?

   (A) Acidic; the strong acid dissociates more than the weak base
   (B) Basic; the only ion present at equilibrium is the conjugate base
   (C) Basic; the higher concentration of the base is the determining factor
   (D) Neutral; equal moles of both acid and base are present

3. If the formic acid were replaced with a strong acid such as HCl at the same concentration (2.0 *M*), how would that change the volume needed to reach the equivalence point?

   (A) The change would reduce the amount as the acid now fully dissociates.
   (B) The change would reduce the amount because the base will be more strongly attracted to the acid.
   (C) The change would increase the amount because the reaction will now go to completion instead of equilibrium.
   (D) Changing the strength of the acid will not change the volume needed to reach equivalance.

4. Which of the following would create a good buffer when dissolved in formic acid?

   (A) $NaCO_2H$
   (B) $HC_2H_3O_2$
   (C) NH_3
   (D) H_2O

5. In which pH range would a buffer made with formic acid be the most effective?

   (A) 0–2
   (B) 2–7
   (C) 7–12
   (D) 12–14

Use the following information to answer questions 6-10.

The following reaction is found to be at equilibrium at 25°C:

$2SO_3(g) \leftrightarrow O_2(g) + 2SO_2(g)$ $\Delta H = -198$ kJ/mol

6. What is the expression for the equilibrium constant, K_c?

(A) $\dfrac{[SO_3]^2}{[O_2][SO_2]^2}$

(B) $\dfrac{2[SO_3]}{[O_2]2[SO_2]}$

(C) $\dfrac{[O_2][SO_2]^2}{[SO_3]^2}$

(D) $\dfrac{[O_2]2[SO_2]}{2[SO_3]}$

7. Which of the following would cause the reverse reaction to speed up?
   (A) Adding more SO_3
   (B) Raising the pressure
   (C) Lowering the temperature
   (D) Removing some SO_2

8. The value for K_c at 25°C is 8.1. What must happen in order for the reaction to reach equilibrium if the initial concentrations of all three species was 2.0 M?
   (A) The rate of the forward reaction would increase, and $[SO_3]$ would decrease.
   (B) The rate of the reverse reaction would increase, and $[SO_2]$ would decrease.
   (C) Both the rate of the forward and reverse reactions would increase, and the value for the equilibrium constant would also increase.
   (D) No change would occur in either the rate of reaction or the concentrations of any of the species.

9. Which of the following would cause a reduction in the value for the equilibrium constant?
   (A) Increasing the amount of SO_3
   (B) Reducing the amount of O_2
   (C) Raising the temperature
   (D) Lowering the temperature

10. Which of the following statements best describes the thermodynamic favorability of the reaction at 25°C?
   (A) The reaction is favored and driven both by enthalpy and entropy changes.
   (B) The reaction is favored but is only driven by enthalpy changes.
   (C) The reaction is favored but is only driven by entropy changes.
   (D) The reaction is not favored.

11. The solubility product, K_{sp}, of AgCl is 1.8×10^{-10}. Which of the following expressions is equal to the solubility of AgCl?

(A) $(1.8 \times 10^{-10})^2$ molar

(B) $\frac{1.8 \times 10^{-10}}{2}$ molar

(C) 1.8×10^{-10} molar

(D) $\sqrt{1.8 \times 10^{-10}}$ molar

12. A 0.1-molar solution of which of the following acids will be the best conductor of electricity?

(A) H_2CO_3
(B) H_2S
(C) HF
(D) HNO_3

13. Which of the following expressions is equal to the K_{sp} of Ag_2CO_3?

(A) $K_{sp} = [Ag^+][CO_3^{2-}]$
(B) $K_{sp} = [Ag^+][CO_3^{2-}]^2$
(C) $K_{sp} = [Ag^+]^2[CO_3^{2-}]$
(D) $K_{sp} = [Ag^+]^2[CO_3^{2-}]^2$

14. If the solubility of BaF_2 is equal to x, which of the following expressions is equal to the solubility product, K_{sp}, for BaF_2?

(A) x^2
(B) $2x^2$
(C) $2x^3$
(D) $4x^3$

Use the following information to answer questions 15-17:

150 mL of saturated SrF_2 solution is present in a 250 mL beaker at room temperature. The molar solubility of SrF_2 at 298 K is 1.0×10^{-3} *M*.

15. What are the concentrations of Sr^{2+} and F^- in the beaker?

(A) $[Sr^{2+}] = 1.0 \times 10^{-3}\ M$ $[F^-] = 1.0 \times 10^{-3}\ M$
(B) $[Sr^{2+}] = 1.0 \times 10^{-3}\ M$ $[F^-] = 2.0 \times 10^{-3}\ M$
(C) $[Sr^{2+}] = 2.0 \times 10^{-3}\ M$ $[F^-] = 1.0 \times 10^{-3}\ M$
(D) $[Sr^{2+}] = 2.0 \times 10^{-3}\ M$ $[F^-] = 2.0 \times 10^{-3}\ M$

16. If some of the solution evaporates overnight, which of the following will occur?

(A) The mass of the solid and the concentration of the ions will stay the same.
(B) The mass of the solid and the concentration of the ions will increase.
(C) The mass of the solid will decrease, and the concentration of the ions will stay the same.
(D) The mass of the solid will increase, and the concentration of the ions will stay the same.

17. How could the concentration of Sr^{2+} ions in solution be decreased?

(A) Adding some NaF(*s*) to the beaker
(B) Adding some $Sr(NO_3)_2(s)$ to the beaker
(C) By heating the solution in the beaker
(D) By adding a small amount of water to the beaker, but not dissolving all the solid

18. A student added 1 liter of a 1.0 *M* KCl solution to 1 liter of a 1.0 *M* $Pb(NO_3)_2$ solution. A lead chloride precipitate formed, and nearly all of the lead ions disappeared from the solution. Which of the following lists the ions remaining in the solution in order of decreasing concentration?

(A) $[NO_3^-] > [K^+] > [Pb^{2+}]$
(B) $[NO_3^-] > [Pb^{2+}] > [K^+]$
(C) $[K^+] > [Pb^{2+}] > [NO_3^-]$
(D) $[K^+] > [NO_3^-] > [Pb^{2+}]$

19. The solubility of PbS in water is 3×10^{-14} molar. What is the solubility product constant, K_{sp}, for PbS?

(A) 9×10^{-7}
(B) 3×10^{-14}
(C) 3×10^{-28}
(D) 9×10^{-28}

20. $2HI(g) + Cl_2(g) \rightleftharpoons 2HCl(g) + I_2(g) + \text{energy}$

A gaseous reaction occurs and comes to equilibrium as shown above. Which of the following changes to the system will serve to increase the number of moles of I_2 present at equilibrium?

(A) Increasing the volume at constant temperature
(B) Decreasing the volume at constant temperature
(C) Increasing the temperature at constant volume
(D) Decreasing the temperature at constant volume

21. A sealed isothermal container initially contained 2 moles of CO gas and 3 moles of H_2 gas. The following reversible reaction occurred:

$$CO(g) + 2H_2(g) \rightleftharpoons CH_3OH(g)$$

At equilibrium, there was 1 mole of CH_3OH in the container. What was the total number of moles of gas present in the container at equilibrium?

(A) 1
(B) 2
(C) 3
(D) 4

22. A 1 *M* solution of $SbCl_5$ in organic solvent shows no noticeable reactivity at room temperature. The sample is heated to 350°C and the following equilibrium reaction is found to occur:

$$SbCl_5 \rightleftharpoons SbCl_3 + Cl_2$$

If the equilibrium concentration of Cl_2 was found to be 0.1 *M*, which of the following best approximated the value of K_{eq} at 350°C?

(A) 1.5
(B) 1.0
(C) 0.1
(D) 0.01

23.

$$2NOBr(g) \rightleftharpoons 2NO(g) + Br_2(g)$$

The reaction above came to equilibrium at a temperature of 100°C. At equilibrium the partial pressure due to NOBr was 4 atmospheres, the partial pressure due to NO was 4 atmospheres, and the partial pressure due to Br_2 was 2 atmospheres. What is the equilibrium constant, K_p, for this reaction at 100°C?

(A) $\frac{1}{4}$

(B) $\frac{1}{2}$

(C) 1

(D) 2

24.

$$Br_2(g) + I_2(g) \leftrightarrow 2IBr(g)$$

At 150°C, the equilibrium constant, K_c, for the reaction shown above has a value of 300. This reaction was allowed to reach equilibrium in a sealed container and the partial pressure due to IBr(*g*) was found to be 3 atm. Which of the following could be the partial pressures due to $Br_2(g)$ and $I_2(g)$ in the container?

	$Br_2(g)$	$I_2(g)$
(A)	0.1 atm	0.3 atm
(B)	0.3 atm	1 atm
(C)	1 atm	1 atm
(D)	1 atm	3 atm

25. $$H_2(g) + CO_2(g) \leftrightarrow H_2O(g) + CO(g)$$

Initially, a sealed vessel contained only $H_2(g)$ with a partial pressure of 6 atm and $CO_2(g)$ with a partial pressure of 4 atm. The reaction above was allowed to come to equilibrium at a temperature of 700 K. At equilibrium, the partial pressure due to $CO(g)$ was found to be 2 atm. What is the value of the equilibrium constant K_p, for the reaction?

(A) $\frac{1}{6}$

(B) $\frac{1}{4}$

(C) $\frac{1}{3}$

(D) $\frac{1}{2}$

26. What is the volume of 0.05 *M* HCl that is required to neutralize 50 mL of 0.10 *M* $Sr(OH)_2$ solution?

(A) 100 mL
(B) 200 mL
(C) 300 mL
(D) 400 mL

27. A laboratory technician wishes to create a buffered solution with a pH of 5. Which of the following acids would be the best choice for the buffer?

(A) $H_2C_2O_4$ — $K_a = 5.9 \times 10^{-2}$
(B) H_3AsO_4 — $K_a = 5.6 \times 10^{-3}$
(C) $H_2C_2H_3O_2$ — $K_a = 1.8 \times 10^{-5}$
(D) HOCl — $K_a = 3.0 \times 10^{-8}$

28. Which of the following species is amphoteric?

(A) H^+
(B) CO_3^{2-}
(C) HCO_3^-
(D) H_2CO_3

29. How many liters of distilled water must be added to 1 liter of an aqueous solution of HCl with a pH of 1 to create a solution with a pH of 2?

(A) 0.1 L
(B) 0.9 L
(C) 2 L
(D) 9 L

30. A 1-molar solution of a very weak monoprotic acid has a pH of 5. What is the value of K_a for the acid?

(A) $K_a = 1 \times 10^{-10}$
(B) $K_a = 1 \times 10^{-7}$
(C) $K_a = 1 \times 10^{-5}$
(D) $K_a = 1 \times 10^{-2}$

31. The value of K_a for HSO_4^- is 1×10^{-2}. What is the value of K_b for SO_4^{2-} ?

(A) $K_b = 1 \times 10^{-12}$
(B) $K_b = 1 \times 10^{-8}$
(C) $K_b = 1 \times 10^{-2}$
(D) $K_b = 1 \times 10^{2}$

32. If 0.630 grams of HNO_3 (molecular weight 63.0) are placed in 1 liter of distilled water at 25°C, what will be the pH of the solution? (Assume that the volume of the solution is unchanged by the addition of the HNO_3.)

(A) 0.01
(B) 0.1
(C) 1
(D) 2

Use the following information to answer questions 33-36.

The following curve is obtained during the titration of 30.0 mL of 1.0 *M* NH_3, a weak base, with a strong acid:

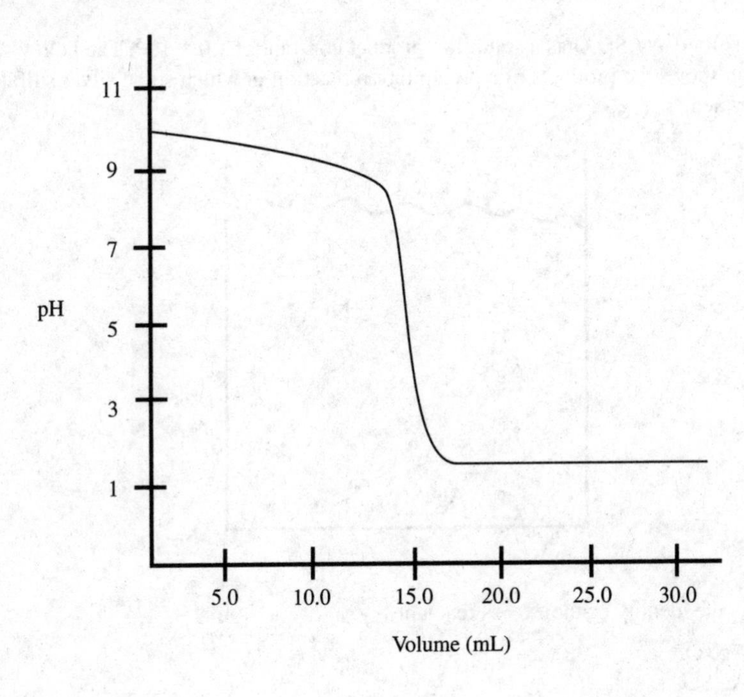

33. Why is the solution acidic at equilibrium?

(A) The strong acid dissociates fully, leaving excess [H^+] in solution.
(B) The conjugate acid of NH_3 is the only ion present at equilibrium.
(C) The water which is being created during the titration acts as an acid.
(D) The acid is diprotic, donating two protons for every unit dissociated.

34. What is the concentration of the acid?

(A) 0.5 *M*
(B) 1.0 *M*
(C) 1.5 *M*
(D) 2.0 *M*

35. What is the pK_b of the base?

(A) 2.5
(B) 4.5
(C) 9.5
(D) 11.5

36. What ions are present in significant amounts during the first buffer region?

(A) NH_3 and NH_4^+
(B) NH_3 and H^+
(C) NH_4^+ and OH^-
(D) H_3O^+ and NH_3

Use the information below to answer questions 37-39.

Silver sulfate, Ag_2SO_4, has a solubility product constant of 1.0×10^{-5}. The below diagram shows the products of a precipitation reaction in which some silver sulfate was formed.

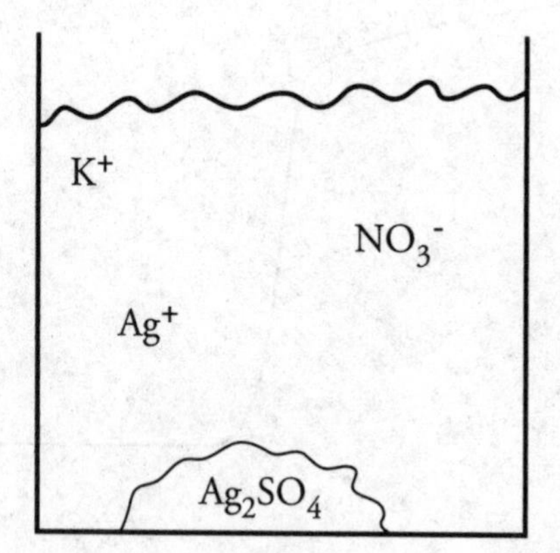

37. What is the identity of the excess reactant?

(A) $AgNO_3$
(B) Ag_2SO_4
(C) $NaNO_3$
(D) Na_2SO_4

38. If the beaker above were left uncovered for several hours:

(A) Some of the Ag_2SO_4 would dissolve.
(B) Some of the spectator ions would evaporate into the atmosphere.
(C) The solution would become electrically imbalanced.
(D) Additional Ag_2SO_4 would precipitate.

39. Which ion concentrations below would have led the precipitate to form?

(A) $[Ag^+] = 0.01\ M$ $[SO_4^{2-}] = 0.01\ M$
(B) $[Ag^+] = 0.10\ M$ $[SO_4^{2-}] = 0.01\ M$
(C) $[Ag^+] = 0.01\ M$ $[SO_4^{2-}] = 0.10\ M$
(D) This is impossible to determine without knowing the total volume of the solution.

40. In a voltaic cell with a Cu (s) | Cu^{2+} cathode and a Pb^{2+} | Pb (s) anode, increasing the concentration of Pb^{2+} causes the voltage to decrease. What is the reason for this?

(A) The value for Q will increase, causing the cell to come closer to equilibrium.
(B) The solution at the anode becomes more positively charged, leading to a reduced electron flow.
(C) The reaction will shift to the right, causing a decrease in favorability.
(D) Cell potential will always decrease anytime the concentration of any aqueous species present increases.

41. Which of the following could be added to an aqueous solution of weak acid HF to increase the percent dissociation?

(A) NaF (*s*)
(B) H_2O (*l*)
(C) NaOH (*s*)
(D) NH_3 (*aq*)

42. A bottle of water is left outside early in the morning. The bottle warms gradually over the course of the day. What will happen to the pH of the water as the bottle warms?

(A) Nothing; pure water always has a pH of 7.00.
(B) Nothing; the volume would have to change in order for any ion concentration to change.
(C) It will increase because the concentration of $[H^+]$ is increasing.
(D) It will decrease because the auto-ionization of water is an endothermic process.

Free-Response Questions

1. The value of the solubility product, K_{sp}, for calcium hydroxide, $Ca(OH)_2$, is 5.5×10^{-6}, at 25°C.
   (a) Write the K_{sp} expression for calcium hydroxide.
   (b) What is the mass of $Ca(OH)_2$ in 500 mL of a saturated solution at 25°C?
   (c) What is the pH of the solution in (b)?
   (d) If 1.0 mole of OH^- is added to the solution in (b), what will be the resulting Ca^{2+} concentration? Assume that the volume of the solution does not change.

2. For sodium chloride, the solution process with water is endothermic.
   (a) Describe the change in entropy when sodium chloride dissociates into aqueous particles.
   (b) Two saturated aqueous NaCl solutions, one at 20°C and one at 50°C, are compared. Which one will have higher concentration? Justify your answer.
   (c) Which way will the solubility reaction shift if the temperature is increased?
   (d) If a saturated solution of NaCl is left out overnight and some of the solution evaporates, how will that affect the amount of solid NaCl present?

3.

$$H_2CO_3 \rightleftharpoons H^+ + HCO_3^- \qquad K_1 = 4.3 \times 10^{-7}$$

$$HCO_3^- \rightleftharpoons H^+ + CO_3^{2-} \qquad K_2 = 5.6 \times 10^{-11}$$

The acid dissociation constants for the reactions above are given at 25°C.

   (a) What is the pH of a 0.050-molar solution of H_2CO_3 at 25°C?
   (b) What is the concentration of CO_3^{2-} ions in the solution in (a)?
   (c) How would the addition of each of the following substances affect the pH of the solution in (a)?
   (i) HCl
   (ii) $NaHCO_3$
   (iii) NaOH
   (iv) NaCl

4. $N_2(g) + 3H_2(g) \rightleftharpoons 2NH_3(g)$ $\Delta H = -92.4$ kJ

When the reaction above took place at a temperature of 570 K, the following equilibrium concentrations were measured:

$[NH_3] = 0.20$ mol/L
$[N_2] = 0.50$ mol/L
$[H_2] = 0.20$ mol/L

(a) Write the expression for K_c and calculate its value.
(b) Calculate ΔG for this reaction.
(c) Describe how the concentration of H_2 at equilibrium will be affected by each of the following changes to the system at equilibrium:
(i) The temperature is increased.
(ii) The volume of the reaction chamber is increased.
(iii) N_2 gas is added to the reaction chamber.
(iv) Helium gas is added to the reaction chamber.

5. In an acidic medium, iron (III) ions will react with thiocyanate (SCN^-) ions to create the following complex ion:

$Fe^{3+}(aq) + SCN^-(aq) \leftrightarrow FeSCN^{2+}(aq)$

Initially, the solution is a light yellow color due to the presence of the Fe^{3+} ions. As the $FeSCN^{2+}$ forms, the solution will gradually darken to a golden yellow. The reaction is not a fast one, and generally after mixing the ions the maximum concentration of $FeSCN^{2+}$ will occur between 2–4 minutes after mixing the solution.

A student creates four solution with varying concentration of $FeSCN^{2+}$ and gathers the following data at 298 K using a spectrophotometer calibrated to 460 nm:

$[FeSCN^{2+}]$	Absorbance
1.1×10^{-4} *M*	0.076
1.6×10^{-4} *M*	0.112
2.2×10^{-4} *M*	0.167
2.5×10^{-4} *M*	0.199

(a) (i) On the axes below, create a Beer's Law calibration plot for $[FeSCN^{2+}]$. Draw a best-fit line through your data points.

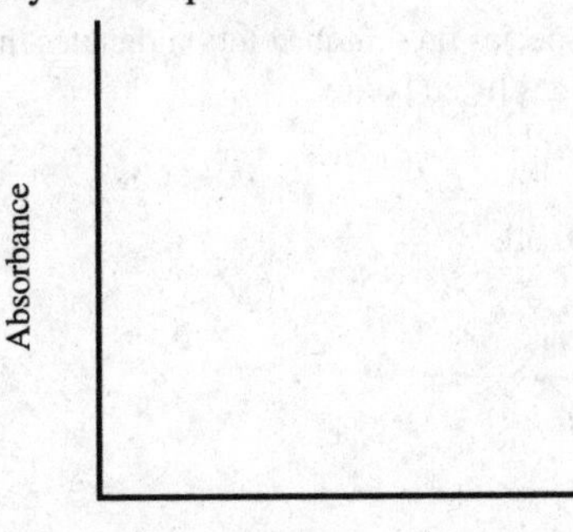

(ii) The slope of the best-fit line for the above set of data points is 879 and the *y*-intercept is –0.024. Write out the equation for this line.

To determine the equilibrium constant for the reaction, a solution is made up in which 5.00 mL of 0.0025 *M* $Fe(NO_3)_3$ and 5.00 mL of 0.0025 *M* KSCN are mixed. After 3 minutes, the absorbance of the solution is found to be 0.134.

(b) (i) Using your Beer's law best-fit line from (a), calculate [$FeSCN^{2+}$] once equilibrium has been established.
(ii) Calculate [Fe^{3+}] and [SCN^-] at equilibrium.
(iii) Calculate K_{eq} for the reaction.

After equilibrium is established, the student heats the solution and observes that it becomes noticeably lighter.

(c) (i) Did heating the mixture increase the equilibrium constant, decrease it, or have no effect on it? Why?
(ii) Is the equilibrium reaction exothermic or endothermic? Justify your answer.

6.

$$HA + OH \rightleftharpoons A^- + H_2O(l)$$

A student titrates a weak acid, HA, with some 1.0 *M* NaOH, yielding the following titration curve:

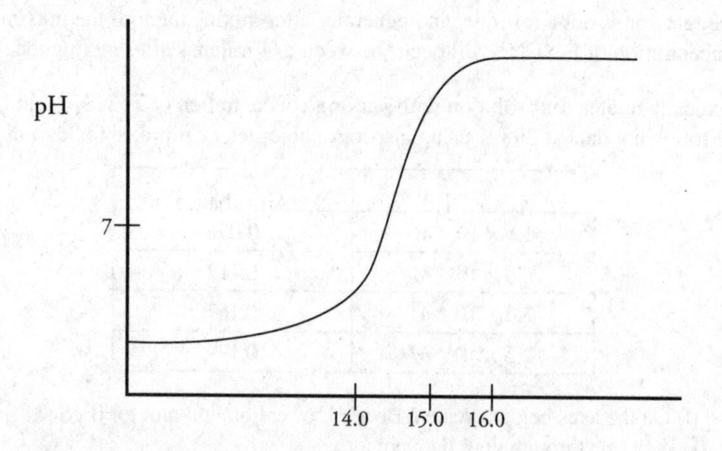

(a) Which chemical species present in solution dictates the pH of the solution in each of the volume ranges listed below?
(i) 1.0 mL–14.0 mL
(ii) 15.0 mL
(iii) 16.0 mL–30.0 mL

(b) At which volumes is:
(i) [HA] > [A^-]?
(ii) [HA] = [A^-]?
(iii) [HA] < [A^-]?

(c) At which point in the titration (if any) would the concentration of the following species be equal to zero? Justify your answers.

(i) HA

(ii) A^-

(d) If the titration were performed again, but this time with 2.0 *M* NaOH, name two things that would change about the titration curve, and explain the reasoning behind your identified changes.

7. A student performs an experiment to determine the concentration of a solution of hypochlorous acid, HOCl ($K_a = 3.5 \times 10^{-8}$). The student starts with 25.00 mL of the acid in a flask and titrates it against a standardized solution of sodium hydroxide with a concentration of 1.47 *M*. The equivalence point is reached after the addition of 34.23 mL of NaOH.

(a) Write the net ionic equation for the reaction that occurs in the flask.

(b) What is the concentration of the HOCl?

(c) What would the pH of the solution in the flask be after the addition of 28.55 mL of NaOH?

(d) The actual concentration of the HOCl is found to be 2.25 *M*. Quantitatively discuss whether or not each of the following errors could have caused the error in the student's results.

(i) The student added additional NaOH past the equivalence point.

(ii) The student rinsed the buret with distilled water but not with the NaOH solution before filling it with NaOH.

(iii) The student measured the volume of acid incorrectly; instead of adding 25.00 mL of HOCl, only 24.00 mL was present in the flask prior to titration.

CHAPTER 8 ANSWERS AND EXPLANATIONS

Multiple-Choice

1. A The equivalence point is defined as the point at which the moles of acid are equal to the moles of base. The amount of NaOH is equal to (1.0 *M*)(0.0200 *L*) = 0.0200 mol. To calculate the volume of acid:

$$2.0\ M = 0.0200\ \text{mol}/V$$

$$V = 0.0100\ \text{L} = 10.0\ \text{mL}$$

2. B Using the ICE chart:

	HCO_2H	OH^-	CO_2H^-	H_2O
Initial	0.020	0.020	0	X
Change	–0.020	–0.020	+ 0.020	X
Equilibrium	0	0	0.020	X

Both the hydroxide ion and the weak acid are not present at equilibrium. The only ionic species left, the O_2CH^-, is basic, and so the solution will be as well.

3. D The number of moles of base is staying the same, and so is the concentration of the acid. Therefore, the same volume of that acid will be needed to get to the equivalence point.

4. A A buffer is made up of an acid and its conjugate base. The conjugate base of HCO_2H is CO_2H^-, which is present as the anion in the $NaCO_2H$ salt.

5. B The ideal buffer will have a pK_a close to the pH of the buffer solution. The K_a of formic acid is 1.8×10^{-4}, meaning its pK_a will be in the 3–4 range.

6. C The equilibrium expression is always products over reactants, so that eliminates (A) and (B). The coefficients in the balanced equation turn into exponents in the expression, leading us to the correct answer.

7. B To speed up the reverse reaction, we are looking to cause a shift to the left via Le Châtelier's principle. If the pressure were to increase, there would be a shift to the side with fewer gas molecules, which in this case means a shift to the left. All other options cause a shift to the right.

8. **A** To determine which way the reaction would shift, the reaction quotient, Q, would need to be calculated. This can be done by plugging the concentrations into the law of mass action. With all of the initial values being 2.0 M:

$$Q = \frac{(2.0)(2.0)^2}{(2.0)^2} = 2.0$$

2.0 < 8.1, so the reaction must shift right in order for the reaction to proceed to equilibrium. This would cause an increase in the rate of the forward reaction, along with a decrease in the $[SO_3]$.

9. **C** Changing the amounts of either reactants or products present will not cause a change in the equilibrium constant (nor would changing the pressure, if that were an option). The only way to actually change the value of the constant is by changing the temperature. As this is an exothermic reaction, adding more heat would cause a shift to the left, increasing the amount of reactants, and thus the denominator in the equilibrium constant expression. This, in turn, reduces the value for K_c.

10. **A** Using $\Delta G = \Delta H - T\Delta S$, we can see the value for ΔH is negative, which promotes a negative ΔG and thus favorability. ΔS would be positive because there are more moles of gas molecules present after the reaction. A positive ΔS leads to a negative $T\Delta S$ term, which also promotes favorability.

11. **D** The solubility of a substance is equal to its maximum concentration in solution.

    For every AgCl in solution, we get one Ag^+ and one Cl^-, so the solubility of AgCl—let's call it x—will be the same as $[Ag^+]$, which is the same as $[Cl^-]$.

    So for AgCl, $K_{sp} = [Ag^+][Cl^-] = 1.8 \times 10^{-10} = x^2$.

    $$x = \sqrt{1.8 \times 10^{-10}}$$

12. **D** The best conductor of electricity (also called the strongest electrolyte) will be the solution that contains the most charged particles.

HNO_3 is the only strong acid listed in the answer choices, so it is the only choice where the acid has dissociated completely in solution into H^+ and NO_3^- ions. So a 0.1-molar HNO_3 solution will contain the most charged particles and, therefore, be the best conductor of electricity.

13. **C** K_{sp} is just the equilibrium constant without a denominator.

When Ag_2CO_3 dissociates, we get the following reaction:

$Ag_2CO_3(s) \rightleftharpoons 2Ag^+ + CO_3^{2-}$

In the equilibrium expression, coefficients become exponents, so we get:

$K_{sp} = [Ag^+]^2[CO_3^{2-}]$

14. **D** For BaF_2, $K_{sp} = [Ba^{2+}][F^-]^2$.

For every BaF_2 that dissolves, we get one Ba^{2+} and two F^-.

So if the solubility of BaF_2 is x, then $[Ba^{2+}] = x$, and $[F^-] = 2x$.

So $K_{sp} = (x)(2x)^2 = (x)(4x^2) = 4x^3$

15. **B** When SrF_2 dissociates, it creates one Sr^{2+} ion and two F^- ions. That means the concentration of fluoride ions will be twice that of strontium ions in a saturated solution.

16. **D** If the solution is saturated, that means the concentrations of Sr^{2+} and F^- are at their maximum values and would remain unchanged even if some water evaporated. If that happens, thus decreasing the volume of the solution, there will not be as much room for the solute ions in the solution. These ions would fall out of solution, increasing the mass of the solid, and maintaining the same concentration of ions in solution.

17. **A** If extra F^- ions are added to the solution via the addition of NaF (which contains an alkali metal cation and is thus fully soluble), they will bond with some of the Sr^{2+} ions that were present to create more $SrF_2(s)$ and reduce the number of Sr^{2+} ions present in solution. This is called the common ion effect. Choice (D) is incorrect because as long as there is still solid on the bottom of the beaker, the equilibrium concentration of the ions will remain unchanged.

18. A At the start, the concentrations of the ions are as follows:

$[K^+] = 1\ M$
$[Cl^-] = 1\ M$
$[Pb^{2+}] = 1\ M$
$[NO_3^-] = 2\ M$

After $PbCl_2$ forms, the concentrations are as follows:

$[K^+] = 1\ M$
$[Cl^-] = 0.5\ M$
$[Pb^{2+}] = 0\ M$
$[NO_3^-] = 2\ M$

So from greatest to least:

$[NO_3^-] > [K^+] > [Pb^{2+}]$

19. D The solubility of a substance is equal to its maximum concentration in solution. For every PbS in solution, we get one Pb^{2+} and one S^{2-}, so the concentration of PbS, $3 \times 10^{-14}\ M$, will be the same as the concentrations of Pb^{2+} and S^{2-}.

$K_{sp} = [Pb^{2+}][S^{2-}]$
$K_{sp} = (3 \times 10^{-14}\ M)(3 \times 10^{-14}\ M) = 9 \times 10^{-28}$

20. D According to Le Châtelier's law, the equilibrium will shift to counteract any stress that is placed on it. If the temperature is decreased, the equilibrium will shift toward the side that produces energy or heat. That's the product side where I_2 is produced.

21. C Because the equation is balanced, the following will occur:

If 1 mole of CH_3OH was created, then 1 mole of CO was consumed and 1 mole of CO remains; and if 1 mole of CH_3OH was created, then 2 moles of H_2 were consumed and 1 mole of H_2 remains. So at equilibrium, there are:

(1 mol CH_3OH) + (1 mol CO) + (1 mol H_2) = 3 moles of gas

22. D Since all the Cl_2 found in solution must have come from $SbCl_5$, we know that at equilibrium

$[Cl_2] = [SbCl_3] = 0.1\ M$, and $[SbCl_5] = (1.0 - 0.1)\ M = 0.99\ M$.

We can then say that $K = (0.1)(0.1)/0.99 = 0.0101$ which is most closely approximated by (D).

23. D $K_p = \dfrac{[NO]^2[Br_2]}{[NOBr]^2} = \dfrac{(4)^2(2)}{(4)^2} = 2$

24. A The equilibrium expression for the reaction is as follows:

$$\frac{P_{IBr}^{\ 2}}{P_{Br_2}P_{I_2}} = 300$$

When all of the values are plugged into the expression, (A) is the only choice that works.

$$\frac{(3)^2}{(0.1)(0.3)} = \frac{9}{0.03} = 300$$

25. D Use a table to see how the partial pressures change. Based on the balanced equation, we know that if 2 atm of $CO(g)$ were formed, then 2 atm of $H_2O(g)$ must also have formed. We also know that the reactants must have lost 2 atm each.

	$H_2(g)$	$CO_2(g)$	$H_2O(g)$	$CO(g)$
Initial	6 atm	4 atm	0	0
Change	−2	−2	+2	+2
Equilibrium	4 atm	2 atm	2 atm	2 atm

Now plug the numbers into the equilibrium expression.

$$K_{eq} = \frac{P_{H_2O}P_{CO}}{P_{H_2}P_{CO_2}} = \frac{(2)(2)}{(4)(2)} = \frac{1}{2}$$

26. **B** Every mole of $Sr(OH)_2$ dissociates to produce 2 moles of OH^- ions, so a 0.10 *M* $Sr(OH)_2$ solution will have a $[OH^-]$ of 0.20 *M*.

The solution will be neutralized when the number of moles of H^+ ions added is equal to the number of OH^- ions originally in the solution.

Moles = (molarity)(volume)

Moles of OH^- = (0.20 *M*)(50 mL) = 10 millimoles = moles of H^+ added

$$\text{Volume} = \frac{\text{moles}}{\text{molarity}}$$

$$\text{Volume of HCl} = \frac{(10\ \text{millimoles})}{(0.05\,M)} = 200\ \text{mL}$$

27. **C** The best buffered solution occurs when pH = pK_a. That happens when the solution contains equal amounts of acid and conjugate base. If you want to create a buffer with a pH of 5, the best choice would be an acid with a pK_a that is as close to 5 as possible. You shouldn't have to do a calculation to see that the pK_a for (C) is much closer to 5 than that of any of the others.

28. **C** An amphoteric species can act either as an acid or a base, gaining or losing a proton.

HCO_3^- can act as an acid, losing a proton to become CO_3^{2-}, or it can act as a base, gaining a proton to become H_2CO_3.

29. **D** We want to change the hydrogen ion concentration from 0.1 *M* (pH of 1) to 0.01 *M* (pH of 2).

The HCl is completely dissociated, so the number of moles of H^+ will remain constant as we dilute the solution.

Moles = (molarity)(volume) = Constant

$(M_1)(V_1) = (M_2)(V_2)$

$(0.1\ M)(1\ \text{L}) = (0.01\ M)(V_2)$

So, V_2 = 10 L, which means that 9 L must be added.

30. **A** A pH of 5 means that $[H^+] = 1 \times 10^{-5}$

$$K_a = \frac{[H^+][A^-]}{[HA]}$$

For every HA that dissociates, we get one H^+ and one A^-, so $[H^+] = [A^-] = 1 \times 10^{-5}$.

The acid is weak, so we can assume that very little HA dissociates and that the concentration of HA remains 1-molar.

$$K_a = \frac{[H^+][A^-]}{[HA]} = \frac{(1\times10^{-5})(1\times10^{-5})}{(1)} = 1 \times 10^{-10}.$$

31. **A** For conjugates, $(K_a)(K_b) = K_w = 1 \times 10^{-14}$

$$K_b = \frac{K_w}{K_a} = \frac{(1\times10^{-14})}{(1\times10^{-2})} = 1 \times 10^{-12}$$

32. **D** Every unit of HNO_3 added to the solution will place 1 unit of H^+ ions in the solution. So first find the moles of HNO_3 added.

$$\text{Moles} = \frac{\text{grams}}{\text{MW}}$$

$$\text{Moles of } H^+ = \frac{0.630 \text{ grams}}{63.0 \text{ g/mole}} = 0.01 \text{ moles}$$

Now it's easy to find the H^+ concentration.

$$\text{Molarity} = \frac{\text{moles}}{\text{liters}}$$

$$[H^+] = \frac{0.01 \text{ moles}}{1 \text{ L}} = 0.01\ M$$

$$\text{pH} = -\log [H^+] = -\log(0.01) = 2$$

33. **B** The reaction here is $NH_3 + H^+ \rightleftharpoons NH_4^+$. At equilibrium, the moles of NH_3 and H^+ would be equal, leaving behind NH_4^+ ions, which will then donate ions to water, creating an acidic medium.

34. **D** At the equivalence point, the moles of acid are equal to the moles of base.

Moles base = $(1.0\ M)(0.030 \text{ L}) = 0.030$ mol base = 0.030 moles acid

It requires 15.0 mL of acid to reach equivalence, so:

$$\frac{0.030 \text{ mol}}{0.015 \text{ L}} = 2.0\ M$$

35. **B** At the half-equivalence point, half of the base has been protonated, so $[NH_3] = [NH_4^+]$. Subbing that into the base version of the Henderson-Hasselbach equation yields

$$pK_b = pOH + \log \frac{[NH_4^+]}{[NH_3]} \qquad \frac{[NH_4^+]}{[NH_3]} = 1, \text{ and } \log 1 = 0.$$

So, at the half-equivalence point, $pK_b = pOH$. The equivalence point is at $V = 15.0$ mL, and at $V = 7.5$ mL the pH is about 9.5. The pOH (and the pK_b) of the base is thus 4.5.

36. **A** The reaction occurring is $NH_3 + H^+ \rightleftharpoons NH_4^+$. During the first buffer region, all added hydrogen ions immediately react with NH_3 to create NH_4^+. NH_3 remains in excess until equilibrium is achieved.

37. **A** There are no sulfate ions in solution, which means that whatever the sulfate was attached to was the limiting reactant. Thus, silver must have been attached to the excess reactant, which is $AgNO_3$. (Remember, Ag_2SO_4 was the product, not a reactant.)

38. **D** Leaving the container uncovered will cause some of the water molecules to evaporate. This allows some Ag+ and SO_4^{2-} ions to "fall" out of solution and combine to create more Ag_2SO_4.

39. **B** For a precipitate to form, $Q > K_{sp}$. In this case, $Q = [Ag^+]^2[SO_4^{2-}]$. The concentrations in (B) lead to a Q of 1.0×10^{-4}, which is greater than the given K_{sp} of 1.0×10^{-5}. The other options lead to a Q value that is equal to or less than K_{sp}.

40. **A** The reaction here is Pb (s) + $Cu^{2+} \rightleftharpoons$ Cu (s) + Pb^{2+}. Increasing the concentration of Pb^{2+} causes Q to increase, and as all voltaic cells have $K > 1$, that brings Q closer to K. For any voltaic cell, the voltage is zero at equilibrium, so bringing the cell closer to equilibrium brings the voltage closer to zero.

41. **B** Percent dissociation is inversely proportional to concentration. Adding some water to the solution will cause the concentration of the HF to decrease, leading to a greater percent dissociation.

42. **D** $H_2O \rightleftharpoons H^+ + OH^-$ is an endothermic process, requiring energy in order for the bonds within the water molecules to break. Thus, heat is a reactant, and as temperature increases the reaction will shift right. Because this is a temperature shift, this causes a permanent increase in the value of K_w. This, in turn, causes the concentrations of H^+ and OH^- to increase, and the –log of an increased concentration leads to a decreased value for pH.

Free-Response

1. (a) The solubility product is the same as the equilibrium expression, but because the reactant is a solid, there is no denominator.

   $K_{sp} = [Ca^{2+}][OH^-]^2$

   (b) Use the solubility product.

   $K_{sp} = [Ca^{2+}][OH^-]^2$

   $5.5 \times 10^{-6} = (x)(2x)^2 = 4x^3$

   $x = 0.01\ M$ for Ca^{2+}

   One mole of calcium hydroxide produces 1 mole of Ca^{2+}, so the concentration of $Ca(OH)_2$ must be 0.01 M.

   Moles = (molarity)(volume)

   Moles of $Ca(OH)_2$ = (0.01 M)(0.500 L) = 0.005 moles

   Grams = (moles)(MW)

   Grams of $Ca(OH)_2$ = (0.005 mol)(74 g/mol) = 0.37 g

   (c) We can find $[OH^-]$ from (b).

   If $[Ca^{2+}] = 0.01\ M$, then $[OH^-]$ must be twice that, so $[OH^-] = 0.02\ M$.

   $pOH = -\log[OH^-] = 1.7$

   $pH = 14 - pOH = 14 - 1.7 = 12.3$

   (d) Find the new $[OH^-]$. The hydroxide already present is small enough to ignore, so we'll use only the hydroxide just added.

   $$\text{Molarity} = \frac{\text{moles}}{\text{liters}}$$

   $$[OH^-] = \frac{(1.0\ \text{mol})}{(0.500\ \text{L})} = 2.0\ M$$

   Now use the K_{sp} expression.

   $K_{sp} = [Ca^{2+}][OH^-]^2$

   $5.5 \times 10^{-6} = [Ca^{2+}](2.0\ M)^2$

   $[Ca^{2+}] = 1.4 \times 10^{-6}\ M$

2. (a) Entropy increases when a salt dissociates because aqueous particles have more randomness than a solid.

(b) Most salt solution processes are endothermic, and endothermic processes are favored by an increase in temperature. Therefore, increasing temperature will increase the solubility of most salts.

(c) In an endothermic reaction, the energy is part of the reactants. Increasing the temperature would thus shift the reaction to the right.

(d) The solution was already saturated, which means $[Na^+]$ and $[Cl^-]$ were already at their maximum possible values. If the volume of water decreases, the number of moles of Na^+ and Cl^- present in solution must also decrease in order for their concentrations to remain constant. In order for that to happen, some of the sodium and chloride ions present will form a precipitate, increasing the mass of NaCl(*s*) present.

3. (a) Use the equilibrium expression.

$$K_1 = \frac{[H^+][HCO_3^-]}{[H_2CO_3]}$$
$$[H^+] = [HCO_3^-] = x$$
$$[H_2CO_3] = (0.050\ M - x)$$

Assume that x is small enough that we can use:
$[H_2CO_3]$ = (0.050 M).

$$4.3 \times 10^{-7} = \frac{x^2}{(0.050)}$$
$$x = [H^+] = 1.5 \times 10^{-4}$$
$$pH = -\log[H^+] = -\log(1.5 \times 10^{-4}) = 3.8$$

(b) Use the equilibrium expression.

$$K_2 = \frac{[H^+][CO_3^{2-}]}{[HCO_3^-]}$$

From (a) we know that $[H^+] = [HCO_3^-] = 1.5 \times 10^{-4}$.

$$5.6 \times 10^{-11} = \frac{(1.5 \times 10^{-4})[CO_3^{2-}]}{(1.5 \times 10^{-4})} = [CO_3^{2-}]$$
$$[CO_3^{2-}] = 5.6 \times 10^{-11}\ M$$

(c) (i) Adding HCl will increase $[H^+]$, lowering the pH.

(ii) From Le Châtelier's law, you can see that adding $NaHCO_3$ will cause the first equilibrium to shift to the left to try to use up the excess HCO_3^-. This will cause a decrease in $[H^+]$, raising the pH.

You may notice that adding $NaHCO_3$ will also cause the second equilibrium to shift toward the right, which should increase $[H^+]$, but because K_2 is much smaller than K_1, this shift is insignificant.

(iii) Adding NaOH will neutralize hydrogen ions, decreasing $[H^+]$ and raising the pH.

(iv) Adding NaCl will have no effect on the pH.

4. (a) $K_c = \frac{[NH_3]^2}{[N_2][H_2]^3}$

$K_c = \frac{(0.20)^2}{(0.50)(0.20)^3} = 10$

(b) $\Delta G = -RT \ln K$

$\Delta G = -(8.31\text{ J/mol} \times \text{K})(570\text{ K}) \ln(10)$

$\Delta G = (-4740)(2.3)$

$\Delta G = -11{,}000\text{ J/mol}$

(c) (i) An increase in temperature will shift the reaction to the left, causing an increase in the H_2 concentration in the short term. The temperature increase also causes an decrease in the value for the equilibrium constant. A smaller equilibrium constant means the species in the denominator of the mass action expression will be present in increased concentrations at equilibrium, so $[H_2]$ will remain increased.

(ii) An increase in volume favors the direction that produces more moles of gas. In this case, that's the reverse direction, so the concentration of H_2 will increase in the short term. However, an increased chamber volume will eventually reduce the concentrations of all species at equilibrium, as the number of moles remains constant and concentration is calculated using moles/volume. Thus, $[H_2]$ will decrease by the time equilibrium re-establishes.

(iii) Adding N_2 will initially shift the reaction right, and the concentration of H_2 will decrease. However, by the time equilibrium is re-established, there will be more moles of all three substances in the container, so ultimately $[H_2]$ will be increased.

(iv) The addition of He, a gas that takes no part in the reaction, will have no effect on the concentration of H_2.

5. (a) (i)

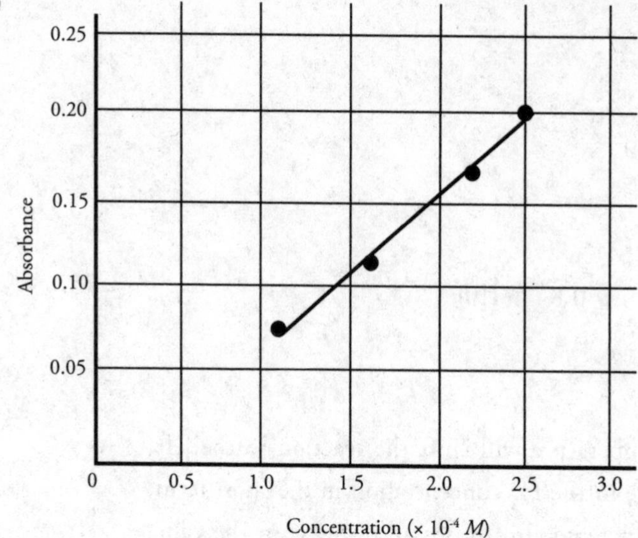

(ii) Slope-intercept form is $y = mx + b$. So:

$$y = 879x - 0.0240$$

It is worth noting that in a perfect world, the y-intercept for this graph would be zero; that is, when there is no $FeSCN^{2+}$ present the solution will not absorb light. Due to the nature of experimental data in the real world being not perfect, though, it is best in this case to include the approximate y-intercept in your best-fit line, as it may account for variations in your equipment.

(b) (i) Using the equation from part (a):

$$y = 879x - 0.0240$$

$$0.134 = 879x - 0.0240$$

$$x = 1.80 \times 10^{-4}\ M$$

(ii) An ICE chart will be very helpful here. To do that, we first need to calculate the initial concentration of the Fe^{3+} and the SCN^{-}. They both start with the same volumes and concentrations, so one calculation will suffice. We'll start with calculating the moles present:

$$0.0025\ M = \frac{n}{0.0050\text{ L}} \qquad n = 1.3 \times 10^{-5}\text{ mol}$$

Then, we have to divide by the total volume of both solutions combined (10.0 mL) in order to find out the concentration of each ion.

$$\frac{1.3 \times 10^{5}\ \text{mol}}{0.010\ \text{L}} = 1.3 \times 10^{-3}\ M$$

Our ICE chart thus looks like:

	$Fe^{3+}(aq)$	+	$SCN^{-}(aq)$	$\leftrightarrow$	$FeSCN^{2+}(aq)$
I	1.3×10^{-3}		1.3×10^{-3}		0
C	-1.80×10^{-4}		-1.80×10^{-4}		$+1.80 \times 10^{-4}$
E	1.1×10^{-3}		1.1×10^{-3}		1.80×10^{-4}

So at equilibrium $[Fe^{3+}] = [SCN^{-}] = 1.1 \times 10^{-3}\ M$.

(iii) $$K_{eq} = \frac{[FeSCN^{2+}]}{[Fe^{3+}][SCN-]} = \frac{1.80 \times 10^{-4}}{(1.1 \times 10^{-3})(1.1 \times 10^{-3})} = 150$$

(c) (i) Changing the temperature of a reaction is the only way to change the equilibrium constant, which is what is happening here. As the solution becomes lighter, the amount of $FeSCN^{2+}$ decreases, which indicates a shift to the left. Doing so causes there to be more products and fewer reactants at equilibrium, which lowers the value of the equilibrium constant.

(ii) Using our trick for temperature-caused shifts, we can see that a shift to the left means that there must have been heat on the products side of the equilibrium. This means the reaction is exothermic.

6. (a) (i) Between these volumes lies the buffer region of the titration, in which both [HA] and [A^-] contribute significantly to the pH of the solution.

(ii) At equivalence, the only species present which affects the pH of the solution is the conjugate base, A^-.

(iii) In this region, the strong base is in excess, and [OH^-] determines the solution.

(b) (i) As the reaction progress, OH^- will take protons from HA to create A^-. There will be more HA in solution until half of it has been deprotonated. The point at which that occurs is the half-equivalence point at 7.50 mL. So, [HA] > [A^-] from 0 to 7.5 mL.

(ii) The only time these values are equal is when exactly half of the HA has been converted to A^-, which occurs at 7.50 mL.

(iii) For the remainder of the titration (7.50 mL+), there will be more conjugate base present than there will be of original weak acid.

(c) (i) and (ii) Neither of these values will ever be equal to zero. Prior to the titration beginning, there is some A^- in solution via the reaction of HA with water, as follows:

$$HA + H_2O \rightleftharpoons A^- + H_3O^+$$

Adding OH^- through titrating the strong base in will cause the HA molecules present to deprotonate fully, but the conjugate base that is created will, in turn, also react with water to create additional HA molecules:

$$A^- + H_2O \rightleftharpoons HA + OH^-$$

As both the reaction of the weak acid with water and that of its conjugate base with water will always be occurring, there will always be some of each species in solution.

(d) (i) Twice as much OH^- is being added per drop, so the pH changes charted will occur over only 15.0 mL instead of 30.0 mL (for instance, equivalence will occur at 7.50 mL instead of 15.0 mL).

(ii) The final pH of the solution will be slightly higher, as a higher concentration of NaOH will lead to a higher pH in the region where [OH^-] dictates the pH of the solution.

7. (a) The hypochlorous acid will donate its proton to the hydroxide. The sodium ions are spectators and would not appear in the net ionic equation.

$HOCl + OH^- \leftrightarrow OCl^- + H_2O(l)$

(b) At the equivalence point, the moles of acid are equal to the moles of base.

Moles of base = (1.47 *M*)(0.03423 L) = 0.0503 mol

Concentration acid = (0.0503 mol/0.02500 L) = 2.01 *M*

(c) This calls for an ICE chart. We will first need the moles of both the acid and base.

Moles HOCl = (2.01 *M*)(0.02500 L) = 0.0503 mol HOCl

Moles OH^- = (1.47 *M*)(0.02855 L) = 0.0420 mol OH^-

Putting those numbers into our ICE chart:

	HOCl	OH^-	OCl^-	H_2O
Initial	0.0503	0.0420	0	X
Change	–0.0420	–0.0420	0.0420	X
Equilibrium	0.0083	0	0.0420	X

Now that we know the number of moles at equilibrium of each species, we need to determine their new concentrations. The total volume of the solution at equilibrium is 25.00 mL + 28.55 mL = 53.55 mL. So the concentrations at equilibrium will be:

[HOCl] = 0.0083 mol/0.05355 L = 0.15 *M*
$[OCl^-]$ = 0.0420 mol/0.05355 L = 0.784 *M*

To finish off we use Henderson-Hasselbalch:

$$pH = pK_a + \log \frac{[OCl^-]}{[HOCl]}$$

$$pH = -\log(3.5 \times 10^{-8}) + \log \frac{(0.784)}{(0.15)}$$

$$pH = 7.5 + \log(5.2)$$

$$pH = 7.5 + 0.72$$

$$pH = 8.2$$

(d) (i) If additional NaOH is added, that would mean a larger number of moles of NaOH added to the flask, and thus more apparent moles of HOCl would be present at the equivalence point. More apparent moles of HOCl would lead to a larger numerator in the acid molarity calculation and thus a larger apparent molarity. As the student's calculated molarity was too low, this could not have caused the error.

(ii) If the buret was not rinsed with the NaOH prior to filling it, the concentration of the NaOH would be diluted by the water inside the buret. The student would then have to add a greater volume of NaOH to reach equivalence. However, in the calculations, using the original concentration of NaOH multiplied by the higher volume would lead to an artificially high number of moles of NaOH and thus, more apparent moles HOCl at equivalence. More apparent moles of HOCl would lead to a larger numerator in the acid molarity calculation and thus a larger apparent molarity. As the student's calculated molarity was too low, this could not have caused the error.

(iii) If the student did not add enough acid to the flask, that would cause the denominator in the molarity of the acid calculation to be artificially high. This, in turn, would make the calculated acid molarity be artificially low. This matches up with the student's results and could be an acceptable source of error.

Chapter 9
Laboratory Overview

INTRODUCTION

The AP Chemistry Exam will test your knowledge of basic lab techniques, as well as your understanding of accuracy and precision and your ability to analyze potential sources of error in a lab. In this section, we will discuss safety and accuracy precautions, laboratory equipment, and laboratory procedures.

SAFETY

Here are some basic safety rules that might turn up in test questions.

- Don't put chemicals in your mouth. You were told this when you were four years old, and it still holds true for the AP Chemistry Exam.
- When diluting an acid, always add the acid to the water. This is to avoid the spattering of hot solution.
- Always work with good ventilation; many common chemicals are toxic.
- When heating substances, do it slowly. When you heat things too quickly, they can spatter, burn, or explode.

ACCURACY

Here are some rules for ensuring the accuracy of experimental results.

- When titrating, rinse the buret with the solution to be used in the titration instead of with water. If you rinse the buret with water, you might dilute the solution, which will cause the volume added from the buret to be too large.
- Allow hot objects to return to room temperature before weighing. Hot objects on a scale create convection currents that may make the object seem lighter than it is.
- Don't weigh reagents directly on a scale. Use a glass or porcelain container to prevent corrosion of the balance pan.
- When collecting a gas over water, remember to take into account the pressure and volume of the water vapor.
- Don't contaminate your chemicals. Never insert another piece of equipment into a bottle containing a chemical. Instead you should always pour the chemical into another clean container. Also, don't let the inside of the stopper for a bottle containing a chemical touch another surface.
- When mixing chemicals, stir slowly to ensure even distribution.
- Be conscious of significant figures when you record your results. The number of significant figures you use should indicate the accuracy of your results.
- Be aware of the difference between accuracy and precision. A measurement is accurate if it is close to the accepted value. A series of measurements is precise if the values of all of the measurements are close together.

SIGNIFICANT FIGURES

When taking measurements in lab, your measurement will always have a certain number of significant figures. For instance, if you are using a balance to measure the weight of an object to the hundredths place, you might get 23.15 g. That number has four significant figures. The balance is no more accurate than that; you can not say the mass of the object is 23.15224 g.

It's important to be able to identify the number of significant figures in any number given to you on the AP Exam. There's a plethora of information on ways to count significant figures out there, but we've reduced it to two simple rules.

1. For numbers without a decimal place, you count every number except for trailing zeroes (those which appear after all non-zeroes). So, 100 mL only has one significant figure, and 250 mL has two. On the other hand, 105 g has three significant figures—the zero in that measurement does not trail all other numbers.
2. For numbers with a decimal place, you count every number except for leading zeroes (those which appear before all non-zeroes). The number 0.052 has 2 significant figures, but 0.0520 g would have three (trailing zeroes DO count in numbers with decimals). Most (but not all) values you get on the AP Exam will have decimal places.

This leads to numbers that otherwise might look very strange. For instance, it is not unusual to see numbers with a decimal at the end but no numbers past it. That's because a number like 100. g has three significant figures, but 100 g only has one. In science, 100. g is not the same as 100 g, which is also not the same as 100.0 g. All of those values have different numbers of significant figures, and that implies different levels of accuracy. Significant figures only apply to measurements with units; pure unitless numbers (such as those you find in math) do not follow these rules. Nonetheless, many math books do have units on practice problems and then ignore significant figures. Yes, it's wrong to do that, but don't tell your math teacher.

When doing calculations, it is important that any calculated value cannot be more accurate than the measurements used in the calculation. Essentially, significant figures can tell you how to round your answers correctly. Again, this can be divided into two categories:

1. When multiplying and dividing, your answer cannot have more significant figures than your least accurate measurement. For instance:

   2.50 g / 12 cm^3 = 0.20833 g/cm^3

   is wrong. Your two measurements have three significant figures and two significant figures, respectively. Your answer cannot have five. It has to have only two because that's how many figures your least accurate measurement had. The correct answer is 0.21 g/cm^3.

2. When adding and subtracting, your answer cannot have more figures after the decimal place than your value with the least number of figures after its decimal place. For instance:

   1.435 cm + 12.1 cm = 13.535 cm

   is wrong. Your values have three figures after the decimal and one figure after the decimal, respectively. Your answer cannot have more than one figure past the decimal, because that's how many figures your least accurate measurement had. The correct answer is 13.5 cm.

On a side note, counting numbers are considered to have an infinite number of significant figures. If you are doing a molar mass conversion and you identify the conversion as 22.99 g of sodium = 1 mol of sodium, your eventual answer will have four significant figures. That 1 mol represents *exactly* 1 mol of sodium, which is essentially a 1 with an infinite number of zeroes past the decimal place. This will usually come up when dealing with stoichiometry—the coefficients in a balanced equation are counting numbers of moles and don't count when considering the number of significant figures your answer should have.

You should get into the habit of making sure your calculations always have the correct number of significant figures as you do them, both in labs and on any practice problems you do in class. The AP Exam may have questions on significant figures in a lab and they are typically integrated into the free-response section. Additionally, you may lose points for the incorrect number of significant figures when you do calculations. Along with making sure you have the correct units on your number, making sure you do your calculations to the correct number of significant figures is a good laboratory practice.

EXPERIMENTAL DESIGN

The AP Chemistry Exam is heavily focused on students being able to understand the various types of experiments that can be undertaken in order to reinforce the concepts that are taught in the classroom. Some, if not most, of the free response questions will be based on laboratory investigations. Here are the seven types that, based on the current framework for the exam, you should be familiar with. We've listed the test-maker's expectations (in their own words), a brief description of the experiment, and where in this book you can find additional explanations and practice problems.

Spectrophotometry

LO 1.16: The student can design and/or interpret the results of an experiment regarding the absorption of light to determine the concentration of an absorbing species in a solution.

Spectrophotometry (aka colorimetry) is often used to measure the concentration of colored solutions. This type of experiment is often done in relation with reaction rates and/or equilibrium calculations.

Conceptual Explanation: Page 206

Free-Response Example: Question 8 (page 220) and Question 5 (page 301)

Gravimetric Analysis

LO 1.19: The student can design, and/or interpret data from, an experiment that uses gravimetric analysis to determine the concentration of an analyte in a solution.

Gravimetric analysis is intrinsically tied in with precipitation reactions and mass percent calculations.

Conceptual Explanation: Page 160

Free-Response Example: Question 1 (page 180)

Acid-Base Titrations

LO 1.20: The student can design, and/or interpret data from, an experiment that uses titration to determine the concentration of an analyte in a solution.

Acid-base titrations are by far the most common form of this learning objective that you will see on the exam. Acid-base titrations typically use indicators to cause the color to change at or near the equivalence point.

Conceptual Explanation: Pages 288–290

Free-Response Example: Question 7 (page 303)

Component Separation

LO 2.10: The student can design and/or interpret the results of a separation experiment (filtration, paper chromatography, column chromatography, or distillation) in terms of the relative strength of interactions among and between the components.

Understanding the best way to separate unique substances out of a mixture ties directly in with concepts of intermolecular forces and bonding. Out of the seven types of experiments, this one is generally the least mathematical and the most conceptual.

Conceptual Explanation: Page 128

Free-Response Example: Question 4 (page 138)

Redox Titrations

LO 3.9: The student is able to design and/or interpret the results of an experiment involving a redox titration.

As the name indicates, these are titrations that are focused on oxidation-reduction reactions. These do not always use indicators (although they can), but instead some of the solutions involved are colored themselves.

Conceptual Explanation: Page 166

Free-Response Example: Question 2 (page 180)

Kinetics

LO 4.1: The student is able to design and/or interpret the results of an experiment regarding the factors (i.e., temperature, concentration, surface area) that may influence the rate of a reaction.

Determining the rate of reaction can usually only be done experimentally. As such, that makes kinetics an easy topic to study via experimental design. Graphical analysis is particularly common in these experiments.

Conceptual Explanation: Pages 200–205

Free-Response Example: Question 6 (page 218)

Calorimetry

LO 5.7 The student is able to design and/or interpret the results of an experiment in which calorimetry is used to determine the change in enthalpy of a chemical process (heating/cooling, phase transition, or chemical reaction) at constant pressure.

The most common experimental data you will deal with involving any kind of thermodynamics calculations begins with calorimetry. Most thermodynamic data is gathered from existing data tables/research, but calorimetry is something that any high school student can experience during lab.

Conceptual Explanation: Pages 240–242

Free-Response Example: Question 6 (page 254)

There will certainly be other types of experiments tested on the AP Exam—there are a countless number of ways to set up and design experiments to test various chemistry concepts. However, focusing on these types is an excellent place to start.

LABORATORY EQUIPMENT

The following charts show some standard chemistry lab equipment.

Equipment	Use
 Beaker	Used to hold and pour liquids—also your favorite muppet.
 Buret	Used to add small but precisely measured volumes of liquid to a solution. Burets are used frequently in titration experiments.
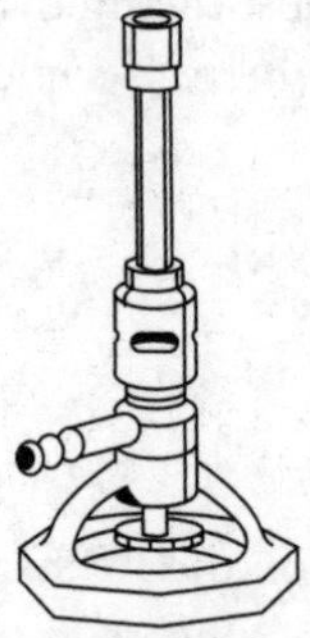 Burner	Used to apply heat and to wake up sleeping AP Chemistry students.
 Crucible tongs	Used to handle objects that are too hot to touch (careful though, they can break test tubes!).
 Dropper pipette	Used to add small amounts of liquid to a solution, but only when a precise volume is not needed, because the drops themselves should be consistent only for one particular dropper (for example, adding an indicator, comparing the amount of drops of strong acid or base needed for a pH change in different solutions).

 Erlenmeyer flask	A flask used for heating liquids. The conic shape allows stirring.
 Evaporating dish	Used to hold liquids for evaporation. The wide mouth allows vapor to escape.
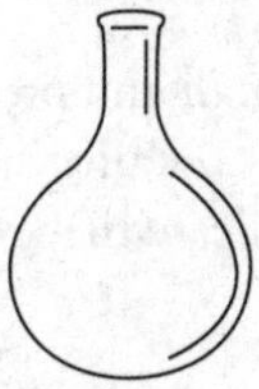 Florence flask	Used for boiling of liquids. The small neck prevents excessive evaporation and splashing.
 Forceps	A fancy name for tweezers.
 Funnel	Used to get liquids into a smaller container. Doubles as a cute hat when inverted.

Graduated cylinder	Used for measuring precisely a volume of liquid to be poured all at once (rather than dripped from a buret).
Graduated pipette	Used to transfer small and precise volumes of liquid from one container to another (the gradations indicate the volume).
Metal spatula	Used to scoop and transport powders.
Mortar and pestle	Used to grind solids into powders suitable for dissolving or mixing.
Pipette bulb	Rubber bulb used to draw liquid into pipette. In a pinch, could be used as a miniature clown nose.

Platform balance (triple beam)	A very precise scale operated by moving a set of three weights (typically corresponding to 100, 10, and 1 gram increments). A measurement will proceed like this: Rear weight is in the notch reading 30 g. Middle weight is in the notch reading 200 g. Front beam weight reads 3.86 g. The sample weighs 200 + 30 + 3.86 = 233.86 g.
Ring clamp	Used to hold funnels or other vessels in conjunction with a stand.
Rubber policeman	A hard tipped rubber scraper used to transfer precipitate.
Safety goggles	Used by chemists to protect their eyes during all laboratory experiments.

Test tube	Used to contain samples, especially when heating. Sometimes babies come out of them.
Thermometer	Measures temperature of a solution (don't use these to measure your body temperature, you don't know where they've been).
Volumetric pipette	Used to transfer small amounts of liquid. The big difference between a graduated pipette and a volumetric pipette is that the volumetric type is suited for only one particular volume. Because of this, they are extremely accurate.

Part VI
Practice Test 2

AP® Chemistry Exam

SECTION I: Multiple-Choice Questions

DO NOT OPEN THIS BOOKLET UNTIL YOU ARE TOLD TO DO SO.

At a Glance

Total Time
1 hour and 30 minutes
Number of Questions
60
Percent of Total Grade
50%
Writing Instrument
Pencil required

Instructions

Section I of this examination contains 60 multiple-choice questions. Fill in only the ovals for numbers 1 through 60 on your answer sheet.

CALCULATORS MAY NOT BE USED IN THIS PART OF THE EXAMINATION.

Indicate all of your answers to the multiple-choice questions on the answer sheet. No credit will be given for anything written in this exam booklet, but you may use the booklet for notes or scratch work. After you have decided which of the suggested answers is best, completely fill in the corresponding oval on the answer sheet. Give only one answer to each question. If you change an answer, be sure that the previous mark is erased completely. Here is a sample question and answer.

Sample Question	Sample Answer
Chicago is a (A) state (B) city (C) country (D) continent	Ⓐ ● Ⓒ Ⓓ

Use your time effectively, working as quickly as you can without losing accuracy. Do not spend too much time on any one question. Go on to other questions and come back to the ones you have not answered if you have time. It is not expected that everyone will know the answers to all the multiple-choice questions.

About Guessing

Many candidates wonder whether or not to guess the answers to questions about which they are not certain. Multiple-choice scores are based on the number of questions answered correctly. Points are not deducted for incorrect answers, and no points are awarded for unanswered questions. Because points are not deducted for incorrect answers, you are encouraged to answer all multiple-choice questions. On any questions you do not know the answer to, you should eliminate as many choices as you can, and then select the best answer among the remaining choices.

Disclaimer

This test is an approximation of the test that you will take. For up-to-date information, please remember to check the AP Students website.

GO ON TO THE NEXT PAGE.

CHEMISTRY
SECTION I
Time—1 hour and 30 minutes

INFORMATION IN THE TABLE BELOW AND ON THE FOLLOWING PAGES MAY BE USEFUL IN ANSWERING THE QUESTIONS IN THIS SECTION OF THE EXAMINATION.

DO NOT DETACH FROM BOOK.

PERIODIC TABLE OF THE ELEMENTS

1 **H** 1.008																	2 **He** 4.00
3 **Li** 6.94	4 **Be** 9.01											5 **B** 10.81	6 **C** 12.01	7 **N** 14.01	8 **O** 16.00	9 **F** 19.00	10 **Ne** 20.18
11 **Na** 22.99	12 **Mg** 24.30											13 **Al** 26.98	14 **Si** 28.09	15 **P** 30.97	16 **S** 32.06	17 **Cl** 35.45	18 **Ar** 39.95
19 **K** 39.10	20 **Ca** 40.08	21 **Sc** 44.96	22 **Ti** 47.90	23 **V** 50.94	24 **Cr** 52.00	25 **Mn** 54.94	26 **Fe** 55.85	27 **Co** 58.93	28 **Ni** 58.69	29 **Cu** 63.55	30 **Zn** 65.39	31 **Ga** 69.72	32 **Ge** 72.59	33 **As** 74.92	34 **Se** 78.96	35 **Br** 79.90	36 **Kr** 83.80
37 **Rb** 85.47	38 **Sr** 87.62	39 **Y** 88.91	40 **Zr** 91.22	41 **Nb** 92.91	42 **Mo** 95.94	43 **Tc** (98)	44 **Ru** 101.1	45 **Rh** 102.91	46 **Pd** 106.42	47 **Ag** 107.87	48 **Cd** 112.41	49 **In** 114.82	50 **Sn** 118.71	51 **Sb** 121.75	52 **Te** 127.60	53 **I** 126.91	54 **Xe** 131.29
55 **Cs** 132.91	56 **Ba** 137.33	57 ***La** 138.91	72 **Hf** 178.49	73 **Ta** 180.95	74 **W** 183.85	75 **Re** 186.21	76 **Os** 190.2	77 **Ir** 192.2	78 **Pt** 195.08	79 **Au** 196.97	80 **Hg** 200.59	81 **Tl** 204.38	82 **Pb** 207.2	83 **Bi** 208.98	84 **Po** (209)	85 **At** (210)	86 **Rn** (222)
87 **Fr** (223)	88 **Ra** 226.02	89 **†Ac** 227.03	104 **Rf** (261)	105 **Db** (262)	106 **Sg** (266)	107 **Bh** (264)	108 **Hs** (277)	109 **Mt** (268)	110 **Ds** (271)	111 **Rg** (272)							

*Lanthanide Series	58 **Ce** 140.12	59 **Pr** 140.91	60 **Nd** 144.24	61 **Pm** (145)	62 **Sm** 150.4	63 **Eu** 151.97	64 **Gd** 157.25	65 **Tb** 158.93	66 **Dy** 162.50	67 **Ho** 164.93	68 **Er** 167.26	69 **Tm** 168.93	70 **Yb** 173.04	71 **Lu** 174.97
†Actinide Series	90 **Th** 232.04	91 **Pa** 231.04	92 **U** 238.03	93 **Np** (237)	94 **Pu** (244)	95 **Am** (243)	96 **Cm** (247)	97 **Bk** (247)	98 **Cf** (251)	99 **Es** (252)	100 **Fm** (257)	101 **Md** (258)	102 **No** (259)	103 **Lr** (262)

GO ON TO THE NEXT PAGE.

ADVANCED PLACEMENT CHEMISTRY EQUATIONS AND CONSTANTS

Throughout the test the following symbols have the definitions specified unless otherwise noted.

L, mL = liter(s), milliliter(s)
g = gram(s)
nm = nanometer(s)
atm = atmosphere(s)
mm Hg = millimeters of mercury
J, kJ = joule(s), kilojoule(s)
V = volt(s)
mol = mole(s)

ATOMIC STRUCTURE

$E = h\nu$
$c = \lambda\nu$

E = energy
ν = frequency
λ = wavelength

Planck's constant, $h = 6.626 \times 10^{-34}$ J s
Speed of light, $c = 2.998 \times 10^{8}$ m s^{-1}
Avogadro's number = 6.022×10^{23} mol^{-1}
Electron charge, $e = -1.602 \times 10^{-19}$ coulombs

EQUILIBRIUM

$K_c = \frac{[C]^c[D]^d}{[A]^a[B]^b}$, where $a\,A + b\,B \rightleftarrows c\,C + d\,D$

$K_p = \frac{(P_C)^c(P_D)^d}{(P_A)^a(P_B)^b}$

$K_a = \frac{[H^+][A^-]}{[HA]}$

$K_b = \frac{[OH^-][HB^+]}{[B]}$

$K_w = [H^+][OH^-] = 1.0 \times 10^{-14}$ at 25°C
$= K_a \times K_b$

$pH = -\log[H^+]$, $pOH = -\log[OH^-]$

$14 = pH + pOH$

$pH = pK_a + \log\frac{[A^-]}{[HA]}$

$pK_a = -\log K_a$, $pK_b = -\log K_b$

Equilibrium Constants

K_c (molar concentrations)
K_p (gas pressures)
K_a (weak acid)
K_b (weak base)
K_w (water)

KINETICS

$\ln[A]_t - \ln[A]_0 = -kt$

$\frac{1}{[A]_t} - \frac{1}{[A]_0} = kt$

$t_{1/2} = \frac{0.693}{k}$

k = rate constant
t = time
$t_{1/2}$ = half-life

GO ON TO THE NEXT PAGE.

GASES, LIQUIDS, AND SOLUTIONS

$$PV = nRT$$

$$P_A = P_{total} \times X_A, \text{ where } X_A = \frac{\text{moles A}}{\text{total moles}}$$

$$P_{total} = P_A + P_B + P_C + \ldots$$

$$n = \frac{m}{M}$$

$$K = {}^\circ C + 273$$

$$D = \frac{m}{V}$$

$$KE \text{ per molecule} = \frac{1}{2}mv^2$$

Molarity, M = moles of solute per liter of solution

$$A = abc$$

P = pressure
V = volume
T = temperature
n = number of moles
m = mass
M = molar mass
D = density
KE = kinetic energy
v = velocity
A = absorbance
a = molar absorptivity
b = path length
c = concentration

Gas constant, $R = 8.314 \text{ J mol}^{-1}\text{K}^{-1}$
$= 0.08206 \text{ L atm mol}^{-1}\text{K}^{-1}$
$= 62.36 \text{ L torr mol}^{-1}\text{K}^{-1}$
1 atm = 760 mm Hg
= 760 torr
STP = 0.00°C and 1.000 atm

THERMOCHEMISTRY/ ELECTROCHEMISTRY

$$q = mc\Delta T$$

$$\Delta S^\circ = \sum S^\circ \text{ products} - \sum S^\circ \text{ reactants}$$

$$\Delta H^\circ = \sum \Delta H_f^\circ \text{ products} - \sum \Delta H_f^\circ \text{ reactants}$$

$$\Delta G^\circ = \sum \Delta G_f^\circ \text{ products} - \sum \Delta G_f^\circ \text{ reactants}$$

$$\Delta G^\circ = \Delta H^\circ - T\Delta S^\circ$$

$$= -RT \ln K$$

$$= -nFE^\circ$$

$$I = \frac{q}{t}$$

q = heat
m = mass
c = specific heat capacity
T = temperature
S° = standard entropy
H° = standard enthalpy
G° = standard free energy
n = number of moles
E° = standard reduction potential
I = current (amperes)
q = charge (coulombs)
t = time (seconds)

Faraday's constant, F = 96,485 coulombs per mole of electrons

$$1 \text{ volt} = \frac{1 \text{ joule}}{1 \text{ coulomb}}$$

GO ON TO THE NEXT PAGE.

1. A compound is made up of entirely silicon and oxygen atoms. If there are 14.0 g of silicon and 32.0 g of oxygen present, what is the empirical formula of the compound?

(A) SiO_2
(B) SiO_4
(C) Si_2O
(D) Si_2O_3

2.

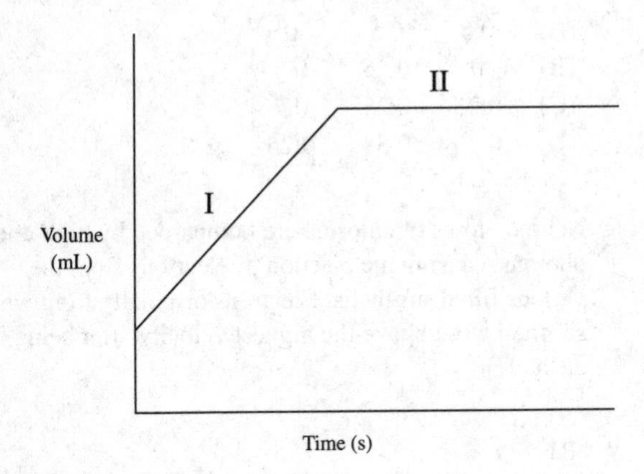

The volume of a gas is charted over time, giving the above results. Which of the following options provides a possible explanation of what was happening to the gas during each phase of the graph?

(A) During phase I, the temperature decreased while the pressure increased. During phase II, the temperature was held constant as the pressure decreased.
(B) During phase I, the temperature increased while the pressure was held constant. During phase II, the temperature and pressure both decreased.
(C) During phase I, the temperature was held constant while the pressure increased. During phase II, the temperature and pressure both decreased.
(D) During phase I, the temperature and pressure both increased. During phase II, the temperature was held constant while the pressure decreased.

3. A solution of sulfurous acid, H_2SO_3, is present in an aqueous solution. Which of the following represents the concentrations of three different ions in solution?

(A) $[SO_3^{2-}] > [HSO_3^-] > [H_2SO_3]$
(B) $[H_2SO_3] > [HSO_3^-] > [SO_3^{2-}]$
(C) $[H_2SO_3] > [HSO_3^-] = [SO_3^{2-}]$
(D) $[SO_3^{2-}] = [HSO_3^-] > [H_2SO_3]$

4. $$2NO(g) + Br_2(g) \leftrightarrow 2NOBr(g)$$

The above experiment was performed several times, and the following data was gathered:

Trial	$[NO]_{init}$ (M)	$[Br_2]_{init}$ (M)	Initial Rate of Reaction (M/min)
1	0.20 M	0.10 M	5.20×10^{-3}
2	0.20 M	0.20 M	1.04×10^{-2}
3	0.40 M	0.10 M	2.08×10^{-2}

What is the rate law for this reaction?

(A) Rate = $k[NO][Br_2]^2$
(B) Rate = $k[NO]^2[Br_2]^2$
(C) Rate = $k[NO][Br_2]$
(D) Rate = $k[NO]^2[Br_2]$

5. $SF_4(g) + H_2O(l) \rightarrow SO_2(g) + 4HF(g)$ $\Delta H = -828$ kJ/mol

Which of the following statements accurately describes the above reaction?

(A) The entropy of the reactants exceeds that of the products.
(B) $H_2O(l)$ will always be the limiting reagent.
(C) This reaction is never thermodynamically favored.
(D) The temperature of the surroundings will increase as this reaction progresses.

Use the following information to answer questions 6-9.

20.0 mL of 1.0 M Na_2CO_3 is placed in a beaker and titrated with a solution of 1.0 M $Ca(NO_3)_2$, resulting in the creation of a precipitate.

6. How much $Ca(NO_3)_2$ must be added to reach the equivalence point?

(A) 10.0 mL
(B) 20.0 mL
(C) 30.0 mL
(D) 40.0 mL

GO ON TO THE NEXT PAGE.

7. Which of the following diagrams correctly shows the species present in the solution in significant amounts at the equivalence point?

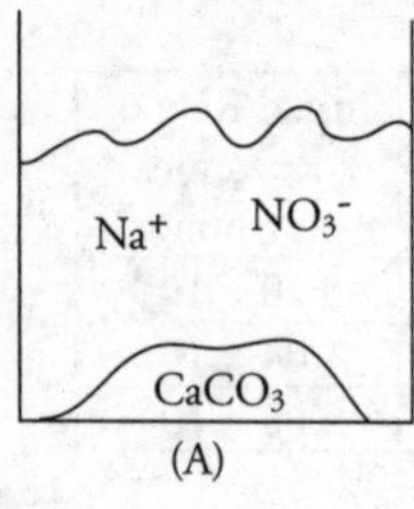

(A)

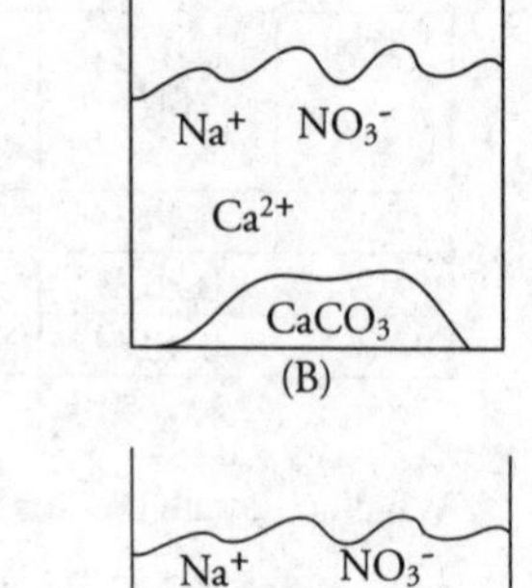

(B)

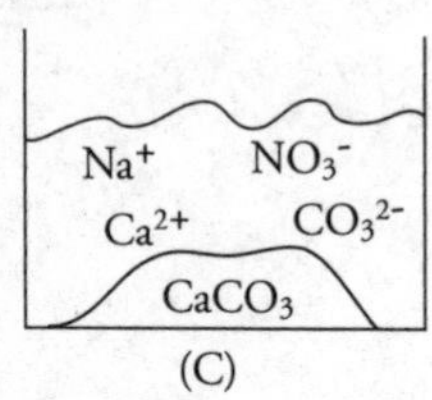

(C)

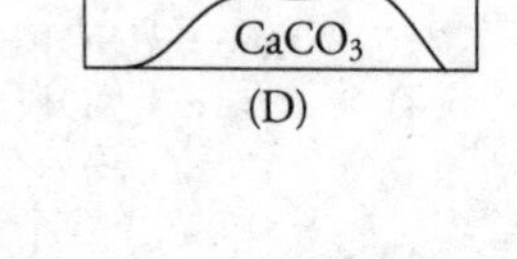

(D)

8. What will happen to the conductivity of the solution after additional $Ca(NO_3)_2$ is added past the equivalence point?

(A) The conductivity will increase as additional ions are being added to the solution.
(B) The conductivity will stay constant as the precipitation reaction has gone to completion.
(C) The conductivity will decrease as the solution will be diluted with the addition of additional $Ca(NO_3)_2$.
(D) The conductivity will stay constant as equilibrium has been established.

9. If the experiment were repeated and the Na_2CO_3 was diluted to 40.0 mL with distilled water prior to the titration, how would that affect the volume of $Ca(NO_3)_2$ needed to reach the equivalence point?

(A) It would be cut in half.
(B) It would decrease by a factor of 1.5.
(C) It would double.
(D) It would not change.

10. $$2CO(g) + O_2(g) \rightarrow 2CO_2(g)$$

2.0 mol of $CO(g)$ and 2.0 mol of $O_2(g)$ are pumped into a rigid, evacuated 4.0-L container, where they react to form $CO_2(g)$. Which of the following values does NOT represent a potential set of concentrations for each gas at a given point during the reaction?

	CO	O_2	CO_2
(A)	0.5	0.5	0
(B)	0	0.25	0.5
(C)	0.25	0.25	0.5
(D)	0.25	0.38	0.25

11. Neutral atoms of chlorine are bombarded by high-energy photons, causing the ejection of electrons from the various filled subshells. Electrons originally from which subshell would have the highest velocity after being ejected?

(A) 1*s*
(B) 2*p*
(C) 3*p*
(D) 3*d*

12. A sample of oxygen gas at 50°C is heated, reaching a final temperature of 100°C. Which statement best describes the behavior of the gas molecules?

(A) Their velocity increases by a factor of two.
(B) Their velocity increases by a factor of four.
(C) Their kinetic energy increases by a factor of 2.
(D) Their kinetic energy increases by a factor of less than 2.

13. The average mass, in grams, of one mole of carbon atoms is equal to

(A) the average mass of a single carbon atom, measured in amus
(B) the ratio of the number of carbon atoms to the mass of a single carbon atom
(C) the number of carbon atoms in one amu of carbon
(D) the mass, in grams, of the most abundant isotope of carbon

GO ON TO THE NEXT PAGE.

14. 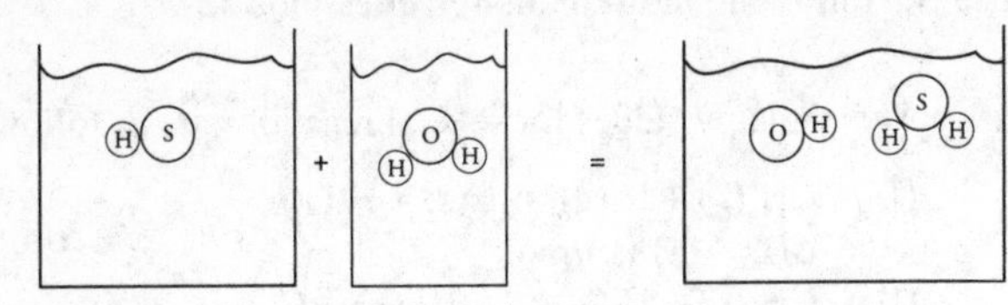

The diagram above best represents which type of reaction?

(A) Acid/base
(B) Oxidation/reduction
(C) Precipitation
(D) Decomposition

15. Which of the following is true for all bases?

(A) All bases donate OH^- ions into solution.
(B) Only strong bases create solutions in which OH^- ions are present.
(C) Only strong bases are good conductors when dissolved in solution.
(D) For weak bases, the concentration of the OH^- ions exceeds the concentration of the base in the solution.

Use the following information to answer questions 16-18.

$14H^+(aq) + Cr_2O_7^{2-}(aq) + 3Ni(s) \rightarrow$
$2Cr^{3+}(aq) + 3Ni^{2+}(aq) + 7H_2O(l)$

In the above reaction, a piece of solid nickel is added to a solution of potassium dichromate.

16. Which species is being oxidized and which is being reduced?

	Oxidized	Reduced
(A)	$Cr_2O_7^{2-}(aq)$	$Ni(s)$
(B)	$Cr^{3+}(aq)$	$Ni^{2+}(aq)$
(C)	$Ni(s)$	$Cr_2O_7^{2-}(aq)$
(D)	$Ni^{2+}(aq)$	$Cr^{3+}(aq)$

17. How many moles of electrons are transferred when 1 mole of potassium dichromate is mixed with 3 mol of nickel?

(A) 2 moles of electrons
(B) 3 moles of electrons
(C) 5 moles of electrons
(D) 6 moles of electrons

18. How does the pH of the solution change as the reaction progresses?

(A) It increases until the solution becomes basic.
(B) It increases, but the solution remains acidic.
(C) It decreases until the solution becomes basic.
(D) It decreases, but the solution remains acidic.

19. A sample of an unknown chloride compound was dissolved in water, and then titrated with excess $Pb(NO_3)_2$ to create a precipitate. After drying, it is determined there are 0.0050 mol of precipitate present. What mass of chloride is present in the original sample?

(A) 0.177 g
(B) 0.355 g
(C) 0.522 g
(D) 0.710 g

20. A photoelectron spectra for which of the following atoms would show peaks at exactly three different binding energies?

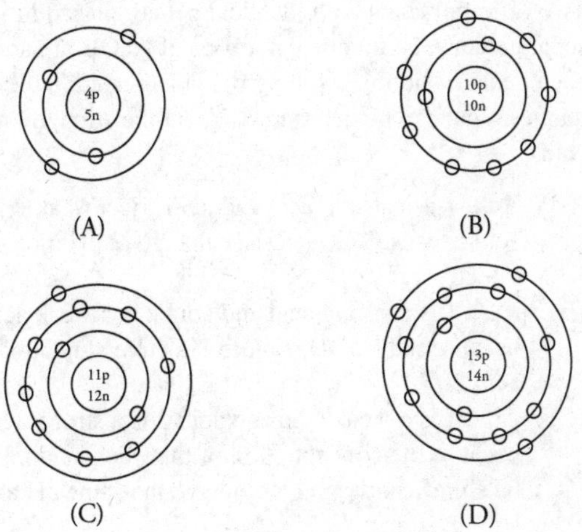

21. The bond length between any two nonmetal atoms is achieved under which of the following conditions?

(A) Where the energy of interaction between the atoms is at its minimum value
(B) Where the nuclei of each atom exhibits the strongest attraction to the electrons of the other atom
(C) The point at which the attractive and repulsive forces between the two atoms are equal
(D) The closest point at which a valence electron from one atom can transfer to the other atom

GO ON TO THE NEXT PAGE.

22. Hydrogen fluoride, HF, is a liquid at 15°C. All other hydrogen halides (represented by HX, where X is any other halogen) are gases at the same temperature. Why?

(A) Fluorine has a very high electronegativity; therefore the H–F bond is stronger than any other H–X bond.
(B) HF is smaller than any other H–X molecule; therefore it exhibits stronger London dispersion forces.
(C) The dipoles in a HF molecule exhibit a particularly strong attraction force to the dipoles in other HF molecules.
(D) The H–F bond is the most ionic in character compared to all other hydrogen halides.

23.

	Initial pH	pH after NaOH addition
Acid 1	3.0	3.5
Acid 2	3.0	5.0

Two different acids with identical pH are placed in separate beakers. Identical portions of NaOH are added to each beaker, and the resulting pH is indicated in the table above. What can be determined about the strength of each acid?

(A) Acid 1 is a strong acid and acid 2 is a weak acid because acid 1 resists change in pH more effectively.
(B) Acid 1 is a strong acid and acid 2 is a weak acid because the NaOH is more effective at neutralizing acid 2.
(C) Acid 1 is a weak acid and acid 2 is a strong acid because the concentration of the weak acid must be significantly greater to have the same pH as the strong acid.
(D) Acid 1 is a weak acid and acid 2 is a strong acid because the concentration of the hydrogen ions will be greater in acid 2 after the NaOH addition.

24. A stock solution of 12.0 *M* sulfuric acid is made available. What is the best procedure to make up 100.0 mL of 4.0 *M* sulfuric acid using the stock solution and water prior to mixing?

(A) Add 33.3 mL of water to the flask, and then add 66.7 mL of 12.0 *M* acid.
(B) Add 33.3 mL of 12.0 *M* acid to the flask, and then dilute it with 66.7 mL of water.
(C) Add 67.7 mL of 12.0 *M* acid to the flask, and then dilute it with 33.3 mL of water.
(D) Add 67.7 mL of water to the flask, and then add 33.3 mL of 12.0 *M* acid.

Use the following data to answer questions 25-29.

The enthalpy values for several reactions are as follows:

(I) $CH_4(g) + H_2(g) \rightarrow C(s) + H_2O(g)$
$\Delta H_{rxn} = -131$ kJ/mol
(II) $CH_4(g) + H_2O(g) \rightarrow 3H_2(g) + CO(g)$
$\Delta H_{rxn} = 206$ kJ/mol
(III) $CO(g) + H_2O(g) \rightarrow CO_2(g) + H_2(g)$
$\Delta H_{rxn} = -41$ kJ/mol
(IV) $CH_4(g) + 2O_2(g) \rightarrow CO_2(g) + H_2O(l)$
$\Delta H_{rxn} = -890$ kJ/mol

25. In which of the reactions does the amount of energy released by the formation of bonds in the products exceed the amount of energy necessary to break the bonds of the reactants by the greatest amount?

(A) Reaction I
(B) Reaction II
(C) Reaction III
(D) Reaction IV

26. In which of the reactions is the value for ΔS the most positive?

(A) Reaction I
(B) Reaction II
(C) Reaction III
(D) Reaction IV

27. Regarding reaction I, how would the addition of a catalyst affect the enthalpy and entropy changes for this reaction?

	Enthalpy	Entropy
(A)	Decrease	Decrease
(B)	Decrease	No Change
(C)	No Change	Decrease
(D)	No Change	No Change

GO ON TO THE NEXT PAGE.

28.

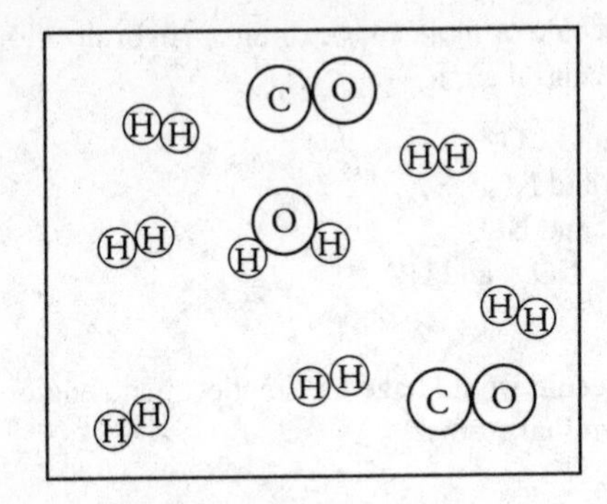

Regarding reaction II, to achieve the products present in the above diagram how many moles of each reactant must be present prior to the reaction?

(A) 1.0 mol of CH_4 and 2.0 mol of H_2O
(B) 2.0 mol of CH_4 and 2.0 mol of H_2O
(C) 2.0 mol of CH_4 and 3.0 mol of H_2O
(D) 3.0 mol of CH_4 and 2.0 mol of H_2O

29. Regarding reaction IV, how much heat is absorbed or released when 2.0 mol of $CH_4(g)$ reacts with 2.0 mol of $O_2(g)$?

(A) 890 kJ of heat is released.
(B) 890 kJ of heat is absorbed.
(C) 1780 kJ of heat is released.
(D) 1780 kJ of heat is absorbed.

---

30. London dispersion forces are caused by

(A) temporary dipoles created by the position of electrons around the nuclei in a molecule
(B) the three-dimensional intermolecular bonding present in all covalent substances
(C) the uneven electron-to-proton ratio found on individual atoms of a molecule
(D) the electronegativity differences between the different atoms in a molecule

31. What is the general relationship between temperature and entropy for diatomic gases?

(A) They are completely independent of each other; temperature has no effect on entropy.
(B) There is a direct relationship, because at higher temperatures there is an increase in energy dispersal.
(C) There is an inverse relationship, because at higher temperatures substances are more likely to be in a gaseous state.
(D) It depends on the specific gas and the strength of the intermolecular forces between individual molecules.

32. Which of the following pairs of ions would make the best buffer with a pH between 6 and 7?

K_a for $HC_3H_2O_2 = 1.75 \times 10^{-5}$
K_a for $HPO_4^{2-} = 4.8 \times 10^{-13}$

(A) H_2SO_4 and H_3PO_4
(B) HPO_4^{2-} and Na_3PO_4
(C) $HC_3H_2O_2$ and $NaC_3H_2O_2$
(D) NaOH and $HC_2H_3O_2$

33. A solution contains a mixture of four different compounds: $KCl(aq)$, $Fe(NO_3)_3(aq)$, $MgSO_4(aq)$, and $N_2H_4(aq)$. Which of these compounds would be easiest to separate via distillation?

(A) $KCl(aq)$
(B) $Fe(NO_3)_3(aq)$
(C) $MgSO_4(aq)$
(D) $N_2H_4(aq)$

34.

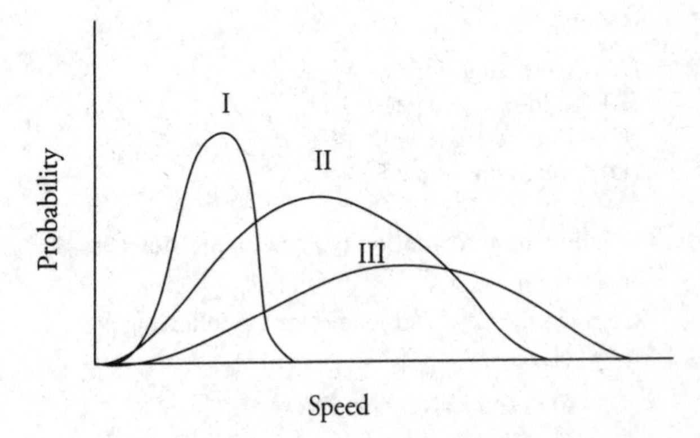

Identify the three gases represented on the Maxwell-Boltzmann diagram above. Assume all gases are at the same temperature.

	I	II	III
(A)	H_2	N_2	F_2
(B)	H_2	F_2	N_2
(C)	F_2	N_2	H_2
(D)	N_2	F_2	H_2

GO ON TO THE NEXT PAGE.

35. A sample of solid $MgCl_2$ would be most soluble in which of the following solutions?

(A) $LiOH(aq)$
(B) $CBr_4(aq)$
(C) $Mg(NO_3)_2(aq)$
(D) $AlCl_3(aq)$

36. Most transition metals share a common oxidation state of +2. Which of the following best explains why?

(A) Transition metals all have a minimum of two unpaired electrons.
(B) Transition metals have unstable configurations and are very reactive.
(C) Transition metals tend to gain electrons when reacting with other elements.
(D) Transition metals will lose their outermost *s*-block electrons when forming bonds.

37. $2Ag^+(aq) + Fe(s) \rightarrow 2Ag(s) + Fe^{2+}(aq)$

Which of the following would cause an increase in potential in the voltaic cell described by the above reaction?

(A) Increasing $[Fe^{2+}]$
(B) Adding more $Fe(s)$
(C) Decreasing $[Fe^{2+}]$
(D) Removing some $Fe(s)$

Use the following information to answer questions 38-41.

Consider the Lewis structures for the following molecules:

CO_2, CO_3^{2-}. NO_2^-, and NO_3^-

38. Which molecule would have the shortest bonds?

(A) CO_2
(B) CO_3^{2-}
(C) NO_2^-
(D) NO_3^-

39. Which molecules are best represented by multiple resonance structures?

(A) CO_2 and CO_3^{2-}
(B) NO_2^- and NO_3^-
(C) CO_3^{2-} and NO_3^-
(D) CO_3^{2-}, NO_2^-, and NO_3^-

40. Which molecule or molecules exhibit sp^2 hybridization around the central atom?

(A) CO_2 and CO_3^{2-}
(B) NO_2^- and NO_3^-
(C) CO_3^{2-} and NO_3^-
(D) CO_3^{2-}, NO_2^-, and NO_3^-

41. Which molecule would have the smallest bond angle between terminal atoms?

(A) CO_2
(B) CO_3^{2-}
(C) NO_2^-
(D) NO_3^-

---

42. $NH_4^+(aq) + NO_2^-(aq) \rightarrow N_2(g) + 2H_2O(l)$

Increasing the temperature of the above reaction will increase the rate of reaction. Which of the following is NOT a reason that increased temperature increases reaction rate?

(A) The reactants will be more likely to overcome the activation energy.
(B) The number of collisions between reactant molecules will increase.
(C) A greater distribution of reactant molecules will have high velocities.
(D) Alternate reaction pathways become available at higher temperatures.

43. Which of the following diagrams best represents what is happening on a molecular level when NaCl dissolves in water?

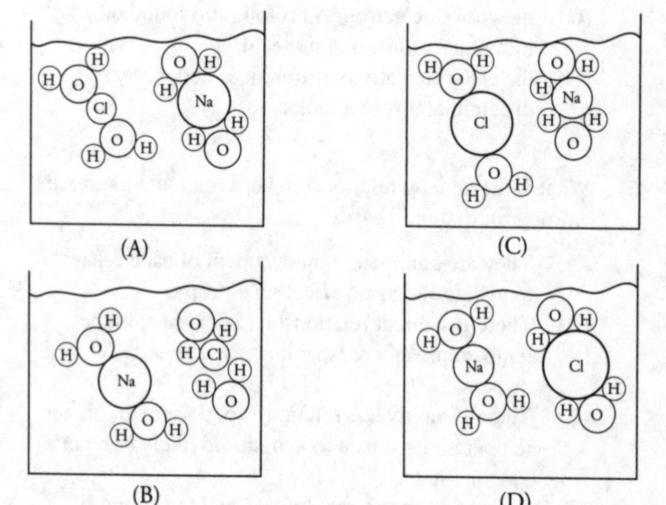

GO ON TO THE NEXT PAGE.

44. Nitrous acid, HNO_2, has a pK_a value of 3.3. If a solution of nitrous acid is found to have a pH of 4.2, what can be said about the concentration of the conjugate acid/base pair found in solution?

(A) $[HNO_2] > [NO_2^-]$
(B) $[NO_2^-] > [HNO_2]$
(C) $[H_2NO_2^+] > [HNO_2]$
(D) $[HNO_2] > [H_2NO_2^+]$

45. Which of the following processes is an irreversible reaction?

(A) $CH_4(g) + O_2(g) \rightarrow CO_2(g) + H_2O(l)$
(B) $HCN(aq) + H_2O(l) \rightarrow CN^-(aq) + H_3O^+(aq)$
(C) $Al(NO_3)_3(s) \rightarrow Al^{3+}(aq) + 3NO_3^-(aq)$
(D) $2Ag^+(aq) + Ti(s) \rightarrow 2Ag(s) + Ti^{2+}(aq)$

Use the following information to answer questions 46-50.

A sample of H_2S gas is placed in an evacuated, sealed container and heated until the following decomposition reaction occurs at 1000 K:

$2H_2S(g) \rightarrow 2H_2(g) + S_2(g)$ $K_c = 1.0 \times 10^{-6}$

46. Which of the following represents the equilibrium constant for this reaction?

(A) $K_c = \dfrac{[H_2]^2[S_2]}{[H_2S]^2}$

(B) $K_c = \dfrac{[H_2S]^2}{[H_2]^2[S_2]}$

(C) $K_c = \dfrac{2[H_2][S_2]}{2[H_2S]}$

(D) $K_c = \dfrac{2[H_2S]}{2[H_2][S_2]}$

47. Which of the following graphs would best represent the change in concentration of the various species involved in the reaction over time?

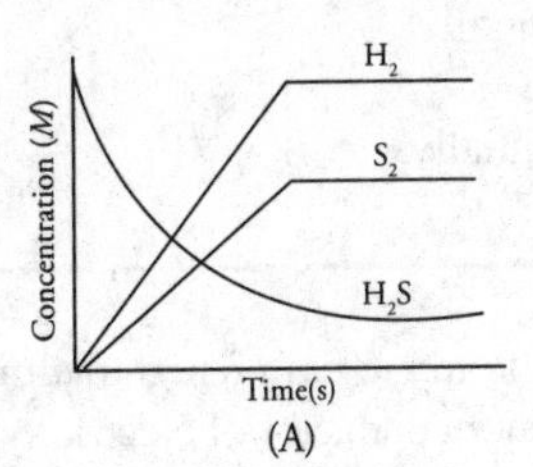

(A)

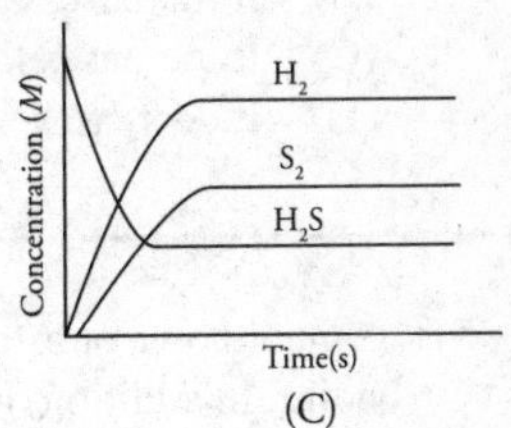

(C)

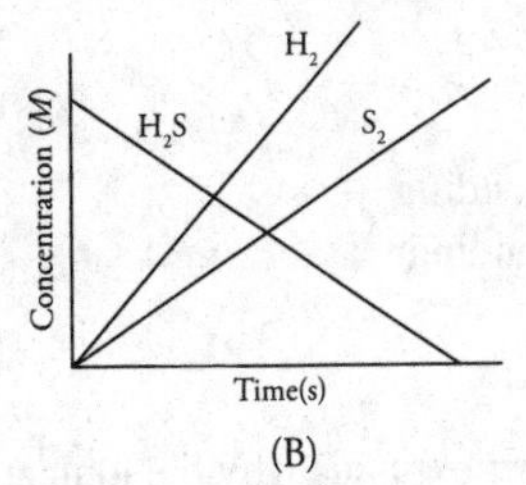

(B)

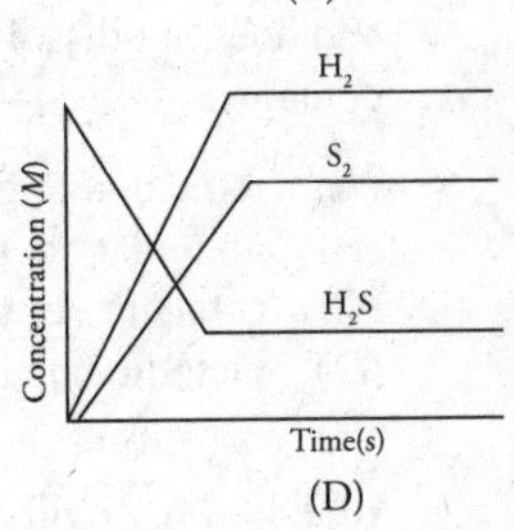

(D)

48. Which option best describes what will immediately occur to the reaction rates if the pressure on the system is increased after it has reached equilibrium?

(A) The rate of both the forward and reverse reactions will increase.
(B) The rate of the forward reaction will increase while the rate of the reverse reaction decreases.
(C) The rate of the forward reaction will decrease while the rate of the reverse reaction increases.
(D) Neither the rate of the forward nor reverse reactions will change.

49. If, at a given point in the reaction, the value for the reaction quotient Q is determined to be 2.5×10^{-8}, which of the following is occurring?

(A) The concentration of the reactant is decreasing while the concentration of the products is increasing.
(B) The concentration of the reactant is increasing while the concentration of the products is decreasing.
(C) The system has passed the equilibrium point and the concentration of all species involved in the reaction will remain constant.
(D) The concentrations of all species involved are changing at the same rate.

GO ON TO THE NEXT PAGE.

50. As the reaction progresses at a constant temperature of 1000 K, how does the value for the Gibbs free energy constant for the reaction change?

(A) It stays constant.
(B) It increases exponentially.
(C) It increases linearly.
(D) It decreases exponentially.

---

51. An unknown substance is found to have a high melting point. In addition, it is a poor conductor of electricity and does not dissolve in water. The substance most likely contains

(A) ionic bonding
(B) nonpolar covalent bonding
(C) covalent network bonding
(D) metallic bonding

52. Which of the following best explains why the ionization of atoms can occur during photoelectron spectroscopy, even though ionization is not a thermodynamically favored process?

(A) It is an exothermic process due to the release of energy as an electron is liberated from the Coulombic attraction holding it to the nucleus.
(B) The entropy of the system increases due to the separation of the electron from its atom.
(C) Energy contained in the light can be used to overcome the Coulombic attraction between electrons and the nucleus.
(D) The products of the ionization are at a lower energy state than the reactants.

53.

$$2H_2O_2(aq) \longrightarrow 2H_2O(l) + O_2(g)$$

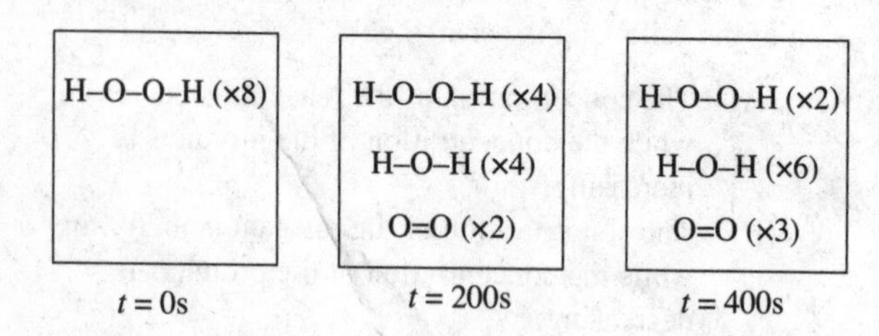

The above diagrams show the decomposition of hydrogen peroxide in a sealed container in the presence of a catalyst. What is the overall order for the reaction?

(A) Zero order
(B) First order
(C) Second order
(D) Third order

54.

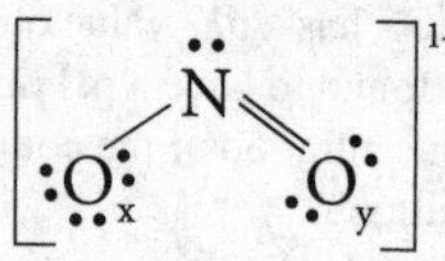

One of the resonance structures for the nitrite ion is shown above. What is the formal charge on each atom?

	O_x	N	O_y
(A)	–1	+1	–1
(B)	+1	–1	0
(C)	0	0	–1
(D)	–1	0	0

Use the following information to answer questions 55-57.

Atoms of four elements are examined: carbon, nitrogen, neon, and sulfur.

55. Atoms of which element would have the strongest magnetic moment?

(A) Carbon
(B) Nitrogen
(C) Neon
(D) Sulfur

56. Atoms of which element are most likely to form a structure with the formula XF_6 (where X is one of the four atoms)?

(A) Carbon
(B) Nitrogen
(C) Neon
(D) Sulfur

57. Which element would have a photoelectron spectra in which the peak representing electrons with the lowest ionization energy would be three times higher than all other peaks?

(A) Carbon
(B) Nitrogen
(C) Neon
(D) Sulfur

GO ON TO THE NEXT PAGE.

58. The diagram below supports which of the following conclusions about the reaction shown below?

$$:\!N \equiv N\!: \; + \; \begin{matrix} H\text{-}H \\ H\text{-}H \\ H\text{-}H \end{matrix} \; \longrightarrow \; \begin{matrix} H - \ddot{N}(H) - H \\ H - \ddot{N}(H) - H \end{matrix}$$

(A) There is an increase in entropy.
(B) Mass is conserved in all chemical reactions.
(C) The pressure increases after the reaction goes to completion.
(D) The enthalpy value is positive.

59. $NO_2 + O_3 \rightarrow NO_3 + O_2$ slow
$NO_3 + NO_2 \rightarrow N_2O_5$ fast

A proposed reaction mechanism for the reaction of nitrogen dioxide and ozone is detailed above. Which of the following is the rate law for the reaction?

(A) Rate = $k[NO_2][O_3]$
(B) Rate = $k[NO_3][NO_2]$
(C) Rate = $k[NO_2]^2[O_3]$
(D) Rate = $k[NO_3][O_2]$

60.

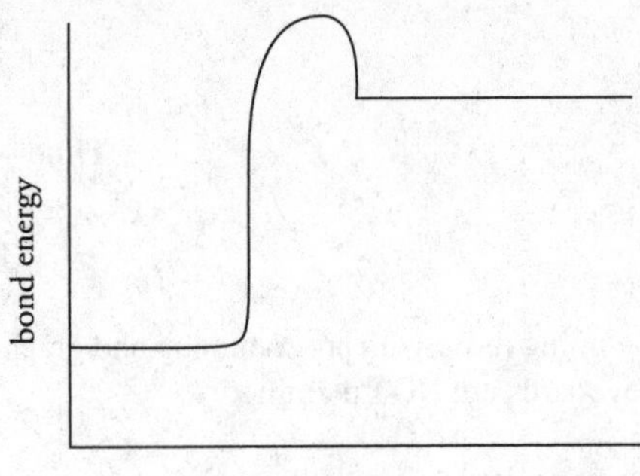

The concentrations of the reactants and products in the reaction represented by the above graph are found to be changing very slowly. Which of the following statements best describes the reaction given that the reaction is exergonic? ($\Delta G < 0$)

(A) The reaction is under kinetic control.
(B) The reaction has reached a state of equilibrium.
(C) The reaction is highly exothermic in nature.
(D) The addition of heat will increase the rate of reaction significantly.

END OF SECTION I

CHEMISTRY
SECTION II
Time—1 hour and 45 minutes

General Instructions

Calculators, including those with programming and graphing capabilities, may be used. However, calculators with typewriter-style (QWERTY) keyboards are NOT permitted.

Pages containing a periodic table and equations commonly used in chemistry will be available for your use.

You may write your answers with either a pen or a pencil. Be sure to write CLEARLY and LEGIBLY. If you make an error, you may save time by crossing it out rather than trying to erase it.

Write all your answers in the essay booklet. Number your answers as the questions are numbered in the examination booklet.

GO ON TO THE NEXT PAGE.

INFORMATION IN THE TABLE BELOW AND ON THE FOLLOWING PAGES MAY BE USEFUL IN ANSWERING THE QUESTIONS IN THIS SECTION OF THE EXAMINATION.

DO NOT DETACH FROM BOOK.

PERIODIC TABLE OF THE ELEMENTS

1 **H** 1.008																	2 **He** 4.00
3 **Li** 6.94	4 **Be** 9.01											5 **B** 10.81	6 **C** 12.01	7 **N** 14.01	8 **O** 16.00	9 **F** 19.00	10 **Ne** 20.18
11 **Na** 22.99	12 **Mg** 24.30											13 **Al** 26.98	14 **Si** 28.09	15 **P** 30.97	16 **S** 32.06	17 **Cl** 35.45	18 **Ar** 39.95
19 **K** 39.10	20 **Ca** 40.08	21 **Sc** 44.96	22 **Ti** 47.90	23 **V** 50.94	24 **Cr** 52.00	25 **Mn** 54.94	26 **Fe** 55.85	27 **Co** 58.93	28 **Ni** 58.69	29 **Cu** 63.55	30 **Zn** 65.39	31 **Ga** 69.72	32 **Ge** 72.59	33 **As** 74.92	34 **Se** 78.96	35 **Br** 79.90	36 **Kr** 83.80
37 **Rb** 85.47	38 **Sr** 87.62	39 **Y** 88.91	40 **Zr** 91.22	41 **Nb** 92.91	42 **Mo** 95.94	43 **Tc** (98)	44 **Ru** 101.1	45 **Rh** 102.91	46 **Pd** 106.42	47 **Ag** 107.87	48 **Cd** 112.41	49 **In** 114.82	50 **Sn** 118.71	51 **Sb** 121.75	52 **Te** 127.60	53 **I** 126.91	54 **Xe** 131.29
55 **Cs** 132.91	56 **Ba** 137.33	57 ***La** 138.91	72 **Hf** 178.49	73 **Ta** 180.95	74 **W** 183.85	75 **Re** 186.21	76 **Os** 190.2	77 **Ir** 192.2	78 **Pt** 195.08	79 **Au** 196.97	80 **Hg** 200.59	81 **Tl** 204.38	82 **Pb** 207.2	83 **Bi** 208.98	84 **Po** (209)	85 **At** (210)	86 **Rn** (222)
87 **Fr** (223)	88 **Ra** 226.02	89 **†Ac** 227.03	104 **Rf** (261)	105 **Db** (262)	106 **Sg** (266)	107 **Bh** (264)	108 **Hs** (277)	109 **Mt** (268)	110 **Ds** (271)	111 **Rg** (272)							

*Lanthanide Series	58 **Ce** 140.12	59 **Pr** 140.91	60 **Nd** 144.24	61 **Pm** (145)	62 **Sm** 150.4	63 **Eu** 151.97	64 **Gd** 157.25	65 **Tb** 158.93	66 **Dy** 162.50	67 **Ho** 164.93	68 **Er** 167.26	69 **Tm** 168.93	70 **Yb** 173.04	71 **Lu** 174.97
†Actinide Series	90 **Th** 232.04	91 **Pa** 231.04	92 **U** 238.03	93 **Np** (237)	94 **Pu** (244)	95 **Am** (243)	96 **Cm** (247)	97 **Bk** (247)	98 **Cf** (251)	99 **Es** (252)	100 **Fm** (257)	101 **Md** (258)	102 **No** (259)	103 **Lr** (262)

GO ON TO THE NEXT PAGE.

ADVANCED PLACEMENT CHEMISTRY EQUATIONS AND CONSTANTS

Throughout the test the following symbols have the definitions specified unless otherwise noted.

L, mL	= liter(s), milliliter(s)	mm Hg	= millimeters of mercury
g	= gram(s)	J, kJ	= joule(s), kilojoule(s)
nm	= nanometer(s)	V	= volt(s)
atm	= atmosphere(s)	mol	= mole(s)

ATOMIC STRUCTURE

$E = h\nu$

$c = \lambda\nu$

E = energy

ν = frequency

λ = wavelength

Planck's constant, $h = 6.626 \times 10^{-34}$ J s

Speed of light, $c = 2.998 \times 10^{8}$ m s^{-1}

Avogadro's number = 6.022×10^{23} mol^{-1}

Electron charge, $e = -1.602 \times 10^{-19}$ coulombs

EQUILIBRIUM

$K_c = \frac{[C]^c[D]^d}{[A]^a[B]^b}$, where $a\,A + b\,B \rightleftarrows c\,C + d\,D$

$K_p = \frac{(P_C)^c(P_D)^d}{(P_A)^a(P_B)^b}$

$K_a = \frac{[H^+][A^-]}{[HA]}$

$K_b = \frac{[OH^-][HB^+]}{[B]}$

$K_w = [H^+][OH^-] = 1.0 \times 10^{-14}$ at 25°C

$= K_a \times K_b$

$pH = -\log[H^+]$, $pOH = -\log[OH^-]$

$14 = pH + pOH$

$pH = pK_a + \log\frac{[A^-]}{[HA]}$

$pK_a = -\log K_a$, $pK_b = -\log K_b$

Equilibrium Constants

K_c (molar concentrations)

K_p (gas pressures)

K_a (weak acid)

K_b (weak base)

K_w (water)

KINETICS

$\ln[A]_t - \ln[A]_0 = -kt$

$\frac{1}{[A]_t} - \frac{1}{[A]_0} = kt$

$t_{1/2} = \frac{0.693}{k}$

k = rate constant

t = time

$t_{1/2}$ = half-life

GO ON TO THE NEXT PAGE.

GASES, LIQUIDS, AND SOLUTIONS

$$PV = nRT$$

$$P_A = P_{total} \times X_A, \text{ where } X_A = \frac{\text{moles A}}{\text{total moles}}$$

$$P_{total} = P_A + P_B + P_C + \ldots$$

$$n = \frac{m}{M}$$

$$K = °C + 273$$

$$D = \frac{m}{V}$$

$$KE \text{ per molecule} = \frac{1}{2}mv^2$$

Molarity, M = moles of solute per liter of solution

$$A = abc$$

P = pressure
V = volume
T = temperature
n = number of moles
m = mass
M = molar mass
D = density
KE = kinetic energy
v = velocity
A = absorbance
a = molar absorptivity
b = path length
c = concentration

Gas constant, $R = 8.314 \text{ J mol}^{-1} \text{ K}^{-1}$
$= 0.08206 \text{ L atm mol}^{-1} \text{ K}^{-1}$
$= 62.36 \text{ L torr mol}^{-1} \text{ K}^{-1}$
1 atm = 760 mm Hg
= 760 torr
STP = 0.00°C and 1.000 atm

THERMOCHEMISTRY/ ELECTROCHEMISTRY

$$q = mc\Delta T$$

$$\Delta S° = \sum S° \text{ products} - \sum S° \text{ reactants}$$

$$\Delta H° = \sum \Delta H_f° \text{ products} - \sum \Delta H_f° \text{ reactants}$$

$$\Delta G° = \sum \Delta G_f° \text{ products} - \sum \Delta G_f° \text{ reactants}$$

$$\Delta G° = \Delta H° - T\Delta S°$$

$$= -RT \ln K$$

$$= -nFE°$$

$$I = \frac{q}{t}$$

q = heat
m = mass
c = specific heat capacity
T = temperature
$S°$ = standard entropy
$H°$ = standard enthalpy
$G°$ = standard free energy
n = number of moles
$E°$ = standard reduction potential
I = current (amperes)
q = charge (coulombs)
t = time (seconds)

Faraday's constant, F = 96,485 coulombs per mole of electrons

$$1 \text{ volt} = \frac{1 \text{ joule}}{1 \text{ coulomb}}$$

GO ON TO THE NEXT PAGE.

CHEMISTRY

Section II

(Total time—105 minutes)

YOU MAY USE YOUR CALCULATOR IN THIS SECTION.

THE METHODS USED AND THE STEPS INVOLVED IN ARRIVING AT YOUR ANSWERS MUST BE SHOWN CLEARLY. It is to your advantage to do this since you may obtain partial credit if you do, and you will receive little or no credit if you do not. Attention should be paid to significant figures.

1. A stock solution of 0.100 *M* cobalt (II) chloride is used to create several solutions, indicated in the data table below:

Sample	Volume $CoCl_2$ (mL)	Volume H_2O (mL)
1	20.00	0
2	15.00	5.00
3	10.00	10.00
4	5.00	15.00

(a) In order to achieve the degree of accuracy shown in the table above, select which of the following pieces of laboratory equipment could be used when measuring out the $CoCl_2$:

150-mL beaker
50-mL buret
400-mL beaker
50-mL graduated cylinder
250-mL Erlenmeyer flask
100-mL graduated cylinder

(b) Calculate the concentration of the $CoCl_2$ in each sample.

The solutions are then placed in cuvettes before being inserted into a spectrophotometer calibrated to 560 nm and their values are measured, yielding the data below:

Sample	Absorbance
1	0.485
2	0.364
3	0.243
4	0.121

(c) If gloves are not worn when handling the cuvettes, how might this affect the absorbance values gathered?
(d) If the path length of the cuvette is 1.00 cm, what is the molar absorptivity value for $CoCl_2$ at 560 nm?
(e) On the axes on the next page, plot a graph of absorbance vs. concentration. The *y*-axes scale is set, and be sure to scale the *x*-axes appropriately

GO ON TO THE NEXT PAGE.

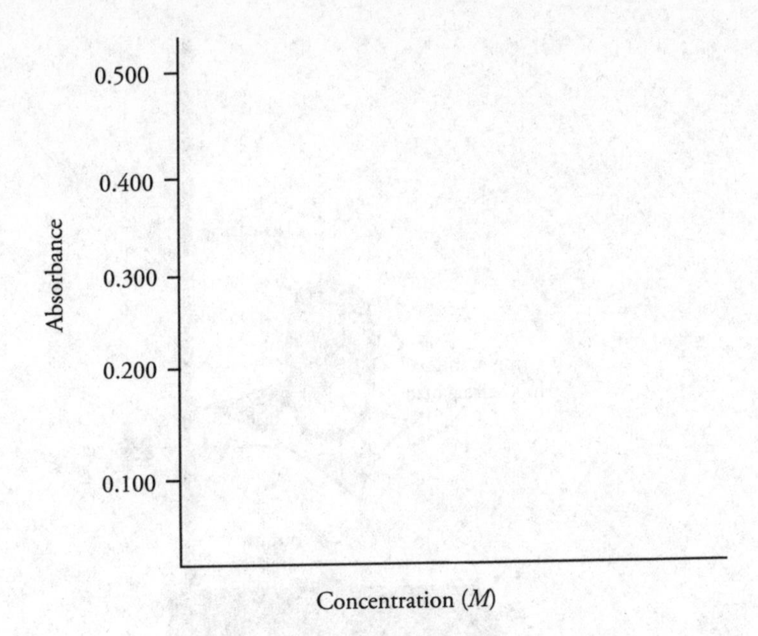

(f) What would the absorbance values be for $CoCl_2$ solutions at the following concentrations?

(i) 0.067 *M*

(ii) 0.180 *M*

GO ON TO THE NEXT PAGE.

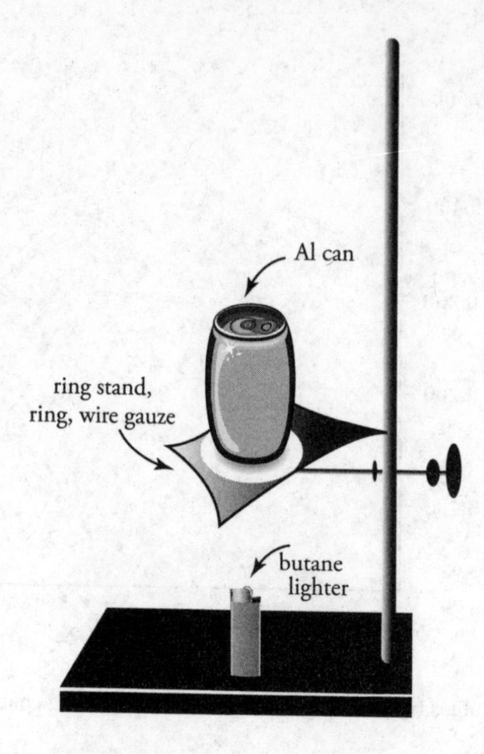

2. A sample of liquid butane (C_4H_{10}) in a pressurized lighter is set up directly beneath an aluminum can, as show in the diagram above. The can contains 100.0 mL of water, and when the butane is ignited the temperature of the water inside the can increases from 25.0°C to 82.3°C. The total mass of butane ignited is found to be 0.51 g, the specific heat of water is 4.18 J/g·°C, and the density of water is 1.00 g/mL.

(a) Write the balanced chemical equation for the combustion of one mole of butane in air.

(b) (i) How much heat did the water gain?

(ii) What is the experimentally determined heat of combustion for butane based on this experiment? Your answer should be in kJ/mol.

(c) Given butane's density of 0.573 g/mL at 25°C, calculate how much heat would be emitted if 5.00 mL of it were combusted at that temperature.

(d) The overall combustion of butane is an exothermic reaction. Explain why this is in terms of bond energies.

(e) One of the major sources of error in this experiment comes from the heat that is absorbed by the air. Why, then, might it not be a good idea to perform this experiment inside a sealed container to prevent the heat from leaving the system?

GO ON TO THE NEXT PAGE.

3. $$2N_2O_5(g) \rightarrow 4NO_2(g) + O_2(g)$$

The data below was gathered for the decomposition of N_2O_5 at 310 K via the equation above.

Time (s)	$[N_2O_5]$ (M)
0	0.250
500.	0.190
1000.	0.145
2000.	0.085

(a) How does the rate of appearance of NO_2 compare to the rate of disappearance of N_2O_5? Justify your answer.

(b) The reaction is determined to be first order overall. On the axes below, create a graph of some function of concentration vs. time that will produce a straight line. Label and scale your axes appropriately.

(c) (i) What is the rate constant for this reaction? Include units.

(ii) What would the concentration of N_2O_5 be at $t = 1500$ s?

(iii) What is the half-life of N_2O_5?

(d) Would the addition of a catalyst increase, decrease, or have no effect on the following variables? Justify your answers.

(i) Rate of disappearance of N_2O_5

(ii) Magnitude of the rate constant

(iii) Half-life of N_2O_5

GO ON TO THE NEXT PAGE.

4. Consider the Lewis structures for the following four molecules:

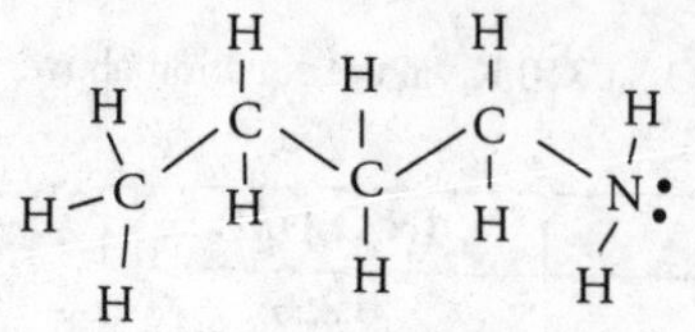

n-Butylamine

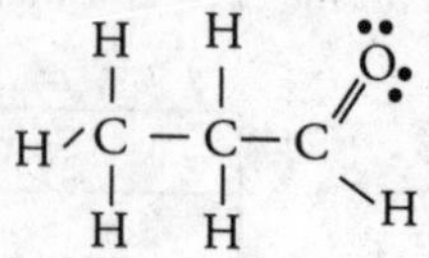

Propanal

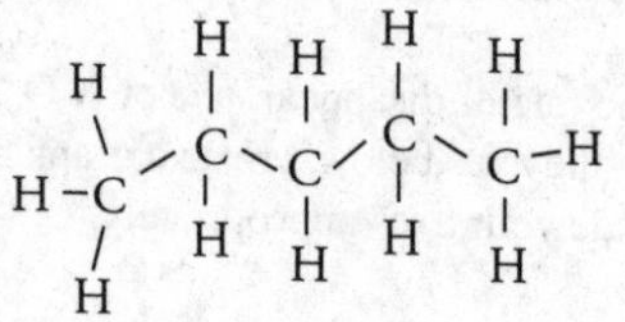

Pentane

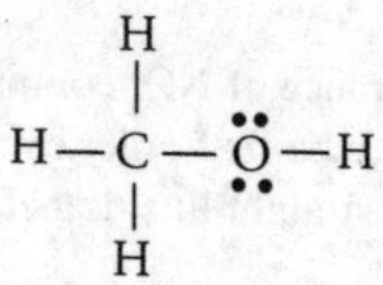

Methanol

(a) All of the substances are liquids at room temperature. Organize them from high to low in terms of boiling points, clearly differentiating between the intermolecular forces in each substance.

(b) On the methanol diagram reproduced below, draw the locations of all dipoles.

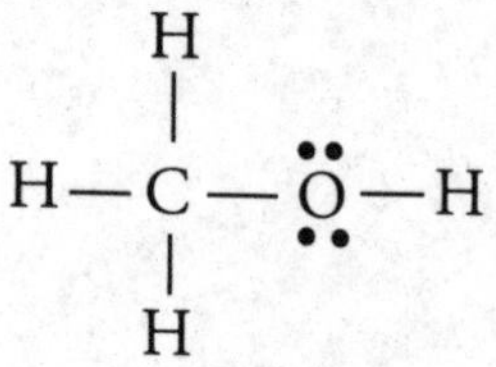

Methanol

(c) n-Butylamine is found to have the lowest vapor pressure at room temperature out of the four liquids. Justify this observation in terms of intermolecular forces.

GO ON TO THE NEXT PAGE.

5. Current is run through an aqueous solution of nickel (II) fluoride, and a gas is evolved at the right-hand electrode, as indicated by the diagram below:

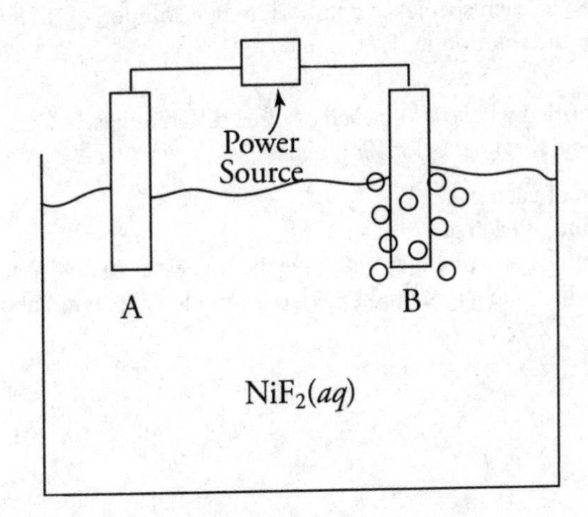

The standard reduction potential for several reactions is given in the following table:

Half-cell	E°_{red}
$F_2(g) + 2e^- \rightarrow 2F^-$	+2.87 V
$O_2(g) + 4H^+ + 4e^- \rightarrow 2H_2O(l)$	+1.23 V
$Ni^{2+} + 2e^- \rightarrow Ni(s)$	–0.25 V
$2H_2O(l) + 2e^- \rightarrow H_2(g) + 2OH^-$	–0.83 V

(a) Determine which half-reaction is occurring at each electrode:
  (i) Oxidation
  (ii) Reduction

(b) (i) Calculate the standard cell potential of the cell.
  (ii) Calculate the Gibbs free energy value of the cell at standard conditions.

(c) Which electrode in the diagram (A or B) is the cathode, and which is the anode? Justify your answers.

GO ON TO THE NEXT PAGE.

6. Aniline, $C_6H_5NH_2$, is a weak base with $K_b = 3.8 \times 10^{-10}$.
   (a) Write out the reaction that occurs when aniline reacts with water.
   (b) (i) What is the concentration of each species at equilibrium in a solution of 0.25 *M* $C_6H_5NH_2$?
   (ii) What is the pH value for the solution in (i)?

7. A rigid, sealed 12.00 L container is filled with 10.00 g each of three different gases: CO_2, NO, and NH_3. The temperature of the gases is held constant 35.0°C. Assume ideal behavior for all gases.
   (a) (i) What is the mole fraction of each gas?
   (ii) What is the partial pressure of each gas?
   (b) Out of the three gases, molecules of which gas will have the highest velocity? Why?
   (c) Name one circumstance in which the gases might deviate from ideal behavior, and clearly explain the reason for the deviation.

STOP

END OF EXAM

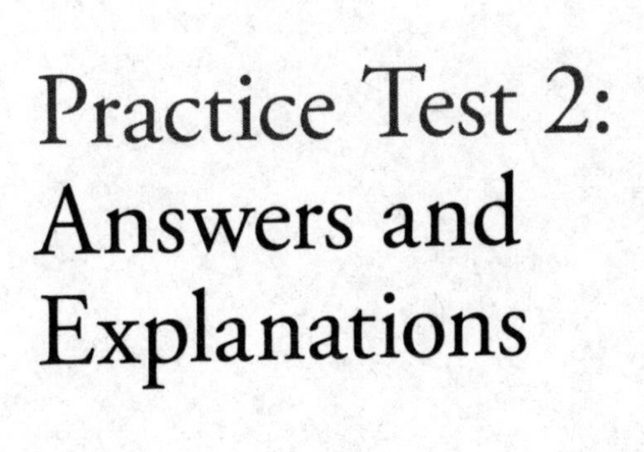

Practice Test 2: Answers and Explanations

PRACTICE TEST 2 MULTIPLE-CHOICE ANSWER KEY

1. B
2. B
3. B
4. D
5. D
6. B
7. A
8. A
9. D
10. C
11. C
12. D
13. A
14. A
15. C
16. C
17. D
18. B
19. B
20. B
21. A
22. C
23. C
24. D
25. D
26. B
27. D
28. C
29. A
30. A
31. B
32. B
33. D
34. C
35. A
36. D
37. C
38. A
39. D
40. D
41. C
42. D
43. D
44. B
45. A
46. A
47. C
48. C
49. A
50. A
51. C
52. C
53. B
54. D
55. B
56. D
57. C
58. B
59. A
60. A

Section I—Multiple-Choice Answers and Explanations

1. **B** 14.0 g of Si is 0.50 mol, and 32.0 g of oxygen is 2.0 mol. Converting that to a whole number ratio gives us 1 mol of Si for every 4 moles of O.

2. **B** In phase I, an increased temperature means the molecules are moving faster and will spread out more, leading to an increased volume. For the volume to remain constant in phase II, either both pressure and volume have to remain constant, or they both have to increase or decrease together, as they are inversely proportional.

3. **B** Weak acids have a very low dissociation value, and for polyprotic weak acids the second dissociation is always weaker than the first.

4. **D** Between trial 1 and 3, the concentration of NO doubled while the concentration of Br_2 held constant, and the rate went up by a factor of four. So, the reaction is second order with respect to NO. Between trial 1 and 2, the concentration of NO was held constant while the concentration of Br_2 doubled, and the rate went up by a factor of two. Thus, the reaction is first order with respect to Br_2.

5. **D** A negative enthalpy change indicates that energy is lost to the surroundings as the reaction occurs.

6. **B** Equivalence is reached when the conductivity of the solution reaches its minimum value, meaning as many of the precipitate ions are out of solution as possible. As the formula of the precipitate is $CaCO_3$, for every mole of sodium carbonate in the beaker, one mole of calcium nitrate will be needed. As both solutions have equal molarity, equal volumes of each would be necessary to ensure the moles are equal.

7. **A** At the equivalence point, the moles of Ca^{2+} and CO_3^{2-} in solution will be negligible.

8. **A** The Ca^{2+} ions will have nothing to react with, so they will just remain in solution (as will the extra NO_3^- ions being added), increasing the conductivity.

9. **D** The equivalence point is reached when there are equal moles of each reactant present. Diluting the sodium carbonate solution will not change the number of moles of Na_2CO_3 present in solution.

10. **C** For every two moles of CO that react, one mole of O_2 will also react. If one mole of CO reacts, half a mole of O_2 will react. At that point, the concentration of CO will be (1.0 mol/4.0 L = 0.25 *M*) and the concentration of the O_2 will be (1.5 mol/4.0 L = 0.38 *M*), rendering (C) impossible.

11. **C** The farther away an electron is from the nucleus, the less ionization energy that is required to eject it, and as a result, the electron will have more kinetic energy after it is ejected. The 3*p* subshell is the farthest one that a neutral chlorine atom would have electrons in. Beware of (D); chlorine does not have a 3*d* subshell.

12. **D** Temperature in Kelvins is a measure of kinetic energy. When that temperature doubles, so does the kinetic energy. However, this is not a direct correlation, so while a doubling of the temperature in Celsius would increase the amount of kinetic energy, it would not double it.

13. **A** This is a straightforward concept to understand. The average atomic mass of an element on the periodic table measures both the average mass of a mole of atoms of that element in grams, as well as being the average mass of a single atom of that element in amus.

14. **A** A proton is transferred from the water to the HS^- ion, making the reaction an acid/base reaction.

15. **C** Strong bases dissociate to donate hydroxide ions into a solution and weak bases cause the creation of hydroxide ions by taking protons from water. However, weak bases do not protonate strongly, meaning there will not be nearly as many ions in solution for a weak base as there are for a strong base. Thus, weak bases are not good conductors.

16. **C** The $Cr_2O_7^{2-}$(*aq*) gains electrons to change the oxidation state on Cr from +6 to +3. The Ni(*s*) loses electrons to change the oxidation state on Ni from 0 to +2.

17. **D** In one mole of dichromate, two moles of chromium are reduced, each requiring three moles of electrons. Three moles of nickel are being oxidized, each requiring two moles of electrons. Thus, six moles of electrons are being transferred.

18. **B** As the reaction progresses, the H^+ ions are converting to H_2O molecules. This would raise the pH of the solution. As the reaction progresses, the solution will become less acidic, but there is no mechanism for it to become basic.

19. **B** The precipitate is $PbCl_2$, and in every mole of precipitate there are two moles of chlorine ions. Thus, there are 0.010 mol of chlorine present in the precipitate, and through conservation of mass, the same number of moles of chlorine were present in the original sample: $0.010 \text{ mol} \times 35.5 \text{ g} = 0.355 \text{ g}$.

20. **B** Diagram B represents an atom of neon, which has three subshells: 1*s*, 2*s*, and 2*p*. Electrons from each subshell would have a different binding energy, yielding three peaks on a PES.

21. **A** The bond length always corresponds to the point where the potential bond energy (a balance of the attraction and repulsion forces between the two atoms) is at its minimum value.

22. **C** The IMFs between HF molecules are hydrogen bonds, which are very strong dipoles created due to the high electronegativity value of fluorine. No other hydrogen halide exhibits hydrogen bonding, and thus they would have weaker intermolecular forces (and lower boiling points) than HF.

23. **C** Weak acids resist changes in pH more effectively than strong acids because so many molecules of weak acid are undissociated in solution. The base must cause those molecules to dissociate before affecting the pH significantly.

24. D Using $M_1V_1 = M_2V_2$, we can calculate the amount of stock solution needed. $(12.0)V_1 = (4.0)(100.0)$, so $V_1 = 33.3$ mL. When making solutions, always add acid to water (in order to more effectively absorb the heat of the exothermic reaction).

25. D When calculating enthalpy, the total energy change in the reaction is always calculated by the following:

Bonds broken (reactants) – Bonds formed (products)

The more negative this value is, the more energy that was released in the bond formation of the products compared to the amount of energy necessary to break the bonds in the reactants.

26. B There are two moles of gaseous reactants, and four moles of gaseous products. That means the disorder increased, yielding a positive value for ΔS.

27. D The addition of a catalyst speeds up the reaction but will not affect the enthalpy or the entropy values.

28. C 2.0 moles of CH_4 would react with 3.0 moles of H_2O, leaving 1.0 mole left. It would also create 6.0 moles of H_2 and 2.0 moles of CO.

29. A When the listed amounts react, the O_2 will be the limiting reactant, as it is used up twice as quickly (this is determined by the coefficients in the reaction).

$$\text{So: } 2.0 \text{ mol } O_2 \times \frac{1 \text{ mol}_{rxn}}{2.0 \text{ mol } O_2} \times \frac{-890 \text{ kJ}}{1 \text{ mol}_{rxn}} = -890 \text{ kJ}$$

That gives us the magnitude of the energy change, and the negative sign means heat will be released.

30. A London dispersion forces are created by temporary dipoles due to the constant motion of electrons in an atom or molecule.

31. B The higher the temperature of any substance, the larger the range of velocities the molecules of that substance can have, and thus the more disorder the substance can have. A Maxwell-Boltzmann diagram represents this in a graphical form.

32. B Strong acids and bases do not make good buffers, so that eliminates (A) and (D). Of the remaining choices, a solution containing $H_2PO_4^-$ (with a K_a in the 10^{-13} range) would have a $[H^+]$ of $\sqrt{10^{-13}}$. That would put $[H^+]$ between 10^{-6} and 10^{-7}, yielding a pH in the required range.

33. D Distillation involves the boiling off of substances with different boiling points. The other three compounds are all ionic, meaning that in solution they are free ions. If we were to boil off all of the N_2H_4 and the water, the remaining ions would all mix together to form multiple precipitates. However, by heating the solution to a boiling point greater than that of N_2H_4 and lower than that of water, the N_2H_4 can be collected in a separate flask.

34. C At identical temperatures, the gases would all have identical amounts of kinetic energy. In order for that to happen, the gas with the lowest mass (H_2) would have to have the highest average velocity, and the gas with the highest mass (F_2) would have to have the lowest average velocity.

35. **A** Like dissolves like, so ionic substances dissolve best in ionic compounds, eliminating (B). Both (C) and (D) share an ion with $MgCl_2$. Via the common ion effect and Le Châtelier's principle, that would reduce the solubility of the $MgCl_2$ in those solutions. The best choice is (A).

36. **D** The outermost *s*-block electrons in a transition metal tend to be lost before the *d*-block electrons. Additionally, the other options do not accurately describe the properties of transition metals.

37. **C** Adding or removing a solid would not cause any equilibrium shift. Decreasing the concentration of the Fe^{2+} causes a shift to the right, which would increase the potential of the cell.

38. **A** The following Lewis structures are necessary to answer questions 38–41:

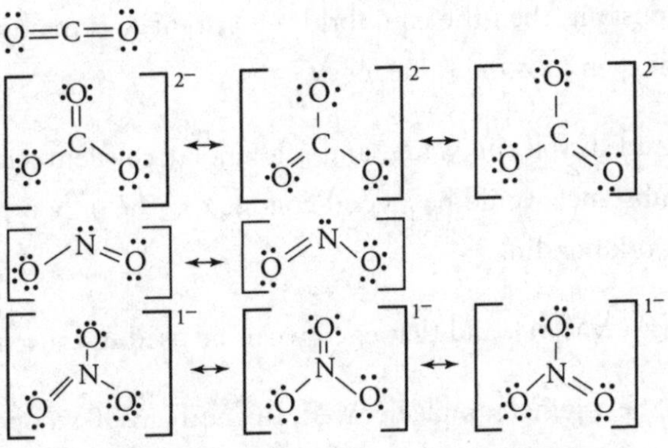

CO_2 has a bond order of 2, which exceeds the order of the other structures. Remember that a higher bond order corresponds with shorter and stronger bonds.

39. **D** Except for CO_2, the other molecules all display resonance.

40. **D** All three of those structures have three electron domains, and thus sp^2 hybridization.

41. **C** The bond angle of NO_2^- would be less than that of NO_3^- or CO_3^{2-} because the unbonded pair of electrons on the nitrogen atom reduces the overall bond angle.

42. **D** Even if this were true, any pathways that become available at higher temperatures would be less likely to be taken than the original pathway.

43. **D** The Na^+ cations would be attracted to the negative (oxygen) end of the water molecules and the Cl^- anions would be attracted to the positive (hydrogen) end. Additionally, Na^+ ions are smaller than Cl^- ions as they have less-filled energy levels.

44. **B** In this case, the pH is greater than the pK_a. This means that there will be more conjugate base present in solution than the original acid. The conjugate base of HNO_2 is NO_2^-.

45. **A** Combustion reactions are irreversible, while the other reactions are examples of acid-base, dissolution, and oxidation-reduction, all of which are reversible in this case.

46. A Equilibrium is always products over reactants, and coefficients in a balanced equilibrium reaction become exponents in the equilibrium expression.

47. C The concentration of the H_2S will decrease exponentially until it reaches a constant value. The concentrations of the two products will increase exponentially (the H_2 twice as quickly as the S_2) until reaching equilibrium.

48. C Increasing the pressure causes a shift toward the side with fewer gas molecules—in this case, a shift to the left. This means the reverse reaction rate increases while the forward reaction rate decreases.

49. A The reaction will progress until $Q = K_c$. If $Q < K_c$, the numerator of the expression (the products) will continue to increase while the denominator (the reactant) decreases until equilibrium is established.

50. A If the temperature is constant, then the equilibrium constant K is unchanged. Via $\Delta G = -RT \ln K$, if K and T are both constant then so is the value for ΔG.

51. C An ionic substance would dissolve in water, and a nonpolar covalent substance would have a low melting point. A metallic substance would be a good conductor. The only type of bonding that meets all the criteria is covalent network bonding.

52. C Light contains energy (via $E = hv$), and that energy can be used to cause a reaction.

53. B In 200 seconds, half of the original sample decayed. In another 200 seconds, half of the remaining sample decayed. This demonstrates a first order reaction.

54. D O_x has 6 valence electrons and 7 assigned electrons: $6 - 7 = -1$. Both O_y and the N atoms have the same number of valence and assigned electrons, making their formal charges zero.

55. B The strength of an atom's magnetic moment increases with an increase in the number of unpaired electrons. Nitrogen has the most unpaired electrons (3), and thus the strongest magnetic moment.

56. D Sulfur is the only element with an empty *d*-block in its outermost energy level, and is thus the only atom of the four which can form an expanded octet.

57. C Neon has six electrons in its subshell with the lowest ionization energy (2*p*), and only two electrons in the other two subshells (1*s* and 2*s*). This means the 2*p* peak will be three times higher than the other two peaks.

58. B The amount of matter is equal on both sides of the reaction. None of the other options are supported by the diagram.

59. A The overall rate law is always equal to the rate law of the slowest elementary step. The rate law of any elementary step can be determined using the coefficients of the reactants in that step.

60. A Reactions with high activation energies that do not proceed at a measurable rate are considered to be under kinetic control—that is, their rate of progress is based on kinetics instead of thermodynamics.

Section II—Free-Response Answers and Explanations

1. A stock solution of 0.100 *M* cobalt (II) chloride is used to create several solutions, indicated in the data table below:

Sample	Volume $CoCl_2$ (mL)	Volume H_2O (mL)
1	20.00	0
2	15.00	5.00
3	10.00	10.00
4	5.00	15.00

(a) In order to achieve the degree of accuracy shown in the table above, select which of the following pieces of laboratory equipment could be used when measuring out the $CoCl_2$.

150-mL beaker | 400-mL beaker | 250-mL Erlenmeyer flask
50-mL buret | 50-mL graduated cylinder | 100-mL graduated cylinder

The 50-mL buret is what is needed here in order to get measurements that are accurate to the hundredths place. Graduated cylinders are generally accurate to the tenths, and using flasks or beakers to measure out volume is highly inaccurate.

(b) Calculate the concentration of the $CoCl_2$ in each sample.

The solutions are then placed in cuvettes before being inserted into a spectrophotometer calibrated to 560 nm and their values are measured, yielding the data below:

Sample	Absorbance
1	0.485
2	0.364
3	0.243
4	0.121

Sample 1: 0.100 *M* (no dilution occurred)

For samples 2–4, the first step is to calculate the number of moles of $CoCl_2$, using the Molarity = moles/volume formula. Then, divide the number of moles by the new volume (volume of $CoCl_2$ + volume H_2O) to determine the new concentration.

Sample 2: 0.100 M = n/0.015 L
n = 0.0015 mol
0.0015 mol/0.020 L = 0.075 M

Sample 3: 0.100 M = n/0.010 L
n = 0.0010 mol
0.0010 mol/0.020 L = 0.050 M

Sample 4: 0.100 M = n/0.0050 L
n = 0.00050 mol
0.00050 mol/0.202 L = 0.025 M

(c) If gloves are not worn when handling the cuvettes, how might this affect the absorbance values gathered?

If gloves are not worn when handling the cuvettes, fingertip oils may be deposited on the surface of the cuvette. These oils may absorb some light, increasing the overall absorbance values.

(d) If the path length of the cuvette is 1.00 cm, what is the molar absorptivity value for $CoCl_2$ at 560 nm?

Using Beer's law: $A = abc$ and the data from Sample 1:

$0.485 = a(1.00\text{ cm})(0.100\ M)$ $\quad a = 4.85\ M^{-1}\text{cm}^{-1}$

Values from any sample can be used with identical results.

(e) On the axes below, plot a graph of absorbance vs. concentration. The *y*-axes scale is set, and be sure to scale the *x*-axes appropriately.

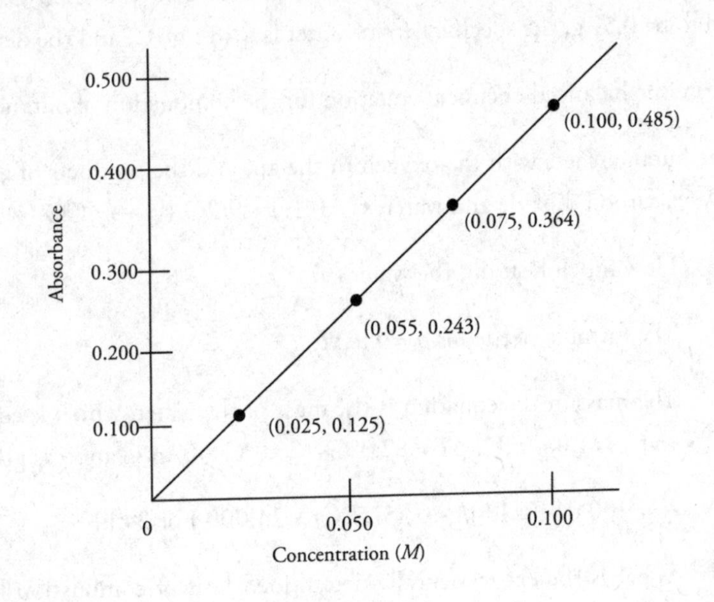

(f) What would the absorbance values be for $CoCl_2$ solutions at the following concentrations?

While the absorbance values for (i) can be estimated from a correctly-drawn graph, the absorbance value for (ii) would be off the chart. The more accurate method to determine the absorbances of both samples would be to determine the slope of the line from the graph in (e) and then plug in the appropriate values. (Note: Estimating the value for (i) is still an acceptable solution.)

Any two points can be taken to determine the slope of the graph, but by examining the Beer's law calculation in (d) we can see it is already in slope-intercept form (the *y*-intercept would be zero, as at 0 concentration there would be no absorbance). So, the slope of the line is equal to $4.85\ M^{-1}\text{cm}^{-1} \times 1.00$ cm $= 4.85\ M^{-1}$

(i) 0.067 *M*

$A = (4.85\ M^{-1})(0.067\ M) = 0.325$

(ii) 0.180 *M*

$A = (4.85\ M^{-1})(0.180\ M) = 0.873$

2. A sample of liquid butane (C_4H_{10}) in a pressurized lighter is set up directly beneath an aluminum can, as show in the diagram above. The can contains 100.0 mL of water, and when the butane is ignited the temperature of the water inside the can increases from 25.0°C to 82.3°C. The total mass of butane ignited is found to be 0.51 g, the specific heat of water is 4.18 J/g·°C, and the density of water is 1.00 g/mL.

(a) Write the balanced chemical equation for the combustion of butane in air.

The butane reacts with the oxygen in the air, and the products of any hydrocarbon combustion are always carbon dioxide and water: $C_4H_{10}(g) + 9/2O_2(g) \rightarrow 4CO_2(g) + 5H_2O(l)$.

(b) (i) How much heat did the water gain?

The formula needed is $q = mc\Delta T$.

The mass in the equation is the mass of the water, which is equal to 100.0 g (as the density of water is 1.0 g/mL). ΔT = 82.3°C – 25.0°C = 57.3°C and c is given as 4.18 J/g°C, so:

q = (100.0 g)(4.18 J/g·°C)(57.3°C) = 24,000 J or 24 kJ

(ii) What is the experimentally determined heat of combustion for butane based on this experiment? Your answer should be in kJ/mol.

0.51 g C_4H_{10} × 1 mol C_4H_{10}/58.14 g C_4H_{10} = 0.0088 mol C_4H_{10}

The sign on the heat calculated in (b)(i) needs to be flipped, as the combustion reaction will generate as much heat as the water gained.

–24 kJ/0.0088 mol C_4H_{10} = –2,700 kJ/mol

(c) Given butane's density of 0.573 g/mL at 25°C, calculate how much heat would be emitted if 5.00 mL of it were combusted at that temperature.

0.573 g/mL = m/5.00 mL m = 2.87 g butane

2.87 g C_4H_{10} × 1 mol C_4H_{10}/58.14 g C_4H_{10} = 0.049 mol C_4H_{10}

0.049 mol C_4H_{10} × 2700 kJ/mol = 130 kJ of heat emitted

(d) The overall combustion of butane is an exothermic reaction. Explain why this is in terms of bond energies.

The overall reaction is exothermic because more energy is released when the products are formed than is required to break the bonds in the reactants.

(e) One of the major sources of error in this experiment comes from the heat which is absorbed by the air. Why, then, might it not be a good idea to perform this experiment inside a sealed container to prevent the heat from leaving the system?

Performing this reaction in a sealed container would limit the amount of oxygen available to react. To get an accurate heat of combustion for butane, it needs to be the limiting reactant, and if the oxygen runs out prior to all the butane combusting, that will not happen.

Alternatively, the pressure of the sealed container could exceed the strength of the container and cause an explosion. There are approximately twice as many gas molecules after the reaction as there are prior to the reaction, and that means the pressure would approximately double if the reaction goes to completion.

3. $2N_2O_5(g) \rightarrow 4NO_2(g) + O_2(g)$

The data below was gathered for the decomposition of N_2O_5 at 310 K via the equation above.

Time (s)	$[N_2O_5]$ (M)
0	0.250
500.	0.190
1000.	0.145
2000.	0.085

(a) How does the rate of appearance of NO_2 compare to the rate of disappearance of N_2O_5? Justify your answer.

Due to the stoichiometric ratios, 4 moles of NO_2 are created for every 2 moles of N_2O_5 that decompose. Therefore, the rate of appearance of NO_2 will be twice the rate of disappearance for N_2O_5. As that rate is constantly changing over the course of the reaction, it is impossible to get exact values, but the ratio of 2:1 will stay constant.

(b) The reaction is determined to be first order overall. On the axes below, create a graph of some function of concentration vs. time that will produce a straight line. Label and scale your axes appropriately.

First order decompositions always create a straight line when plotting the natural log of concentration of the reactant vs. time.

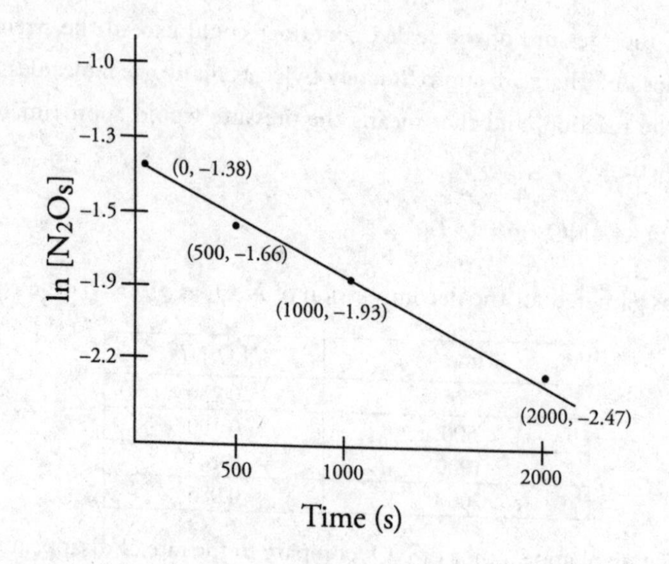

(c) (i) What is the rate constant for this reaction? Include units.

Interpreting the graph using slope-intercept form, we get $\ln[N_2O_5]_t = -kt + \ln[N_2O_5]_0$

Any (non-intercept) values on the graph can be plugged in to determine the rate constant.

Using at $[N_2O_5] = 0.190\ M$ at $t = 500$s:

$\ln(0.190) = -k(500.) + \ln(0.250)$
$-1.66 = -k(500.) + -1.39$
$-.27 = -k(500.)$
$k = 5.40 \times 10^{-4}$

In terms of units, if rate = $k[N_2O_5]$, then via analyzing the units:

$M/\text{s} = k(M)$ $\qquad k = \text{s}^{-1}$

So $k = 5.40 \times 10^{-4}\ \text{s}^{-1}$. Any point along the line will give the same value (as it is equal to the negative slope of the line).

(ii) What would the concentration of N_2O_5 be at t = 1500 s?

Using the equation from (i):

$\ln[N_2O_5]_{1500} = -5.40 \times 10^{-4}\ s^{-1}(1500.\ s) + \ln(0.250)$
$\ln[N_2O_5]_{1500} = -2.20$
$[N_2O_5]_{1500} = 0.111\ M$

(iii) What is the half-life of N_2O_5?

The half life is defined as how long it takes half of the original sample to decay. Thus, at the half-life of N_2O_5 for this reaction, $[N_2O_5] = 0.125\ M$. So:

$\ln(0.125) = -5.40 \times 10^{-4}\ (t) + \ln(0.250)$
$-2.08 = -5.40 \times 10^{-4}\ (t) + -1.39$
$-0.69 = -5.40 \times 10^{-4}\ (t)$
$t = 1280$ s

(d) Would the addition of a catalyst increase, decrease, or have no effect on the following variables? Justify your answers.

(i) Rate of disappearance of N_2O_5

Catalysts speed up the reaction, so adding a cataylst would increase the rate of disappearance of N_2O_5.

(ii) Magnitude of the rate constant

If the reaction is moving faster, the slope of any line graphing concentration vs. time would be steeper. This increases the magnitude of the slope, and thus, increases the magnitude of the rate constant as well.

(iii) Half-life of N_2O_5

If the reaction is proceeding at a faster rate, it will take less time for the N_2O_5 to decompose, and thus the half-life of the N_2O_5 would decrease.

4. Consider the Lewis structures for the following four molecules:

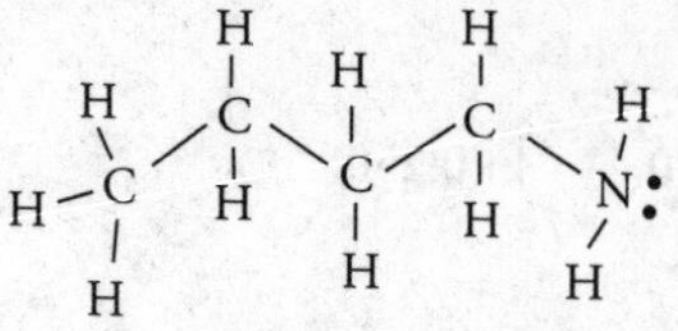

n-Butylamine

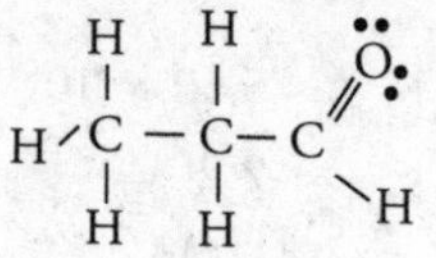

Propanal

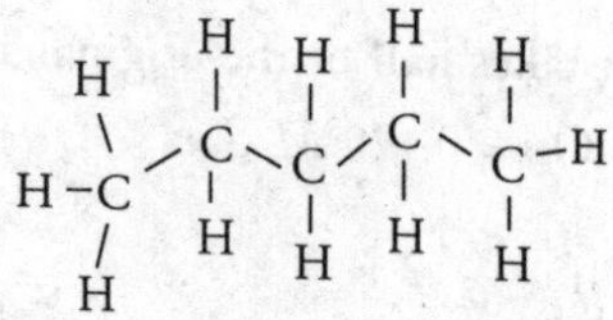

Pentane

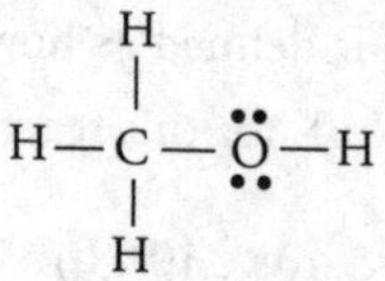

Methanol

(a) All of the substances are liquids at room temperature. Organize them from high to low in terms of boiling points, clearly differentiating between the intermolecular forces in each substance.

The strongest type of intermolecular force present here is hydrogen bonding, which both n-Butylamine and methanol exhibit. Of the two, n-Butylamine would have the stronger London disperion forces as it has more electrons (is more polarizable), and so it has stronger IMFs than methanol. For the remaining two structures, propanal has permanent dipoles while pentane is completely nonpolar. Therefore, propanal would have stronger IMFs than pentane.

So, from high to low boiling point: n-Butylamine > methanol > propanal > pentane

(b) On the methanol diagram reproduced below, draw the locations of all dipoles.

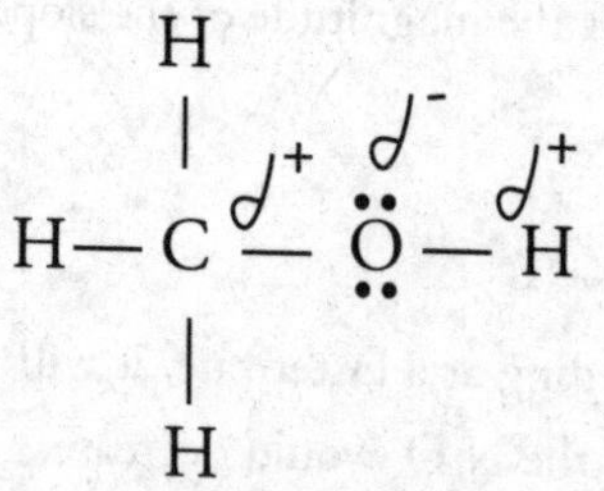

(c) n-Butylamine is found to have the lowest vapor pressure at room temperature out of the four liquids. Justify this observation in terms of intermolecular forces.

Vapor pressure arises from molecules overcoming the intermolecular forces to other molecules and escaping the surface of the liquid to become a gas. Due to its hydrogen bonding and large, highly-polarizable electron cloud, n-Butylamine would have the strongest intermolecular forces and thus its molecules would have the hardest time escaping the surface, leading to a low vapor pressure.

5. Current is run through an aqueous solution of nickel (II) fluoride, and a gas is evolved at the right-hand electrode, as indicated by the diagram below:

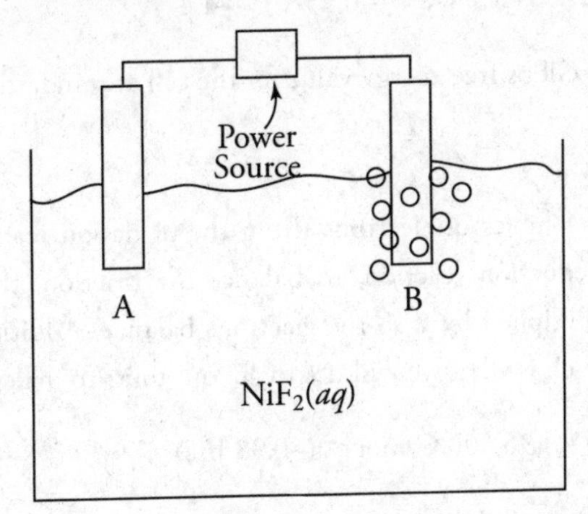

The standard reduction potential for several reactions is given in the following table:

Half-cell	E°_{red}
$F_2(g) + 2e^- \rightarrow 2F^-$	+2.87 V
$O_2(g) + 4H^+ + 4e^- \rightarrow 2H_2O(l)$	+1.23 V
$Ni^{2+} + 2e^- \rightarrow Ni(s)$	–0.25 V
$2H_2O(l) + 2e^- \rightarrow H_2(g) + 2OH^-$	–0.83 V

(a) Determine which half-reaction is occurring at each electrode:

(i) Oxidation:

The two choices for oxidation reductions are $2F^- \rightarrow F_2(g) + 2e^-$ (because fluoride ions are present in the solution) and $2H_2O(l) \rightarrow O_2(g) + 4H^+ + 4e^-$ (as water will lose electrons when oxidized). As we flipped the reactions, we must also flip the signs, so for the fluoride reaction $E^\circ_{ox} = -2.87$ V and for the water one $E^\circ_{ox} = -1.23$ V. The half-reaction with the less negative value is more likely to occur, so the answer is $2H_2O(l) \rightarrow O_2(g) + 4H^+ + 4e^-$.

(ii) Reduction:

Using the same logic as above, the two choices for reduction reaction are $Ni^{2+} + 2e^- \rightarrow Ni(s)$ and $2H_2O(l) + 2e^- \rightarrow H_2(g) + 2OH^-$. Unlike the oxidation reaction, there is no need to flip the signs on these half-reactions. The more positive value in this case belongs to the nickel reaction, so the answer is $Ni^{2+} + 2e^- \rightarrow Ni(s)$.

(b) (i) Calculate the standard cell potential of the cell.

$E^{\circ}_{cell} = E^{\circ}_{ox} + E^{\circ}_{red} = -1.23\text{ V} + (-0.25\text{ V}) = -1.47\text{ V}$

(ii) Calculate the Gibbs free energy value for the cell at standard conditions.

$\Delta G = -nFE$

n is equal to 4 moles of electrons (from the oxidation reaction). Even though $n = 2$ in the unbalanced reduction reaction, to balance the reaction, the reduction half-reaction would need to be multiplied by 2 so the electrons balance. Additionally, a volt is equal to a Joule/Coulomb, which is what we will use to get our units to make sense. So:

$\Delta G = -(4\text{ mol }e^{-})(96{,}500\text{ C/mol }e^{-})(-0.98\text{ J/C})$

$\Delta G = 380{,}000\text{ J or }380\text{ kJ}$

(c) Which electrode in the diagram (A or B) is the cathode, and which is the anode? Justify your answers.

Oxygen gas is evolved in the oxidation reaction, meaning that is occuring at electrode B. Oxidation always occurs at the anode, so electrode A is the cathode and electrode B is the anode.

6. Aniline, $C_6H_5NH_2$, is a weak base with $K_b = 3.8 \times 10^{-10}$.

(a) Write out the reaction which occurs when aniline reacts with water.

The aniline will act as a proton acceptor and take a proton from the water.

$C_6H_5NH_2(aq) + H_2O(l) \leftrightarrow C_6H_5NH_3^{+}(aq) + OH^{-}(aq)$

(b) (i) What is the concentration of each species at equilibrium in a solution of 0.25 M $C_6H_5NH_2$?

For the above equilibrium, $K_b = [C_6H_5NH_3^{+}][OH^{-}]/[C_6H_5NH_2]$ The concentrations of both the conjugate acid and the hydroxide ion will be equal, and the concentration of the aniline itself will be approximately the same, as it is a weak base which has a very low protonation rate. You can do an ICE chart to confirm this, but it is not really necessary if you understand the concepts underlying weak acids and bases.

$3.8 \times 10^{-10} = (x)(x)/(0.25)$
$x^2 = 9.5 \times 10^{-11}$
$x = 9.7 \times 10^{-6}$

$[C_6H_5NH_3^{+}] = [OH^{-}] = 9.7 \times 10^{-6}\ M$ and
$[C_6H_5NH_2] = 0.25\ M$

(ii) What is the pH value for the solution in (b)(i)?

$pOH = -\log[OH^-]$
$pOH = -\log (9.7 \times 10^{-6} M)$
$pOH = 5.01$

$pOH + pH = 14$
$5.01 + pH = 14$
$pH = 8.99$

7. A rigid, sealed 12.00 L container is filled with 10.00 g each of three different gases: CO_2, NO, and NH_3. The temperature of the gases is held constant 35.0°C. Assume ideal behavior for all gases.

(a) (i) What is the mole fraction of each gas?

First, the moles of each gas need calculating:

CO_2 = 10.00 g × 1 mol/44.01 g = 0.227 mol
NO = 10.00 g × 1 mol/30.01 g = 0.333 mol
NH_3 = 10.00g × 1 mol/17.03 g = 0.587 mol

Total moles of gas = 1.147 moles

$X_{CO_2} = 0.227/1.147 = 0.198$
$X_{NO} = 0.333/1.147 = 0.290$
$X_{NH_3} = 0.587/1.147 = 0.512$

(ii) What is the partial pressure of each gas?

Using the ideal gas law, we can calculate the total pressure in the container:

$PV = nRT$
$P\,(12.00\ L) = (1.147\ mol)(0.0821\ atm\cdot L/mol\cdot K)(308\ K)$
$P = 2.42\ atm$

The partial pressure of a gas is equal to the total pressure time the mole fraction of that gas.

$P_{CO_2} = (2.42\ atm)(0.198) = 0.479\ atm$
$P_{NO} = (2.42\ atm)(0.290) = 0.702\ atm$
$P_{NH_3} = (2.42\ atm)(0.512) = 1.24\ atm$

(b) Out of the three gases, molecules of which gas will have the highest velocity? Why?

If all gases are at the same temperature, they have the same amount of kinetic energy. KE has aspects of both mass and velocity, so the gas with the lowest mass would have the highest velocity. Thus, the NH_3 molecules have the highest velocity.

(c) Name one circumstance in which the gases might deviate from ideal behavior, and clearly explain the reason for the deviation.

The most common reason for deviation from ideal behavior is that the intermolecular forces of the gas molecules are acting upon each other. This would occur when the molecules are very close together and/or moving very slowly, so deviations would occur at high pressures and/or low temperatures.

Completely darken bubbles with a No. 2 pencil. If you make a mistake, be sure to erase mark completely. Erase all stray marks.

1.
YOUR NAME: (Print) ______ Last ______ First ______ M.I.

SIGNATURE: ______ DATE: __/__/__

HOME ADDRESS: (Print) ______ Number and Street

______ City ______ State ______ Zip Code

PHONE NO.: ______

IMPORTANT: Please fill in these boxes exactly as shown on the back cover of your test book.

2. TEST FORM

3. TEST CODE

4. REGISTRATION NUMBER

5. YOUR NAME

First 4 letters of last name | FIRST INIT | MID INIT

6. DATE OF BIRTH

Month	Day	Year
JAN		
FEB		
MAR		
APR		
MAY		
JUN		
JUL		
AUG		
SEP		
OCT		
NOV		
DEC		

7. GENDER
MALE
FEMALE

The Princeton Review®

1.	A	B	C	D	21.	A	B	C	D	41.	A B C D
2.	A	B	C	D	22.	A	B	C	D	42.	A B C D
3.	A	B	C	D	23.	A	B	C	D	43.	A B C D
4.	A	B	C	D	24.	A	B	C	D	44.	A B C D
5.	A	B	C	D	25.	A	B	C	D	45.	A B C D
6.	A	B	C	D	26.	A	B	C	D	46.	A B C D
7.	A	B	C	D	27.	A	B	C	D	47.	A B C D
8.	A	B	C	D	28.	A	B	C	D	48.	A B C D
9.	A	B	C	D	29.	A	B	C	D	49.	A B C D
10.	A	B	C	D	30.	A	B	C	D	50.	A B C D
11.	A	B	C	D	31.	A	B	C	D	51.	A B C D
12.	A	B	C	D	32.	A	B	C	D	52.	A B C D
13.	A	B	C	D	33.	A	B	C	D	53.	A B C D
14.	A	B	C	D	34.	A	B	C	D	54.	A B C D
15.	A	B	C	D	35.	A	B	C	D	55.	A B C D
16.	A	B	C	D	36.	A	B	C	D	56.	A B C D
17.	A	B	C	D	37.	A	B	C	D	57.	A B C D
18.	A	B	C	D	38.	A	B	C	D	58.	A B C D
19.	A	B	C	D	39.	A	B	C	D	59.	A B C D
20.	A	B	C	D	40.	A	B	C	D	60	A B C D

Completely darken bubbles with a No. 2 pencil. If you make a mistake, be sure to erase mark completely. Erase all stray marks.

1.

YOUR NAME: (Print) ____________ Last ____________ First ____________ M.I.

SIGNATURE: ____________________ DATE: ___ / ___ / ___

HOME ADDRESS: (Print) ____________________ Number and Street

____________ City ____________ State ____________ Zip Code

PHONE NO.: ____________________

IMPORTANT: Please fill in these boxes exactly as shown on the back cover of your test book.

2. TEST FORM

3. TEST CODE

0 1 2 3 4 5 6 7 8 9

A B C D E F G H I

J K L M N O P Q R

0 1 2 3 4 5 6 7 8 9

4. REGISTRATION NUMBER

0 1 2 3 4 5 6 7 8 9

5. YOUR NAME

First 4 letters of last name | FIRST INIT | MID INIT

A B C D E F G H I J K L M N O P Q R S T U V W X Y Z

6. DATE OF BIRTH

Month: JAN FEB MAR APR MAY JUN JUL AUG SEP OCT NOV DEC

Day: 0 1 2 3 / 0 1 2 3 4 5 6 7 8 9

Year: 0 1 2 3 4 5 6 7 8 9 / 0 1 2 3 4 5 6 7 8 9

7. GENDER

MALE

FEMALE

The Princeton Review®

1. (A) (B) (C) (D)
2. (A) (B) (C) (D)
3. (A) (B) (C) (D)
4. (A) (B) (C) (D)
5. (A) (B) (C) (D)
6. (A) (B) (C) (D)
7. (A) (B) (C) (D)
8. (A) (B) (C) (D)
9. (A) (B) (C) (D)
10. (A) (B) (C) (D)
11. (A) (B) (C) (D)
12. (A) (B) (C) (D)
13. (A) (B) (C) (D)
14. (A) (B) (C) (D)
15. (A) (B) (C) (D)
16. (A) (B) (C) (D)
17. (A) (B) (C) (D)
18. (A) (B) (C) (D)
19. (A) (B) (C) (D)
20. (A) (B) (C) (D)
21. (A) (B) (C) (D)
22. (A) (B) (C) (D)
23. (A) (B) (C) (D)
24. (A) (B) (C) (D)
25. (A) (B) (C) (D)
26. (A) (B) (C) (D)
27. (A) (B) (C) (D)
28. (A) (B) (C) (D)
29. (A) (B) (C) (D)
30. (A) (B) (C) (D)
31. (A) (B) (C) (D)
32. (A) (B) (C) (D)
33. (A) (B) (C) (D)
34. (A) (B) (C) (D)
35. (A) (B) (C) (D)
36. (A) (B) (C) (D)
37. (A) (B) (C) (D)
38. (A) (B) (C) (D)
39. (A) (B) (C) (D)
40. (A) (B) (C) (D)
41. (A) (B) (C) (D)
42. (A) (B) (C) (D)
43. (A) (B) (C) (D)
44. (A) (B) (C) (D)
45. (A) (B) (C) (D)
46. (A) (B) (C) (D)
47. (A) (B) (C) (D)
48. (A) (B) (C) (D)
49. (A) (B) (C) (D)
50. (A) (B) (C) (D)
51. (A) (B) (C) (D)
52. (A) (B) (C) (D)
53. (A) (B) (C) (D)
54. (A) (B) (C) (D)
55. (A) (B) (C) (D)
56. (A) (B) (C) (D)
57. (A) (B) (C) (D)
58. (A) (B) (C) (D)
59. (A) (B) (C) (D)
60 (A) (B) (C) (D)

NOTES

NOTES

NOTES

NOTES

NOTES

NOTES

NOTES

NOTES

International Offices Listing

China (Beijing)
1501 Building A,
Disanji Creative Zone,
No.66 West Section of North 4th Ring Road Beijing
Tel: +86-10-62684481/2/3
Email: tprkor01@chol.com
Website: www.tprbeijing.com

China (Shanghai)
1010 Kaixuan Road
Building B, 5/F
Changning District, Shanghai, China 200052
Sara Beattie, Owner: Email: tprenquiry.sha@sarabeattie.com
Tel: +86-21-5108-2798
Fax: +86-21-6386-1039
Website: www.princetonreviewshanghai.com

Hong Kong
5th Floor, Yardley Commercial Building
1–6 Connaught Road West, Sheung Wan, Hong Kong
(MTR Exit C)
Sara Beattie, Owner: Email: tprenquiry.sha@sarabeattie.com
Tel: +852-2507-9380
Fax: +852-2827-4630
Website: www.princetonreviewhk.com

India (Mumbai)
Score Plus Academy
Office No.15, Fifth Floor
Manek Mahal 90
Veer Nariman Road
Next to Hotel Ambassador
Churchgate, Mumbai 400020
Maharashtra, India
Ritu Kalwani: Email: director@score-plus.com
Tel: + 91 22 22846801 / 39 / 41
Website: www.scoreplusindia.com

India (New Delhi)
South Extension
K–16, Upper Ground Floor
South Extension Part–1,
New Delhi-110049
Aradhana Mahna: aradhana@manyagroup.com
Monisha Banerjee: monisha@manyagroup.com
Ruchi Tomar: ruchi.tomar@manyagroup.com
Rishi Josan: Rishi.josan@manyagroup.com
Vishal Goswamy: vishal.goswamy@manyagroup.com
Tel: +91-11-64501603/ 4, +91-11-65028379
Website: www.manyagroup.com

Lebanon
463 Bliss Street
AlFarra Building–2nd floor
Ras Beirut
Beirut, Lebanon
Hassan Coudsi: Email: hassan.coudsi@review.com
Tel: +961-1-367-688
Website: www.princetonreviewlebanon.com

Korea
945-25 Young Shin Building
25 Daechi-Dong, Kangnam-gu
Seoul, Korea 135-280
Yong-Hoon Lee: Email: TPRKor01@chollian.net
In-Woo Kim: Email: iwkim@tpr.co.kr
Tel: + 82-2-554-7762
Fax: +82-2-453-9466
Website: www.tpr.co.kr

Kuwait
ScorePlus Learning Center
Salmiyah Block 3, Street 2 Building 14
Post Box: 559, Zip 1306, Safat, Kuwait
Email: infokuwait@score-plus.com
Tel: +965-25-75-48-02 / 8
Fax: +965-25-75-46-02
Website: www.scorepluseducation.com

Malaysia
Sara Beattie MDC Sdn Bhd
Suites 18E & 18F
18th Floor
Gurney Tower, Persiaran Gurney
Penang, Malaysia
Email: tprkl.my@sarabeattie.com
Sara Beattie, Owner: Email: tprenquiry.sha@sarabeattie.com
Tel: +604-2104 333
Fax: +604-2104 330
Website: www.princetonreviewKL.com

Mexico
TPR México
Guanajuato No. 242 Piso 1 Interior 1
Col. Roma Norte
México D.F., C.P.06700
registro@princetonreviewmexico.com
Tel: +52-55-5255-4495
+52-55-5255-4440
+52-55-5255-4442
Website: www.princetonreviewmexico.com

Qatar
Score Plus
Villa No. 49, Al Waab Street
Opp Al Waab Petrol Station
Post Box: 39068, Doha, Qatar
Email: infoqatar@score-plus.com
Tel: +974 44 36 8580, +974 526 5032
Fax: +974 44 13 1995
Website: www.scorepluseducation.com

Taiwan
The Princeton Review Taiwan
2F, 169 Zhong Xiao East Road, Section 4
Taipei, Taiwan 10690
Lisa Bartle (Owner): lbartle@princetonreview.com.tw
Tel: +886-2-2751-1293
Fax: +886-2-2776-3201
Website: www.PrincetonReview.com.tw

Thailand
The Princeton Review Thailand
Sathorn Nakorn Tower, 28th floor
100 North Sathorn Road
Bangkok, Thailand 10500
Thavida Bijayendrayodhin (Chairman)
Email: thavida@princetonreviewthailand.com
Mitsara Bijayendrayodhin (Managing Director)
Email: mitsara@princetonreviewthailand.com
Tel: +662-636-6770
Fax: +662-636-6776
Website: www.princetonreviewthailand.com

Turkey
Yeni Sülün Sokak No. 28
Levent, Istanbul, 34330, Turkey
Nuri Ozgur: nuri@tprturkey.com
Rona Ozgur: rona@tprturkey.com
Iren Ozgur: iren@tprturkey.com
Tel: +90-212-324-4747
Fax: +90-212-324-3347
Website: www.tprturkey.com

UAE
Emirates Score Plus
Office No: 506, Fifth Floor
Sultan Business Center
Near Lamcy Plaza, 21 Oud Metha Road
Post Box: 44098, Dubai
United Arab Emirates
Hukumat Kalwani: skoreplus@gmail.com
Ritu Kalwani: director@score-plus.com
Email: info@score-plus.com
Tel: +971-4-334-0004
Fax: +971-4-334-0222
Website: www.scorepluseducation.com

Our International Partners

The Princeton Review also runs courses with a variety of partners in Africa, Asia, Europe, and South America.

Georgia
LEAF American-Georgian Education Center
www.leaf.ge

Mongolia
English Academy of Mongolia
www.nyescm.org

Nigeria
The Know Place
www.knowplace.com.ng

Panama
Academia Interamericana de Panama
http://aip.edu.pa/

Switzerland
Institut Le Rosey
http://www.rosey.ch/

All other inquiries, please email us at
internationalsupport@review.com